Lecture Notes in Mathematics　　　1717

Editors:
A. Dold, Heidelberg
F. Takens, Groningen
B. Teissier, Paris

Springer
Berlin
Heidelberg
New York
Barcelona
Hong Kong
London
Milan
Paris
Singapore
Tokyo

J. Bertoin   F. Martinelli   Y. Peres

Lectures on Probability Theory and Statistics

Ecole d'Eté de Probabilités
de Saint-Flour XXVII – 1997

Editor: Pierre Bernard

Springer

Authors

Jean Bertoin
Laboratoire de Probabilités
Université Pierre et Marie Curie
4, Place Jussieu
75252 Paris Cedex 05, France

Fabio Martinelli
Dipartimento di Matematica
Università di Roma Tre
Via C. Segre 2
00146 Roma, Italy

Yuval Peres
Institute of Mathematics
Hebrew University, Givat Ram
91904 Jerusalem, Israël

Editor

Pierre Bernard
Laboratoire de Mathématiques Appliquées
UMR CNRS 6620
Université Blaise Pascal
Clermont-Ferrand
63177 Aubière Cedex, France

Cataloging-in-Publication Data applied for

Die Deutsche Bibliothek - CIP-Einheitsaufnahme

Lectures on probability theory and statistics / Ecole d'Eté de
Probabilités de Saint-Flour XXVII - 1997. J. Bertoin ; F. Martinelli ;
Y. Peres. Ed.: P. Bernard. - Berlin ; Heidelberg ; New York ;
Barcelona ; Hong Kong ; London ; Milan ; Paris ; Singapore ; Tokyo
: Springer, 2000
(Lecture notes in mathematics ; Vol. 1717)
ISBN 3-540-66593-5

Mathematics Subject Classification (1991): 60-01, 60-06, 60D05, 60G17, 60G18,
60J15, 60J30, 60J55, 60K35, 82-01, 82-02, 82B20, 82B26, 82B44, 82C05

ISSN 0075-8434
ISBN 3-540-66593-5 Springer-Verlag Berlin Heidelberg New York

The use of general descriptive names, registered names, trademarks, etc. in this
publication does not imply, even in the absence of a specific statement, that such
names are exempt from the relevant protective laws and regulations and therefore
free for general use.

Typesetting: Camera-ready T_EX output by the author
SPIN: 10997390   41/3111-54321- Printed on acid-free paper

INTRODUCTION

This volume contains lectures given at the Saint-Flour Summer School of Probability Theory during the period 7th - 23rd July, 1997.

We thank the authors for all the hard work they accomplished. Their lectures are a work of reference in their domain.

The School brought together 81 participants, 37 of whom gave a lecture concerning their research work.

At the end of this volume you will find the list of participants and their papers.

Finally, to facilitate research concerning previous schools we give here the number of the volume of "Lecture Notes" where they can be found :

Lecture Notes in Mathematics

1971 : n° 307 -	1973 : n° 390 -	1974 : n° 480 -	1975 : n° 539 -	1976 : n° 598 -
1977 : n° 678 -	1978 : n° 774 -	1979 : n° 876 -	1980 : n° 929 -	1981 : n° 976 -
1982 : n° 1097 -	1983 : n° 1117 -	1984 : n° 1180 -	1985 - 1986 et	1987 : n° 1362 -
1988 : n° 1427 -	1989 : n° 1464 -	1990 : n° 1527 -	1991 : n° 1541 -	1992 : n° 1581 -
1993 : n° 1608 -	1994 : n° 1648 -	1995 : n° 1690 -	1996 : n° 1665	

Lecture Notes in Statistics

1986 : n° 50

TABLE OF CONTENTS

Fabio MARTINELLI :   « LECTURES ON GLAUBER DYNAMICS
                        FOR DISCRETE SPIN MODELS »

Yuval PERES : « PROBABILITY ON TREES :
AN INTRODUCTORY CLIMB »

SUBORDINATORS :

EXAMPLES AND APPLICATIONS

Jean BERTOIN

Table of Contents

Foreword

A subordinator is an increasing process that has independent and homogeneous increments. Subordinators thus form one of the simplest family of random processes in continuous time. The purpose of this course is two-fold: First to expose salient features of the theory and second to present a variety of examples and applications. The theory mostly concerns the statistical and sample path properties. The applications we have in mind essentially follow from the connection between subordinators and regenerative sets, that can be thought of as the set of times when a Markov process visits some fixed point of the state space. Typically, this enables us to translate certain problems on a given Markov process in terms of some subordinator, and then to use general known results on the latter.

Here is a sketch of the content. The first chapter introduces the basic notions and properties of subordinators, such as the Lévy-Khintchine formula, Itô's decomposition, renewal measures, ranges $\cdots$, and the second presents the correspondence relating subordinators, regenerative sets, and local times and excursions of Markov processes, which is essential to the future applications. More advanced material in that field is developed in chapters 3-5, which concern respectively the asymptotic behaviour of last-passage times in connection with the Dynkin-Lamperti theorem, the smoothness of the local times (law of the iterated logarithm, modulus of continuity) and some geometric properties of regenerative sets including fractal dimensions and the study of the intersection with a given set. Applications are presented in chapters 6-9. First, we describe the law of the solution of the inviscid Burgers equation with Brownian initial velocity in terms of a subordinator, which enables us to investigate its statistical properties. Next, we study the closed subset of $[0, \infty)$ that is left uncovered by open intervals sampled from a Poisson point process, following the ingenious approach of Fitzsimmons *et al.* Then, we turn our attention to two natural regenerative sets associated with a real-valued Lévy process: The set of passage times at a fixed state, and the set of times when a new maximum is achieved. Some applications of Bochner's subordination for Lévy processes are also given. Finally we investigate the class of subordinators that appears in connection with occupation times of a linear Brownian motion, or, equivalently, with the zero set of one-dimensional diffusions, by making use of M. G. Krein's spectral theory of vibrating strings. The choice of the examples discussed here is quite arbitrary; for instance, Marsalle [117] exposes further applications in the same vein, to increase times of stable processes, slow or fast points for local times, and the favorite site of a Brownian motion with drift.

Last but not least, it is my pleasure to thank Marc Yor for his very valuable comments on the first draft of this work.

Chapter 1

Elements on subordinators

The purpose of this chapter is to introduce basic notions on subordinators.

1.1 Definitions and first properties

Let $(\Omega, \mathbb{P})$ denote a probability space endowed with a right-continuous and complete filtration $(\mathcal{F}_t)_{t\geq 0}$. We consider right-continuous increasing adapted processes started from 0 and with values in the extended half-line $[0, \infty]$, where ∞ serves as a cemetery point (i.e. ∞ is an absorbing state). If $\sigma = (\sigma_t, t \geq 0)$ is such a process, we denote its lifetime by

$$\zeta = \inf\{t \geq 0 : \sigma_t = \infty\}$$

and call σ a *subordinator* if it has independent and homogeneous increments on $[0, \zeta)$. That is to say that for every $s, t \geq 0$, conditionally on $\{t < \zeta\}$, the increment $\sigma_{t+s} - \sigma_t$ is independent of $\mathcal{F}_t$ and has the same distribution as σ_s (under $\mathbb{P}$). When the lifetime is infinite a.s., we say that σ is a subordinator in the strict sense. The terminology has been introduced by Bochner [25]; see the forthcoming section 8.4.

Here is a standard example that will be further developed in Section 8.3. Consider a linear Brownian motion $B = (B_t : t \geq 0)$ started at 0, and the first passage times

$$\tau_t = \inf\{s \geq 0 : B_s > t\}, \qquad t \geq 0$$

(it is well-known that $\tau_t < \infty$ for all $t \geq 0$, a.s.). We write $\mathcal{F}_t$ for the complete sigma-field generated by the Brownian motion stopped at time τ_t, viz. $(B_{s\wedge\tau_t} : s \geq 0)$. According to the strong Markov property,

$$B'_s = B_{s+\tau_t} - t, \qquad s \geq 0$$

is independent of $\mathcal{F}_t$ and is again a Brownian motion. Moreover, it is clear that for every $s \geq 0$

$$\tau_{t+s} - \tau_t = \inf\{u \geq 0 : B'_u > s\}.$$

This shows that $\tau = (\tau_t : t \geq 0)$ is an increasing $(\mathcal{F}_t)$-adapted process with independent and homogeneous increments. Its paths are right-continuous and have an infinite lifetime a.s.; and hence τ is a strict subordinator.

We assume henceforth that σ is a subordinator. The independence and homogeneity of the increments immediately yield the (simple) Markov property: For every fixed $t \geq 0$, conditionally on $\{t < \zeta\}$, the process $\sigma' = (\sigma'_s = \sigma_{s+t} - \sigma_t, s \geq 0)$ is independent of $\mathcal{F}_t$ and has the same law as σ. The one-dimensional distributions of σ

$$p_t(dy) = \mathbb{P}(\sigma_t \in dy, t < \zeta), \qquad t \geq 0, y \in [0, \infty)$$

thus give rise to a convolution semigroup $(P_t, t \geq 0)$ by

$$P_t f(x) = \int_{[0,\infty)} f(x+y) p_t(dy) = \mathbb{E}(f(\sigma_t + x), t < \zeta)$$

where f stands for a generic nonnegative Borel function. It can be checked that this semigroup has the Feller property, cf. Proposition I.5 in [11] for details.

The simple Markov property can easily be reinforced, i.e. extended to stopping times:

Proposition 1.1 *If T is a stopping time, then, conditionally on $\{T < \zeta\}$, the process $\sigma' = (\sigma'_t = \sigma_{T+t} - \sigma_T, t \geq 0)$ is independent of $\mathcal{F}_T$ and has the same law as σ (under $\mathbb{P}$).*

Proof: For an elementary stopping time, the statement merely rephrases the simple Markov property. If T is a general stopping time, then there exists a sequence of elementary stopping times $(T_n)_{n \in \mathbb{N}}$ that decrease towards T, a.s. For each integer n, conditionally on $\{T_n < \zeta\}$, the shifted process $(\sigma_{T_n+t} - \sigma_{T_n}, t \geq 0)$ is independent of $\mathcal{F}_{T_n}$ (and thus of $\mathcal{F}_T$), and has the same law as σ. Letting $n \to \infty$ and using the right-continuity of the paths, this entails our assertion. ∎

The law of a subordinator is specified by the Laplace transforms of its one-dimensional distributions. To this end, it is convenient to use the convention that $e^{-\lambda \times \infty} = 0$ for any $\lambda \geq 0$, so that

$$\mathbb{E}(\exp\{-\lambda \sigma_t\}, t < \zeta) = \mathbb{E}(\exp\{-\lambda \sigma_t\}), \qquad t, \lambda \geq 0.$$

The independence and homogeneity of the increments then yield the multiplicative property

$$\mathbb{E}(\exp\{-\lambda \sigma_{t+s}\}) = \mathbb{E}(\exp\{-\lambda \sigma_t\}) \mathbb{E}(\exp\{-\lambda \sigma_s\})$$

for every $s, t \geq 0$. We can therefore express these Laplace transforms in the form

$$\mathbb{E}(\exp\{-\lambda \sigma_t\}) = \exp\{-t\Phi(\lambda)\}, \qquad t, \lambda \geq 0 \tag{1.1}$$

where the function $\Phi : [0, \infty) \to [0, \infty)$ is called the *Laplace exponent* of σ.

Returning to the example of the first passage process τ of a linear Brownian motion, one can use the scaling property of Brownian motion and the reflexion principle to determine the distribution of τ_1. Specifically, for every $t > 0$

$$\mathbb{P}(\tau_1 < t) = \mathbb{P}\left(\sup_{0 \leq s \leq t} B_s > 1\right) = \mathbb{P}\left(\sup_{0 \leq s \leq 1} B_s > 1/\sqrt{t}\right) = \mathbb{P}\left(|B_1| > 1/\sqrt{t}\right)$$

$$= \sqrt{\frac{2}{\pi}} \int_0^t s^{-3/2} e^{-1/2s} ds.$$

It is easy to deduce that the Laplace exponent of τ is

$$\Phi(\lambda) = -\log \mathbb{E}(\exp\{-\lambda\tau_1\}) = \sqrt{2\lambda}.$$

1.2 The Lévy-Khintchine formula

The next theorem gives a necessary and sufficient analytic condition for a function to be the Laplace exponent of a subordinator.

Theorem 1.2 (de Finetti, Lévy, Khintchine)(i) *If Φ is the Laplace exponent of a subordinator, then there exist a unique pair* $(\mathbf{k}, \mathbf{d})$ *of nonnegative real numbers and a unique measure* Π *on* $(0, \infty)$ *with* $\int (1 \wedge x) \, \Pi(dx) < \infty$, *such that for every* $\lambda \geq 0$

$$\Phi(\lambda) = \mathbf{k} + \mathbf{d}\lambda + \int_{(0,\infty)} \left(1 - e^{-\lambda x}\right) \Pi(dx). \tag{1.2}$$

(ii) *Conversely, any function Φ that can be expressed in the form (1.2) is the Laplace exponent of a subordinator.*

Equation (1.2) will be referred to as the *Lévy-Khintchine formula*; one calls $\mathbf{k}$ the *killing rate*, $\mathbf{d}$ the *drift coefficient* and Π the *Lévy measure* of σ. It is sometimes convenient to perform an integration by parts and rewrite the Lévy-Khintchine formula as

$$\Phi(\lambda)/\lambda = \mathbf{d} + \int_0^\infty e^{-\lambda x} \overline{\Pi}(x) dx, \qquad \text{with } \overline{\Pi}(x) = \mathbf{k} + \Pi\left((x, \infty)\right).$$

We call $\overline{\Pi}$ the *tail of the Lévy measure*. Note that the killing rate and the drift coefficient are given by

$$\mathbf{k} = \Phi(0) \quad, \quad \mathbf{d} = \lim_{\lambda \to \infty} \frac{\Phi(\lambda)}{\lambda}.$$

In particular, the lifetime ζ has an exponential distribution with parameter $\mathbf{k} \geq 0$ ($\zeta \equiv \infty$ for $\mathbf{k} = 0$).

Before we proceed to the proof of Theorem 1.2, we present some well-known examples of subordinators. The simplest is the Poisson process with intensity $c > 0$, which corresponds to the Laplace exponent

$$\Phi(\lambda) = c(1 - e^{-\lambda}),$$

that is the killing rate $\mathbf{k}$ and the drift coefficient $\mathbf{d}$ are zero and the Lévy measure $c\delta_1$, where δ_1 stands for the Dirac point mass at 1. Then the so-called standard stable subordinator with index $\alpha \in (0, 1)$ has a Laplace exponent given by

$$\Phi(\lambda) = \lambda^\alpha = \frac{\alpha}{\Gamma(1-\alpha)} \int_0^\infty (1 - e^{-\lambda x}) x^{-1-\alpha} dx.$$

The restriction on the range of the index is due to the requirement $\int (1 \wedge x) \, \Pi(dx) < \infty$. The boundary case $\alpha = 1$ is degenerate since it corresponds to the deterministic process $\sigma_t \equiv t$, and is usually implicitly excluded. A third family of examples is provided by the Gamma processes with parameters $a, b > 0$, for which the Laplace exponent is

$$\Phi(\lambda) = a \log(1 + \lambda/b) = \int_0^\infty (1 - e^{-\lambda x}) a x^{-1} e^{-bx} dx \ ,$$

where the second equality stems from the Frullani integral. We see that the Lévy measure is $\Pi^{(a,b)}(dx) = a x^{-1} e^{-bx} dx$ and the killing rate and the drift coefficient are zero.

Proof of Theorem 1.2: (i) Making use of the independence and homogeneity of the increments in the second equality below, we get from (1.1) that for every $\lambda \geq 0$

$$\Phi(\lambda) = \lim_{n \to \infty} n\left(1 - \exp\{-\Phi(\lambda)/n\}\right) = \lim_{n \to \infty} n\mathbb{E}\left(1 - \exp\{-\lambda \sigma_{1/n}\}\right)$$

$$= \lambda \lim_{n \to \infty} \int_0^\infty e^{-\lambda x} n\mathbb{P}\left(\sigma_{1/n} \geq x\right) dx \ .$$

Write $\overline{\Pi}_n(x) = n\mathbb{P}\left(\sigma_{1/n} \geq x\right)$, so that

$$\frac{\Phi(\lambda)}{\lambda} = \lim_{n \to \infty} \int_0^\infty e^{-\lambda x} \overline{\Pi}_n(x) dx \ .$$

This shows that the sequence of absolutely continuous measures $\overline{\Pi}_n(x) dx$ converges vaguely as $n \to \infty$. As each function $\overline{\Pi}_n(\cdot)$ decreases, the limit has necessarily the form $d\delta_0(dx) + \overline{\Pi}(x) dx$, where $d \geq 0$, $\overline{\Pi} : (0, \infty) \to [0, \infty)$ is a non-increasing function, and δ_0 stands for the Dirac point mass at 0. Thus

$$\frac{\Phi(\lambda)}{\lambda} = d + \int_0^\infty e^{-\lambda x} \overline{\Pi}(x) dx$$

and this yields (1.2) with $k = \overline{\Pi}(\infty)$ and $\Pi(dx) = -d\overline{\Pi}(x)$ on $(0, \infty)$. It is plain that we must have $\int_{(0,1)} x\Pi(dx) < \infty$ since otherwise $\Phi(\lambda)$ would be infinite. Uniqueness is obvious.

(ii) Consider a Poisson point process $\Delta = (\Delta_t, t \geq 0)$ with values in $(0, \infty]$ and with characteristic measure $\Pi + k\delta_\infty$. This means that for every Borel set $B \subseteq (0, \infty]$, the counting process $N^B = \text{Card}\{s \in [0, \cdot] : \Delta_s \in B\}$ is a Poisson process with intensity $\Pi(B) + k\delta_\infty(B)$, and to disjoint Borel sets correspond independent Poisson processes. In particular, the instant of the first infinite point, $\tau_\infty = \inf\{t \geq 0 : \Delta_t = \infty\}$, has an exponential distribution with parameter k ($\tau_\infty \equiv \infty$ if $k = 0$), and is independent of the Poisson point process restricted to $(0, \infty)$. Moreover, the latter is a Poisson point process with characteristic measure Π.

Introduce $\Sigma = (\Sigma_t, t \geq 0)$ by

$$\Sigma_t = dt + \sum_{0 \leq s \leq t} \Delta_s \ .$$

The condition $\int (1 \wedge x) \Pi(dx) < \infty$ ensures that $\Sigma_t < \infty$ whenever $t < \tau_\infty$, a.s. It is plain that Σ is a right-continuous increasing process started at 0, with lifetime τ_∞, and that its increments are stationary and independent on $[0, \tau_\infty)$. In other words, Σ is a subordinator. Finally, the exponential formula for a Poisson point process (e.g. Proposition XII.1.12 in [132]) gives for every $t, \lambda \geq 0$

$$\mathbb{E}\left(\exp\{-\lambda\Sigma_t\}\right) = \exp\left\{-t\left(k + d\lambda + \int_{(0,\infty)} (1 - e^{-\lambda x})\Pi(dx)\right)\right\} \ ,$$

which shows that the Laplace exponent of Σ is given by (1.2). ∎

More precisely, the proof of (ii) contains relevant information on the canonical decomposition of a subordinator as the sum of its continuous part and its jumps.

Proposition 1.3 (Itô [81]) *One has a.s., for every $t \geq 0$:*

$$\sigma_t = \mathrm{d}t + \sum_{0 \leq s \leq t} \Delta_s \,,$$

where $\Delta = (\Delta_s, s \geq 0)$ is a Poisson point process with values in $(0, \infty]$ and characteristic measure $\Pi + \mathrm{k}\delta_\infty$, where δ_∞ stands for the Dirac point mass at ∞. The lifetime of σ is then given by $\zeta = \inf\{t \geq 0 : \Delta_t = \infty\}$.

As a consequence, we see that a subordinator is a step process if its drift coefficient is $\mathrm{d} = 0$ and its Lévy measure has a finite mass, $\Pi((0, \infty)) < \infty$ (this is also equivalent to the boundedness of the Laplace exponent). Otherwise σ is a strictly increasing process. In the first case, we say that σ is a *compound Poisson process*. A compound Poisson process can be identified as a random walk time-changed by an independent Poisson process; and in many aspects, it can be thought of as a process in discrete time. Because we are mostly concerned with 'truly' continuous time problems, it will be more convenient to concentrate on strictly increasing subordinators in the sequel.

Henceforth, the case when σ is a compound Poisson process is implicitly excluded.[1]

1.3 The renewal measure

A subordinator is a transient Markov process; its potential measure $U(dx)$ is called the *renewal measure*. It is given by

$$\int_{[0,\infty)} f(x) U(dx) = \mathbb{E}\left(\int_0^\infty f(\sigma_t) dt\right).$$

The distribution function of the renewal measure

$$U(x) = \mathbb{E}\left(\int_0^\infty \mathbf{1}_{\{\sigma_t \leq x\}} dt\right), \qquad x \geq 0$$

is known as the *renewal function*. If we introduce the continuous inverse of the strictly increasing process σ:

$$L_x = \sup\{t \geq 0 : \sigma_t \leq x\} = \inf\{t > 0 : \sigma_t > x\}, \qquad x \geq 0,$$

we then see that

$$U(x) = \mathbb{E}(L_x) \,;$$

---

[1]Nonetheless, many results presented in this text still hold in the general case.

in particular we obtain by an application of the theorem of dominated convergence that the renewal function is continuous. It is also immediate to deduce from the Markov property that the renewal function is subadditive, that is

$$U(x+y) \leq U(x) + U(y) \qquad \text{for all } x, y \geq 0.$$

Because the Laplace transform of the renewal measure is

$$\mathcal{L}U(\lambda) = \int_{[0,\infty)} e^{-\lambda x} U(dx) = \frac{1}{\Phi(\lambda)}, \qquad \lambda > 0$$

the renewal measure characterizes the law of the subordinator.

We next present useful estimations for the renewal measure in terms of the Laplace exponent and of the tail of the Lévy measure, which follow from the fact that the Laplace transforms of U and $\overline{\Pi}$ admit simple expressions in terms of Φ, and adequate Tauberian theorems. To this end, we first state a general result. When f and g are two nonnegative functions, we use the notation $f \asymp g$ to indicate that there are two positive constants, c and c', such that $cf \leq g \leq c'f$. Introduce the so-called *integrated tail*

$$I(t) = \int_0^t \overline{\Pi}(x) dx = \int_0^t \left(k + \Pi((x, \infty)) \right) dx.$$

Proposition 1.4 *We have*

$$U(x) \asymp 1/\Phi(1/x) \quad and \quad \Phi(x)/x \asymp I(1/x) + \mathrm{d}.$$

Proof: Recall that $1/\Phi$ is the Laplace transform of the renewal measure. As Φ is concave and monotone increasing, the Tauberian theorem of de Haan and Stadtmüller (see [20] on page 118) applies and yields $U(x) \asymp 1/\Phi(1/x)$. The second estimate follows similarly, using the fact that the Laplace transform of the tail of the Lévy measure is $-\mathrm{d} + \Phi(\lambda)/\lambda$ (by the Lévy-Khintchine formula). ∎

Sharper estimates follow from Karamata's Tauberian theorem when one imposes that the Laplace exponent has regular variation. Recall that a measurable function $f : (0, \infty) \to [0, \infty)$ is *regularly varying* at $0+$ (respectively, at ∞) if for every $x > 0$, the ratio $f(\lambda x)/f(\lambda)$ converges as $\lambda \to 0+$ (respectively, $\lambda \to \infty$). The limit is then necessarily x^α for some real number α which is called the index. When $\alpha = 0$, we will simply say that f is slowly varying. We refer to Chapter XIII in Feller [53] for the basic theory, and to Bingham *et al.* [20] for the complete account. We stress that when the Laplace exponent Φ is regularly varying (at $0+$ or at ∞) then, due to the Lévy-Khintchine formula, the index necessarily lies between 0 and 1.

Proposition 1.5 *Suppose that Φ is regularly varying at $0+$ (respectively, at ∞) with index $\alpha \in [0, 1]$. Then,*

$$\Gamma(1+\alpha)U(ax) \sim a^\alpha/\Phi(1/x) \quad as\ x \to \infty\ (respectively,\ as\ x \to 0+),$$

uniformly as a varies on any fixed compact interval of $(0, \infty)$.

Moreover, if $\alpha < 1$, then

$$\Gamma(1-\alpha)\overline{\Pi}(ax) \sim a^{-\alpha}\Phi(1/x) \quad as\ x \to \infty\ (respectively,\ as\ x \to 0+),$$

uniformly as a varies on any fixed compact interval of $(0, \infty)$

11

Proof: The first assertion follows from Karamata's Tauberian theorem and the uniform convergence theorem; cf. Theorems 1.7.1 and 1.5.2 in [20]. The second requires the monotone density theorem; see Theorem 1.7.2 in [20]. ∎

Next, local estimates for the renewal measure in the neighbourhood of ∞ are given by the renewal theorem.

Proposition 1.6 (Renewal theorem) *Put* $\mathbb{E}(\sigma_1) = \mu \in (0, \infty]$. *Then for every* $h > 0$

$$\lim_{x \to \infty} (U(x + h) - U(x)) = h/\mu.$$

This renewal theorem is essentially a consequence of the standard renewal theorem in discrete time (i.e. for so-called renewal processes; see e.g. Feller [53]). Recall that the compound Poisson case has been ruled out, so σ is a 'non-lattice' process. Plainly, it is mostly useful in the finite mean case $\mu < \infty$; we refer to Doney [48] for recent progress in the (discrete) infinite mean case.

There is also an analogue of the renewal theorem in the neighbourhood of $0+$ when the drift coefficient is positive.

Proposition 1.7 (Neveu [122]) *Suppose that* $d > 0$. *Then the renewal measure is absolutely continuous and has a continuous everywhere positive density* $u : [0, \infty) \to (0, \infty)$ *given by*

$$u(x) = d^{-1}\mathbb{P}(\exists t \geq 0 : \sigma_t = x).$$

In particular, $u(0) = 1/d$.

Proof: As $d > 0$, the Laplace transform of the renewal measure has

$$\int_0^\infty e^{-\lambda x} U(dx) = \frac{1}{\Phi(\lambda)} \sim \frac{1}{d\lambda} \qquad \text{as } \lambda \to \infty.$$

By a Tauberian theorem, this entails

$$U(\varepsilon) \sim \varepsilon/d = \varepsilon u(0) \qquad \text{as } \varepsilon \to 0+. \tag{1.3}$$

The Markov property applied at the stopping time $L(x) = \inf\{t \geq 0 : \sigma_t > x\}$ gives

$$
\begin{aligned}
U(x + \varepsilon) - U(x) &= \mathbb{E}\left(\int_{L(x)}^\infty 1_{\{\sigma_t \in (x, x+\varepsilon)\}} dt\right) \\
&= \int_{[x, x+\varepsilon]} \mathbb{P}(\sigma_{L(x)} \in dy) U(x + \varepsilon - y) \\
&= \mathbb{P}\left(\sigma_{L(x)} = x\right) U(\varepsilon) + \int_{(x, x+\varepsilon]} \mathbb{P}(\sigma_{L(x)} \in dy) U(x + \varepsilon - y).
\end{aligned}
$$

The second term in the sum is bounded from above by $\mathbb{P}(\sigma_{L(x)} \in (x, x+\varepsilon]) U(\varepsilon) = o(U(\varepsilon))$. We deduce from (1.3) that

$$d^{-1}\mathbb{P}(\exists t : \sigma_t = x) = d^{-1}\mathbb{P}(\sigma_{L(x)} = x) = \lim_{\varepsilon \to 0+} \frac{U(x + \varepsilon) - U(x)}{\varepsilon}$$

(the first equality stems from the fact that $L(x)$ depends continuously on x). In particular, the renewal measure is absolutely continuous; we henceforth denote by $u(x)$ the version of its density that is specified by the last displayed formula. Note that $u(x) \leq 1/d$ and also, by an immediate application of the Markov property at $L(x)$, that for every $x, y \geq 0$

$$du(x + y) = \mathbb{P}(\exists t : \sigma_t = x + y) \geq \mathbb{P}(\exists t : \sigma_t = x)\mathbb{P}(\exists t : \sigma_t = y) = d^2 u(x)u(y).$$
(1.4)

To prove the continuity of u at $x = 0$, fix $\eta > 0$ and consider the Borel set $B_\eta = \{x \geq 0 : 1/d \leq u(x) + \eta\}$. As u is bounded from above by $1/d$, we see from (1.3) that 0 is a point of density of B_η, in the sense that $m([0, \varepsilon] \cap B_\eta) \sim \varepsilon$ as $\varepsilon \to 0+$, where m stands for the Lebesgue measure. By a standard result of measure theory, this implies that for some $a > 0$ and every $0 < x < a$, we can find $y, y' \in B_\eta$ such that $x = y + y'$. Using (1.4), we deduce

$$u(x) \geq du(y)u(y') \geq d\left(\frac{1}{d} - \eta\right)^2,$$

so that $\lim_{x \to 0+} u(x) = 1/d = u(0)$.

We next prove the continuity at some arbitrary $x > 0$. The same argument as above based on (1.4) yields

$$\limsup_{y \to x-} u(y) \leq u(x) \leq \liminf_{y \to x+} u(y).$$

On the one hand, the right-continuity of the paths shows that if y_n is a sequence that decreases towards x, then

$$\limsup \{\exists t : \sigma_t = y_n\} \subseteq \{\exists t : \sigma_t = x\},$$

so an application of Fatou's lemma gives

$$\limsup_{y \to x+} u(y) \leq u(x).$$

On the other hand, an application of the Markov property as in (1.4) yields that for every $\varepsilon > 0$

$$\mathbb{P}(\exists t : \sigma_t = x) \leq \mathbb{P}(\exists t : \sigma_t = \varepsilon)\mathbb{P}(\exists t : \sigma_t = x - \varepsilon) + \mathbb{P}(\forall t : \sigma_t \neq \varepsilon).$$

We know that the second term in the sum tends to 0 as $\varepsilon \to 0+$, so that

$$\liminf_{y \to x-} u(y) \geq u(x),$$

and the continuity of u is proven. Finally, we know that u is positive in some neighbourhood of 0, and it follows from (1.4) that u is positive everywhere. ∎

To conclude this section, we mention that large deviations estimates for the one-dimensional distributions of σ have been obtained by Jain and Pruitt [87]; see also Fristedt and Pruitt [61] for some more elementary results in that field.

1.4 The range of a subordinator

The range of a subordinator σ is the random closed subset of $[0, \infty)$ defined by

$$\mathcal{R} = \overline{\{\sigma_t : 0 \le t < \zeta\}}.$$

Note that $\mathcal{R}$ is a perfect (i.e. without isolated points) and $0 \in \mathcal{R}$. Because the paths of σ are càdlàg, the range can also be expressed as

$$\mathcal{R} = \{\sigma_t : 0 \le t < \zeta\} \bigcup \{\sigma_{s-} : s \in \mathcal{J}\}$$

where $\mathcal{J} = \{0 \le s \le \zeta : \Delta_s > 0\}$ denotes the set of jump times of σ. To this end, observe that $\{\sigma_{s-} : s \in \mathcal{J}\}$ is precisely the set of points in $\mathcal{R}$ that are isolated on their right. Alternatively, the canonical decomposition of the open set $\mathcal{R}^c = [0, \infty) - \mathcal{R}$, is

$$\mathcal{R}^c = \bigcup_{s \in \mathcal{J}} (\sigma_{s-}, \sigma_s). \tag{1.5}$$

Recall that $L. = \inf\{t \ge 0 : \sigma_t > \cdot\}$ stands for the -continuous- inverse of σ; it should be plain that $\mathcal{R}$ also coincides with the support of the Stieltjes measure dL:

$$\mathcal{R} = \mathrm{Supp}(dL),$$

which provides another useful representation of the range.

We next present basic properties of the range that will be useful in the sequel. First, an interesting problem that frequently arises about random sets, is the evaluation of their sizes. The simplest result in that field for the range of a subordinator concerns its Lebesgue measure. Sharper results involving Hausdorff and packing dimension will be presented in section 5.1.

Proposition 1.8 *We have*

$$m\left(\mathcal{R} \cap [0, t]\right) = m\left(\{\sigma_s : s \ge 0\} \cap [0, t]\right) = \mathrm{d}L_t \qquad a.s. \text{ for all } t \ge 0,$$

where d *is the drift coefficient and* m *the Lebesgue measure on* $[0, \infty)$. *In particular* $\mathcal{R}$ *has zero Lebesgue measure a.s. if and only if* $\mathrm{d} = 0$, *and we then say that* $\mathcal{R}$ *is light. Otherwise we say that* $\mathcal{R}$ *is heavy.*

Proof: The first equality is obvious as $\mathcal{R}$ differs from $\{\sigma_s : s \ge 0\}$ by at most countably many points. Next note that it suffices to treat the case $\mathrm{k} = 0$ (i.e. $\zeta = \infty$ a.s.), because the case $\mathrm{k} > 0$ then will then follow by introducing a killing at some independent time.

Recall that the canonical decomposition of the complementary set $\mathcal{R}^c$ is given by (1.5). In particular, for every fixed $t \ge 0$, the Lebesgue measure of $\mathcal{R}^c \cap [0, \sigma_t]$ is $\sum_{s \le t} \Delta_s$, and the latter quantity equals $\sigma_t - \mathrm{d}t$ by virtue of Proposition 1.3. This gives $m\left([0, \sigma_t] \cap \mathcal{R}\right) = \mathrm{d}t$ for all $t \ge 0$, a.s. Because the quantity on the right depends continuously on t, this entails by an argument of monotonicity that

$$m\left([0, \sigma_t] \cap \mathcal{R}\right) = m\left([0, \sigma_{t-}] \cap \mathcal{R}\right) = \mathrm{d}t.$$

Replacing t by L_t and recalling that $t \in [\sigma_{L_t-}, \sigma_{L_t}]$ completes the proof. ∎

We then specify the probability that $x \in \mathcal{R}$ for any fixed $x > 0$.

Proposition 1.9 (i) (Kesten [96]) *If the drift is* $\mathrm{d} = 0$, *then* $\mathbb{P}(x \in \mathcal{R}) = 0$ *for every* $x > 0$.

(ii) (Neveu [122]) *If* $\mathrm{d} > 0$, *then the function* $u(x) = \mathrm{d}^{-1}\mathbb{P}(x \in \mathcal{R})$ *is the version of the renewal density* $dU(x)/dx$ *that is continuous and everywhere positive on* $[0, \infty)$.

Proof: (i) An application of Fubini's theorem gives

$$\int_0^\infty \mathbb{P}(x \in \mathcal{R}) \, dx = \mathbb{E}(m(\mathcal{R})) \,,$$

where $m(\mathcal{R})$ stands for the Lebesgue measure of $\mathcal{R}$. We know from Proposition 1.8 that the latter is zero as $\mathrm{d} = 0$. In other words, $\mathbb{P}(x \in \mathcal{R}) = 0$ for almost every $x \geq 0$. That we may drop "almost" in the last sentence is easily seen when the renewal measure is absolutely continuous. More precisely, let τ be an independent random time with an exponential distribution with parameter 1. For every fixed $q > 0$, we have for any Borel set A

$$\mathbb{P}\left(\sigma_{\tau/q} \in A\right) = q \int_0^\infty e^{-qt} \mathbb{P}(\sigma_t \in A) dt \leq qU(A) \,,$$

which implies by virtue of the Radon-Nikodym theorem that the distribution of $\sigma_{\tau/q}$ is also absolutely continuous. Applying the Markov property at time τ/q, we deduce that

$$\mathbb{P}\left(\sigma_{\tau/q+t} = x \text{ or } \sigma_{\tau/q+t-} = x \text{ for some } t > 0\right) = \int_0^\infty \mathbb{P}(\sigma_{\tau/q} \in dy) \mathbb{P}(x - y \in \mathcal{R}) \,,$$

and the right-hand side equals zero as $\mathbb{P}(a \in \mathcal{R}) = 0$ for almost every $a > 0$. Letting q go to ∞, we deduce that $\mathbb{P}(x \in \mathcal{R}) = 0$ for every $x > 0$.

The same holds true even when the renewal measure is not absolutely continuous. This requires a more delicate analysis; we refer to the proof of Theorem III.4 in [11] for details.

(ii) By Proposition 1.7, all that is needed is to check that

$$\mathbb{P}(\exists t > 0 : \sigma_{t-} = x < \sigma_t) = 0 \,.$$

By the compensation formula for Poisson point processes, we have for every $\varepsilon > 0$

$$\begin{aligned}
\mathbb{P}(\exists t > 0 : \sigma_{t-} = x < \sigma_t - \varepsilon) &= \mathbb{E}\left(\sum_{t \geq 0} \mathbf{1}_{\{\sigma_{t-}=x\}} \mathbf{1}_{\{\Delta_t > \varepsilon\}}\right) \\
&= \overline{\Pi}(\varepsilon) \mathbb{E}\left(\int_0^\infty \mathbf{1}_{\{\sigma_t=x\}} dt\right) \,,
\end{aligned}$$

and the ultimate quantity is zero as the renewal measure has no atom. As ε is arbitrary, this completes the proof. $\blacksquare$

We next turn our attention to the left and right extremities of $\mathcal{R}$ as viewed from a fixed point $t \geq 0$:

$$g_t = \sup\{s < t : s \in \mathcal{R}\} \quad \text{and} \quad D_t = \inf\{s > t : s \in \mathcal{R}\} \,.$$

We call $(D_t : t \geq 0)$ and $(g_t : t > 0)$ the processes of first-passage and last-passage in $\mathcal{R}$, respectively. The use of an upper-case letter (respectively, of a lower-case letter) refers to the right-continuity (respectively, the left-continuity) of the sample paths. We immediately check that these processes can be expressed in terms of σ and its inverse L as follows :

$$g_t = \sigma(L_t-) \quad \text{and} \quad D_t = \sigma(L_t) \qquad \text{for all } t \geq 0, \text{ a.s.} \tag{1.6}$$

We present an useful expression for the distribution of the pair (g_t, D_t) in terms of the renewal function and the tail of the Lévy measure.

Lemma 1.10 *For every real numbers a, b, t such that $0 \leq a < t \leq a + b$, we have*

$$\mathbb{P}(g_t \in da, D_t - g_t \in db) = \Pi(db)U(da) \quad , \quad \mathbb{P}(g_t \in da, D_t = \infty) = kU(da).$$

In particular, we have for $a \in [0, t)$

$$\mathbb{P}(g_t \in da) = \overline{\Pi}(t-a)U(da).$$

Proof: Recall from (1.6) the identities $g_t = \sigma_{L_t-}$ and $D_t - g_t = \Delta_{L_t}$. Then observe that for any $u > 0$

$$\sigma_{L_t-} < a \quad \text{and} \quad L_t = u \iff \sigma_{u-} < a \quad \text{and} \quad \sigma_u \geq t.$$

Using the canonical expression of σ given in Proposition 1.3, we see that

$$\mathbb{P}(g_t < a, D_t - g_t \geq b) = \mathbb{E}\left(\sum 1_{\{\sigma_{u-} < a\}} 1_{\{\Delta_u \geq (t - \sigma_{u-}) \vee b\}}\right),$$

where the sum in the right-hand side is taken over all the instants when the point process Δ jumps. The process $u \to \sigma_{u-}$ is left continuous and hence predictable, so the compensation formula (see e.g. Proposition XII.1.10 in [132]) entails that the right-hand-side in the last displayed formula equals

$$\mathbb{E}\left(\int_0^\infty 1_{\{\sigma_u < a\}} \overline{\Pi}\left(((t - \sigma_u) \vee b) -\right) du\right) = \int_{[0,a)} \overline{\Pi}\left(((t - x) \vee b) -\right) U(dx).$$

This shows that for $0 \leq a < t < a + b$

$$\mathbb{P}(g_t \in da, D_t - g_t \in db) = \Pi(db)U(da).$$

Integrating this when b ranges over $[t-a, \infty]$ yields $\mathbb{P}(g_t \in da) = \overline{\Pi}((t-a)-)U(da)$. Since the renewal measure has no atom and the tail of the Lévy measure has at most countably many discontinuities, we may replace $\overline{\Pi}((t-a)-)$ by $\overline{\Pi}(t-a)$. ∎

A possible drawback of Lemma 1.10 is that it is not expressed explicitly in terms of the Laplace exponent Φ. Considering a double Laplace transform easily yields the following formula.

Lemma 1.11 *For every $\lambda, q > 0$*

$$\int_0^\infty e^{-qt} \mathbb{E}\left(\exp\{-\lambda g_t\}\right) dt = \frac{\Phi(q)}{q\Phi(\lambda + q)}.$$

Proof: It is immediately seen from Lemma 1.10 that $\mathbb{P}(g_t < t = D_t) = 0$ for every $t > 0$; it follows that $\mathbb{P}(g_t = t) = \mathbb{P}(t \in \mathcal{R})$. Using Proposition 1.9 and the fact that the Laplace transform of the renewal measure is $1/\Phi$, we find

$$\int_0^\infty e^{-qt} \mathbb{P}(g_t = t) dt = \frac{\mathrm{d}}{\Phi(q)}.$$

We then obtain from Lemma 1.10

$$\begin{aligned}
\int_0^\infty e^{-qt} \mathbb{E}\left(\exp\{-\lambda g_t\}\right) dt &= \int_0^\infty e^{-qt} \left(e^{-t\lambda}\mathbb{P}(g_t = t) + \int_{[0,t)} e^{-\lambda s}\overline{\Pi}(t-s)U(ds)\right) dt \\
&= \frac{\mathrm{d}}{\Phi(q+\lambda)} + \int_0^\infty dt \int_{[0,t)} U(ds)\, e^{-q(t-s)}\overline{\Pi}(t-s)e^{-(\lambda+q)s} \\
&= \frac{\mathrm{d}}{\Phi(q+\lambda)} + \mathcal{L}U(q+\lambda)\mathcal{L}\overline{\Pi}(q) \\
&= \frac{\mathrm{d}}{\Phi(q+\lambda)} + \frac{1}{\Phi(q+\lambda)}\left(\frac{\Phi(q)}{q} - \mathrm{d}\right).
\end{aligned}$$

This establishes our claim. ∎

One should note that Lemma 1.11 entails that the law of a subordinator is essentially characterized by that of g_τ, where τ is an independent exponential time. Specifically, if $\sigma^{(1)}$ and $\sigma^{(2)}$ are two subordinators such that, in the obvious notation, $g_\tau^{(1)}$ and $g_\tau^{(2)}$ have the same law, then there is a constant $c > 0$ such that $\Phi^{(1)} = c\Phi^{(2)}$. This observation will be quite useful in the sequel.

Chapter 2

Regenerative property

This chapter is mostly expository; its purpose is to stress the correspondence between a regenerative set, the range of a subordinator, and the set of times when a Markov process visits a fixed point. We refer to Blumenthal [21], Blumenthal and Getoor [23], Dellacherie *et al.* [44, 45], Kingman [100] and Sharpe [141] for background and much more on this topic.

2.1   Regenerative sets

The Markov property of a subordinator has a remarkable consequence on its range. First, note that for every $s \geq 0$, $L_s = \inf\{t \geq 0 : \sigma_t > s\}$ is an $(\mathcal{F}_t)$-stopping time, and the sigma-fields $(\mathcal{M}_s = \mathcal{F}_{L_s})_{s \geq 0}$ thus form a filtration. Because L is a continuous $(\mathcal{M}_s)$-adapted process that increases exactly on $\mathcal{R}$, the latter is an $(\mathcal{M}_s)$-progressive set. Then fix $s \geq 0$. An application of the Markov property at L_s shows that, conditionally on $\{L_s < \infty\}$, the shifted subordinator $\sigma' = \{\sigma_{L_s+t} - \sigma_{L_s}, t \geq 0\}$ is independent of $\mathcal{M}_s$ and has the same law as σ. Recall also from (1.6) that

$$\sigma(L_s) = D_s = \inf\{t > s : t \in \mathcal{R}\}$$

is the first-passage time in $\mathcal{R}$ after s. We thus see that conditionally on $\{D_s < \infty\}$, the shifted range

$$\mathcal{R} \circ \theta_{D_s} = \{v \geq 0 : v + D_s \in \mathcal{R}\} = \overline{\{\sigma'_t : t \geq 0\}}$$

is independent of $\mathcal{M}_s$ and is distributed as $\mathcal{R}$. This is usually referred to as the *regenerative property* of the range. We stress that the regenerative property of $\mathcal{R}$ does not merely hold at the first passage times D_s, but more generally at any $(\mathcal{M}_s)$-stopping time S which takes values in the subset of points in $\mathcal{R}$ which are not isolated on their right, a.s. on $\{S < \infty\}$. In that case, one can express S in the form $S = \sigma_T$, where $T = L_S$ is an $(\mathcal{F}_t)$-stopping time. Then conditionally on $\{L_S < \infty\}$, the shifted range $\mathcal{R} \circ \theta_S = \{v \geq 0 : v + S \in \mathcal{R}\}$ is independent of $\mathcal{M}_S = \mathcal{F}_T$ and is distributed as $\mathcal{R}$.

The regenerative property of the range of a subordinator motivates the definition of a regenerative set, that has been studied in particular by Krylov and Yushkevich,

Kingman, Hoffmann-Jørgensen and Maisonneuve. We refer to Fristedt [60] for a detailed survey including a connection with related concepts and a comprehensive list of references.

Consider a probability space endowed with a complete filtration $(\mathcal{M}_t)_{t \geq 0}$. Let S be a progressively measurable closed subset of $[0, \infty)$ which contains 0 and has no isolated point. We say that S is a perfect[1] *regenerative* set if for every $s \geq 0$, conditionally on $D_s := \inf \{t > s : t \in S\} < \infty$, the right-hand portion $S \circ \theta_{D_s}$ of S as viewed from D_s, is independent of $\mathcal{M}_{D_s}$ and has the same distribution S.

We have seen above that the range of a subordinator is a regenerative set; here is the converse.

Theorem 2.1 (Hoffmann-Jørgensen [74], Maisonneuve [109]) *Let S be a regenerative set.*

(i) *There is a subordinator σ such that $S = \mathcal{R} = \overline{\{\sigma_t : 0 \leq t < \zeta\}}$ a.s., and the inverse L of σ is an $(\mathcal{M}_t)$-adapted process.*

(ii) *If $\tilde{\sigma}$ is a second subordinator with range S, then there is a real number $c > 0$ such that $\tilde{\sigma}_t = \sigma_{ct}$ for all $t \geq 0$, a.s.*

We refer to Maisonneuve [109] or Chapter XX in Dellacherie *et al.* [45] for the proof.

With regards to Theorem 2.1, it will be convenient to use henceforth the notation $\mathcal{R}$ instead of S to designate a regenerative set. Plainly, if $\tilde{\sigma}$ is as in Theorem 2.1(ii), then $\tilde{\Phi}(\lambda) = c\Phi(\lambda)$ in the obvious notation. Hence, among the one-parameter family of subordinators having the range $\mathcal{R}$, there is a unique one for which Laplace exponent satisfies the -arbitrary- normalizing condition

$$\Phi(1) = 1 . \tag{2.1}$$

We refer to Φ as *the* Laplace exponent of $\mathcal{R}$

The inverse L of the subordinator σ is called the *local time* on $\mathcal{R}$, it can be constructed explicitly as a function of $\mathcal{R}$ as follows. Recall first from Proposition 1.8 that $\mathcal{R}$ is called heavy if it has a positive Lebesgue measure (or equivalently if the drift coefficient of σ is positive) and light otherwise. In the heavy case, one can express the local time as

$$L_t = \mathrm{d}^{-1} m \left([0, t] \cap \mathcal{R}\right) , \qquad t \geq 0$$

where m stands for the Lebesgue measure. In the light case, Fristedt and Pruitt [61] have obtained a remarkable analogue of Proposition 1.8. Specifically, they have been able to exhibit a deterministic measure m_H on $[0, \infty)$ (which is the Hausdorff measure associated with some increasing function) such that

$$L_t = m_H \left([0, t] \cap \mathcal{R}\right) , \qquad t \geq 0 .$$

---

[1] The qualification 'perfect' refers to the absence of isolated points and will be frequently omitted in the sequel in the sense that, for us, a regenerative set has no isolated points. This squares with the fact that compound Poisson processes have been ruled out in this text. For completeness, we mention that a closed random set that has the regenerative property and possesses at least one isolated point with positive probability, is in fact discrete a.s. and can be identified as the range of a compound Poisson process.

We refer to Greenwood and Pitman [68] and Fristedt and Taylor [64] for alternative constructions of the local time on a regenerative set.

We also stress the *additive property* of the local time: If S is an $(\mathcal{M}_s)$-stopping time which takes values in points in $\mathcal{R}$ with are not isolated on their right, then on $\{S < \infty\}$, the local time L' on $\mathcal{R}' = \mathcal{R} \circ \theta_S$ is given by

$$L'_t = L_{S+t} - L_S \qquad \text{for all } t \geq 0, \ a.s.$$

2.2 Connection with Markov processes

Consider some Polish space E and write $\mathcal{D}$ for the space of càdlàg paths valued in E, endowed with Skorohod's topology. Let $X = (\Omega, \mathcal{M}, \mathcal{M}_t, X_t, \theta_t, \mathbf{P}^x)$ be a strong Markov process with sample paths in $\mathcal{D}$. As usual, $\mathbf{P}^x$ refers to its law started at x, θ_t for the shift operator and $(\mathcal{M}_t)_{t \geq 0}$ for the filtration.

A point r of the state space is *regular for itself* if

$$\mathbf{P}^r \left(T_r = 0 \right) = 1 \,,$$

where $T_r = \inf\{t > 0 : X_t = r\}$ is the first hitting time of r. In words, r is regular for itself if the Markov process started at r, returns to r at arbitrarily small times, a.s. Applying the Markov property at the first return-time to r after a fixed time s, we see that the closure of set of times when X visits r,

$$\mathcal{R} = \overline{\{t \geq 0 : X_t = r\}}$$

is regenerative for $(\Omega, \mathcal{M}, \mathcal{M}_t, \mathbf{P}^r)$. (Conversely, it can be proved that any -perfect-regenerative set can be viewed as the closed set of times when some Markov process visits a regular point, see Horowitz [76].)

According to Theorem 2.1, $\mathcal{R}$ can thus be viewed as the range of some subordinator σ. The inverse L of σ is a continuous increasing process which increases exactly when X passes at r, in the sense that $\mathrm{Supp}\,(dL) = \mathcal{R}$, $\mathbf{P}^r$-a.s. One calls $L = (L_t, t \geq 0)$ the local time of X at r; its existence has been established originally by Blumenthal and Getoor [23], following the pioneering contribution of Lévy in the Brownian case.

The killing rate of the inverse local time has an obvious probabilistic interpretation in terms of the Markov process. One says that r is a *transient state* if $\mathcal{R}$ is bounded a.s., so that

$$r \text{ is a transient state} \iff \mathrm{k} > 0 \iff L_\infty < \infty \text{ a.s.} \tag{2.2}$$

More precisely, L_∞ has then an exponential distribution with parameter k. In the opposite case, $\mathcal{R}$ is unbounded a.s., and we say that r is a *recurrent state*.

We next present a simple criterion to decide whether a point is regular for itself, and in that case, give an explicit expression for the Laplace exponent of the inverse local time. This requires some additional assumption of duality type on the Markov process. Typically, suppose that $X = (\Omega, \mathcal{M}, \mathcal{M}_t, X_t, \theta_t, \mathbf{P}^x)$ and $\widehat{X} = \left(\Omega, \widehat{\mathcal{M}}, \widehat{\mathcal{M}}_t, \widehat{X}_t, \widehat{\theta}_t, \widehat{\mathbf{P}}^x \right)$

are two standard Markov processes with state space E. For every $\lambda > 0$, the λ-resolvent operators of X and $\widehat{X}$ are given by

$$V^\lambda f(x) = \mathbf{E}^x \left(\int_0^\infty f(X_t)e^{-\lambda t}dt \right) \quad , \quad \widehat{V}^\lambda f(x) = \widehat{\mathbf{E}}^x \left(\int_0^\infty f(\widehat{X}_t)e^{-\lambda t}dt \right) , \quad x \in E ,$$

where $f \geq 0$ is a generic measurable function on E. We recall that $f \geq 0$ is called λ-excessive with respect to $\{V^\alpha\}$ if $\alpha V^{\alpha+\lambda}f \leq f$ for every $\alpha > 0$ and $\lim_{\alpha \to \infty} \alpha V^\alpha f = f$ pointwise.

We suppose that X and $\widehat{X}$ are in duality with respect to some sigma-finite measure ξ. That is, the resolvent operators can be expressed in the form

$$V^\lambda f(x) = \int_E v^\lambda(x,y)f(y)\xi(dy) \quad , \quad \widehat{V}^\lambda f(x) = \int_E v^\lambda(y,x)f(y)\xi(dy) .$$

Here, $v^\lambda : E \times E \to [0,\infty]$ stands for the version of the resolvent density such that, for every $x \in E$, the function $v^\lambda(\cdot,x)$ is λ-excessive with respect to the resolvent $\{V^\alpha\}$, and the function $v^\lambda(x,\cdot)$ is λ-excessive with respect to the resolvent $\{\widehat{V}^\alpha\}$. Under a rather mild hypothesis on the resolvent density, one has the following simple necessary and sufficient condition for a point to be regular for itself (see e.g. Proposition 7.3 in [24]).

Proposition 2.2 *Suppose that for every $\lambda > 0$ and $y \in E$, the function $x \to v^\lambda(x,y)$ is lower-semicontinuous. Then, for each fixed $r \in E$ and $\lambda > 0$, the following assertions are equivalent:*

(i) *r is regular for itself.*

(ii) *For every $x \in E$, $v^\lambda(x,r) \leq v^\lambda(r,r) < \infty$.*

(iii) *The function $x \to v^\lambda(x,r)$ is bounded and continuous at $x = r$.*

Finally, if these assertions hold, then the Laplace exponent Φ of the inverse local time at r is given by

$$\Phi(\lambda) = v^1(r,r)/v^\lambda(r,r), \qquad \lambda > 0.$$

In the case when the semigroup of X is absolutely continuous with respect to ξ, the resolvent density can be expressed in the form

$$v^\lambda(x,y) = \int_0^\infty e^{-\lambda t}p_t(x,y)dt .$$

As the Laplace transform of the renewal measure U of the inverse local time at r is $1/\Phi(\lambda)$, a quantity that is proportional to $v^\lambda(r,r)$ by Proposition 2.2, we see by Laplace inversion that U is absolutely continuous with respect to the Lebesgue measure, with density u given by

$$u(t) = cp_t(r,r), \qquad t > 0.$$

Observe also that in this framework, Proposition 1.9 entails that for each fixed $t > 0$, the probability that $t \in \mathcal{R}$, that is that $X_t = r$, is proportional to $p_t(r,r)$ in the heavy case, and is zero in the light case. Of course, this easy fact can be also checked directly.

Suppose for instance that X is a real-valued Brownian motion. The resolvent density (with respect to the Lebesgue measure) is

$$v^\lambda(x,y) = \int_0^\infty e^{-\lambda t} \frac{1}{\sqrt{2\pi t}} \exp\left(-\frac{(x-y)^2}{2t}\right) dt = \frac{1}{\sqrt{2\lambda}} \exp\left\{-\sqrt{2\lambda}|x-y|\right\}.$$

This quantity depends symmetrically on x and y, so the dual process is simply $\widehat{X} = X$. Proposition 2.2 applies and shows that any $r \in \mathbb{R}$ is a regular point for itself, and the Laplace exponent of the inverse local time is always $\Phi(\lambda) = \sqrt{\lambda}$. More generally, when X is a so-called Bessel process of dimension $d \in (0,2)$ (see chapter XI in Revuz and Yor [132]), then $r = 0$ is a regular point and the inverse local time at 0 is a stable subordinator with index $\alpha = 1 - d/2$. Making use of the results of chapter 5, we see for instance that the fractal dimension (both lower and upper) of the zero set of a d-dimensional Bessel process is $1 - d/2$. Alternatively, when X is a stable Lévy process with index $\beta \in (1,2]$, then any $r \in \mathbb{R}$ is a regular point for itself and the inverse local time is always a stable subordinator with index $\alpha = 1 - 1/\beta$ (see the forthcoming Proposition 8.1).

We next turn our attention to one of the most important applications of the notion of local time to Markov processes, namely Itô's theory of excursions. This is a vast topic and we shall merely recall the basic result of Itô and refer to the literature for developments (cf. in particular Blumenthal [21], chapter XII in Revuz and Yor [132], chapter 8 in Rogers and Williams [137] and also Rogers [136] for an elementary approach).

Call excursion intervals the maximal open time-intervals on which $X \neq r$. In other words, the excursion intervals are those that appear in the canonical decomposition of the open set $[0,\infty) - \mathcal{R}$. We have already pointed out that those open intervals are precisely of the type $(\sigma(t-), \sigma(t))$ for the t's such that $\Delta_t > 0$. Itô used this observation and defined the excursion process $(e_t, t \geq 0)$ of X away from r, which is a process valued in the path-space $\mathcal{D}$ given by

$$e_t(s) = \begin{cases} X_{\sigma(t-)+s} & \text{if } 0 \leq s < \sigma(t) - \sigma(t-) \\ r & \text{otherwise} \end{cases}$$

Recall that a point process $(\xi_t : t \geq 0)$ with values in some metric-complete separable space is called a Poisson point process with characteristic measure μ if for every Borel set B, the counting process $N^B = \text{Card}\{t \in [0,\cdot] : \xi_t \in B\}$ is a Poisson process with intensity $\mu(B)$, and to disjoint Borel sets correspond independent counting processes. We are now able to state Itô's description of the excursions of a Markov process away from a point r; we focus for the sake of simplicity on the case when r is a regular recurrent state.

Theorem 2.3 (Itô [82]) *When r is a regular recurrent state, the excursion process $(e_t, t \geq 0)$ is a Poisson point process under $\mathbf{P}^r$. Its characteristic measure n is called Itô's excursion measure of X away from r.*

We henceforth suppose that r is a regular recurrent point. The excursion measure yields a very useful expression for the (essentially unique) invariant measure of X, which is well-known in the context of Markov chains. Specifically, let $\epsilon \in \mathcal{D}$ be a generic path; write $\rho(\epsilon) = \inf\{t > 0 : \epsilon(t) = r\}$ for its first-return time to r. The sigma-finite measure μ related to the occupation measure under Itô's excursion measure n by

$$\int f d\mu = \mathrm{d} f(r) + n \left(\int_0^{\rho(\epsilon)} f(\epsilon(t)) dt \right) ,$$

where d is the drift coefficient of the inverse local time σ and $f \geq 0$ any measurable function, is an invariant measure for X. We refer to Getoor [66] or to section XIX.46 in [45] for a proof, and to Maisonneuve [112] for some applications.

Recall that a recurrent Markov process is called *positive recurrent* if there is an invariant probability measure, and *null recurrent* otherwise. In our setting, we see that positive recurrence is equivalent to the integrability of the first-return time to r, ρ, under Itô's excursion measure. On the other hand, the very definition of the excursion process implies that $\rho(e_t) = \Delta_t$, i.e. the durations of the excursion process coincide with the lengths of the jumps of the inverse local time σ. In particular, the comparison between Theorem 2.3 and Proposition 1.3 shows that the distribution of ρ under n can be identified as the Lévy measure Π of σ:

$$n (\rho \in dt) = \Pi(dt).$$

In conclusion, we have the equivalence:

$$X \text{ is positive recurrent} \iff \mathbb{E}(\sigma_1) < \infty \iff \int_0^\infty \overline{\Pi}(x) \, dx < \infty. \qquad (2.3)$$

We now end this chapter by presenting a brief dictionary in which the main connections between subordinators, local times of Markov processes and regenerative sets are summarized.

Subordinator	$\sigma_t = L_t^{-1} = \inf\{s : L_s > t\}$
Local time	$L_t = \inf\{s : \sigma_s > t\}$
Lifetime	$\zeta = L_\infty$
Regenerative set	$\mathcal{R} = \overline{\{t \geq 0 : X_t = r\}} = \overline{\{\sigma_s : s \in [0, \zeta)\}} = \mathrm{Supp}(dL_t)$
First passage time	$D_t = \inf\{s > t : s \in \mathcal{R}\} = \sigma(L_t)$
Last passage time	$g_t = \sup\{s < t : s \in \mathcal{R}\} = \sigma(L_t-)$
Probability	$\mathbb{P} = \mathbf{P}^r$
Filtration	$\mathcal{F}_t = \mathcal{M}_{\sigma_t}$

Chapter 3

Asymptotic behaviour of last passage times

We are concerned with the process $(g_t : t > 0)$ of the last passage times in a regenerative set $\mathcal{R}$. When $\mathcal{R}$ is self-similar, $t^{-1}g_t$ always has a generalized arcsine law. In the general case, we consider the asymptotic behaviour of $t^{-1}g_t$ as t goes to ∞, first in distribution, and then pathwise. Special properties of the jump process of a subordinator play a key part in this study.

3.1 Asymptotic behaviour in distribution

3.1.1 The self-similar case

We say that a regenerative set $\mathcal{R}$ is *self-similar* if for every $k > 0$, it has the same distribution as $k\mathcal{R}$. If we think of $\mathcal{R}$ as the range of a subordinator σ, this is equivalent to the condition that the Laplace exponent of σ is proportional to that of $k\sigma$, i.e. $\Phi(\lambda) = c_k\Phi(k\lambda)$ for every $\lambda \geq 0$, where $c_k > 0$ is some constant that depends only on k. Due to the normalization $\Phi(1) = 1$, this holds if and only if $\Phi(\lambda) = \lambda^\alpha$ for some $\alpha \in [0,1]$, that is if σ is a standard stable subordinator of index α. The cases $\alpha = 0$ and $\alpha = 1$ are somewhat degenerate, as they corresponds to the situation where $\mathcal{R} = \{0\}$ and $\mathcal{R} = [0, \infty)$ a.s., respectively; we shall exclude them in the sequel.

Recall that $g_t = \sup\{s < t : s \in \mathcal{R}\}$ denotes the last passage time in $\mathcal{R}$ before time $t > 0$. When $\mathcal{R}$ is self-similar, the distribution of $t^{-1}g_t$ does not depend on $t > 0$ and can be given explicitly in terms of α.

Proposition 3.1 *Suppose that* $\Phi(\lambda) = \lambda^\alpha$ *for some* $0 < \alpha < 1$. *Then* g_1 *has the so-called generalized arcsine law, that is*

$$\mathbb{P}(g_1 \in ds) = \frac{s^{\alpha-1}(1-s)^{-\alpha}}{\Gamma(\alpha)\Gamma(1-\alpha)}ds = \frac{\sin \alpha\pi}{\pi}s^{\alpha-1}(1-s)^{-\alpha}\,ds \qquad (0 < s < 1).$$

For instance, when $\mathcal{R} = \{t : B_t = 0\}$ is the zero set of a one-dimensional Brownian motion B started at 0, we have $\Phi(\lambda) = \sqrt{\lambda}$ (the absence of the usual factor $\sqrt{2}$ is due to the normalization (2.1) of the local time) and one gets

$$\mathbb{P}(g_1 \le t) = \frac{2}{\pi} \arcsin \sqrt{t} \qquad (t \in [0,1]).$$

This is the celebrated first arcsine theorem of Paul Lévy; see e.g. Exercise III.3.20 in [132] for a direct proof. Proposition 3.1 also applies to the particular cases when one replaces the Brownian motion B by a Bessel process of dimension $d \in (0,2)$ (then $\alpha = 1 - d/2$), or a stable Lévy process with index $\beta \in (1,2]$ (then $\alpha = 1 - 1/\beta$).

We now proceed to the proof of Proposition 3.1.

Proof: The Laplace transform of the renewal measure is given by

$$\int_0^\infty e^{-\lambda x} U(dx) = \lambda^{-\alpha} = \frac{1}{\Gamma(\alpha)} \int_0^\infty e^{-\lambda x} x^{\alpha-1} dx$$

and that of the tail of the Lévy measure by

$$\int_0^\infty e^{-\lambda x} \overline{\Pi}(x) dx = \lambda^{\alpha-1} = \frac{1}{\Gamma(1-\alpha)} \int_0^\infty e^{-\lambda x} x^{-\alpha} dx$$

We conclude by Laplace inversion and Lemma 1.10. ∎

The distribution of the first-passage time $D_t = \inf\{s > t : s \in \mathcal{R}\}$ readily follows from Proposition 3.1 (still in the case when $\Phi(\lambda) = \lambda^\alpha$). Specifically, for every $0 < s < t$, we have

$$g_t \le s \iff \mathcal{R} \cap (s,t) = \emptyset \iff D_s \ge t.$$

An application of the scaling property then yields for $t > 1$

$$\mathbb{P}(D_1 \ge t) = \mathbb{P}(D_{1/t} \ge 1) = \mathbb{P}(g_1 \le 1/t) = \frac{\sin \alpha \pi}{\pi} \int_0^{1/t} s^{\alpha-1}(1-s)^{-\alpha} ds,$$

and we deduce that the distribution of D_1 is given by

$$\mathbb{P}(D_1 \in dt) = \frac{\sin \alpha \pi}{\pi} t^{-1} (t-1)^{-\alpha} dt, \qquad t > 1.$$

Finally we refer to Pitman and Yor [126, 127, 129] and the references therein for further recent remarkable results about the interval partitions of $[0,\infty)$ induced by self-similar regenerative sets.

3.1.2 The Dynkin-Lamperti theorem

We next turn our attention to the asymptotic behaviour of the last passage time in the case when $\mathcal{R}$ is not necessarily self-similar. Informally, the rescaled set $t^{-1}\mathcal{R}$ is the range of the subordinator $t^{-1}\sigma$; its Laplace exponent is thus $\Phi_t(q) = \Phi(q/t)/\Phi(1/t)$, due to (2.1). This quantity converges as $t \to \infty$ if and only if Φ is regularly varying at 0+, and then the limit is q^α for some $\alpha \in [0,1]$, that is the Laplace exponent of a stable subordinator with index α. In view of Proposition 3.1, one naturally expects that $t^{-1}g_t$ should then converge in distribution towards the generalized arcsine law with parameter α. The Dynkin-Lamperti theorem not only provides a rigorous setting to this informal argument, but also states a converse.

Theorem 3.2 (Dynkin [50], Lamperti [106]) *The following assertions are equivalent:*

(i) $t^{-1}g_t$ *converges in law as* $t \to \infty$.

(ii) $\lim_{t \to \infty} t^{-1}\mathbb{E}(g_t) = \alpha \in [0,1]$.

(iii) $\lim_{q \to 0+} q\Phi'(q)/\Phi(q) = \alpha \in [0,1]$.

(iv) Φ *is regularly varying at* 0+ *with index* $\alpha \in [0,1]$.

Moreover, when these assertions hold, then the limit distribution of $t^{-1}g_t$ *is the Dirac point mass at* 0 *(respectively, at* 1*) for* $\alpha = 0$ *(respectively,* $\alpha = 1$*); and for* $0 < \alpha < 1$, *the generalized arcsine law of parameter* α *that appears in Proposition 3.1.*

There is also a similar result for small times; more precisely a true statement is obtained after exchanging the rôles of 0+ and ∞. We also mention that, more generally, the limit behaviour in distribution of the pair (g_t, D_t) can be studied, using essentially the same arguments as below.

Proof: (i) $\Longrightarrow$ (ii) is obvious as $g_t/t \leq 1$.

(ii) $\Longrightarrow$ (iii) On the one hand, we know from Lemma 1.11 that

$$\int_0^\infty e^{-qt}\mathbb{E}(g_t)dt = \frac{\Phi'(q)}{q\Phi(q)}.$$

On the other hand, we see by an Abelian theorem that (ii) entails

$$\lim_{q \to 0+} q^2 \int_0^\infty e^{-qt}\mathbb{E}(g_t)dt = \alpha.$$

(iii) $\Longrightarrow$ (iv) When (iii) holds, the logarithmic derivative of $t \to t^{-\alpha}\Phi(t)$ can be expressed as $t \to \varepsilon(t)/t$, with $\lim_{t \to 0+} \varepsilon(t) = 0$. That is

$$t^{-\alpha}\Phi(t) = c \exp\left\{\int_t^1 \frac{\varepsilon(s)}{s} ds\right\}.$$

According to the representation theorem of slowly varying functions (see e.g. [20]), this shows that $t \to t^{-\alpha}\Phi(t)$ is slowly varying at 0+, and hence Φ is regularly varying at 0+ with index α.

(iv) $\Longrightarrow$ (i) Suppose first that (iv) holds with $0 < \alpha < 1$. According to Proposition 1.5, we have

$$\lim_{t \to \infty} U(tx)\Phi(1/t) = \frac{x^\alpha}{\Gamma(1+\alpha)} \qquad \text{uniformly for } x \in K, \tag{3.1}$$

and

$$\lim_{t \to \infty} \frac{\overline{\Pi}(tx)}{\Phi(1/t)} = \frac{x^{-\alpha}}{\Gamma(1-\alpha)} \qquad \text{uniformly for } x \in K, \tag{3.2}$$

where K stands for a generic compact subset on $(0, \infty)$.

Next, fix $0 < a < b < 1$. According to Lemma 1.10, we have

$$\mathbb{P}(at \leq g_t < bt) = \int_{[at,bt)} \overline{\Pi}(t-s)dU(s) = \int_{[a,b)} \overline{\Pi}(t(1-u))dU(tu)$$

$$= \int_{[a,b)} \frac{\overline{\Pi}(t(1-u))}{\Phi(1/t)}d\left(U(tu)\Phi(1/t)\right).$$

Applying (3.1) and (3.2), we deduce that

$$\lim_{t \to \infty} \mathbb{P}(at \leq g_t < bt) = \int_a^b \frac{(1-s)^{-\alpha} s^{\alpha-1}}{\Gamma(\alpha)\Gamma(1-\alpha)} ds.$$

In words, $t^{-1}g_t$ converges in distribution to the generalized arcsine law with parameter α.

An easy variation of this argument applies for $\alpha = 0$, but not for $\alpha = 1$ (the quantity $\Gamma(1-\alpha)$ in (3.2) is then infinite). So suppose that $\alpha = 1$, take any $a \in (0,1)$ and observe from Lemma 1.10 that

$$\mathbb{P}(t^{-1}g_t < a) = \int_{[0,ta)} \overline{\Pi}(t-u)U(du) \leq \overline{\Pi}(t(1-a))U(ta).$$

A Tauberian theorem applied to the Lévy-Khintchine formula now gives

$$I(s) \sim s\Phi(1/s) \qquad \text{as } s \to \infty, \tag{3.3}$$

where I is the integrated tail of the Lévy measure. In particular I is slowly varying. The inequality

$$I(s) - I(s/2) = \int_{s/2}^s \overline{\Pi}(t)dt \geq s\overline{\Pi}(s)/2$$

and the fact that I is slowly varying entail that $\overline{\Pi}(s) = o(I(s)/s) = o(\Phi(1/s))$. Using Proposition 1.4 and (3.3) gives

$$\lim_{t \to \infty} \mathbb{P}(t^{-1}g_t < a) = 0,$$

and the proof of Theorem 3.2 is complete. ∎

Theorem 3.2 is essentially an application of the estimates of Proposition 1.4 for the tail of the Lévy measure and the renewal measure. In the same vein, the renewal theorem readily yields the following well-known limit theorem.

Proposition 3.3 *Suppose that* $\mathbb{E}(\sigma_1) = \mu < \infty$. *Then*

$$\lim_{t \to \infty} \mathbb{P}(t - g_t \in ds) = \frac{1}{\mu}\overline{\Pi}(s)ds, \qquad s > 0,$$

and

$$\lim_{t \to \infty} \mathbb{P}(t - g_t = 0) = \mathrm{d}/\mu.$$

Proof: This is an easy application of the renewal theorem (see Proposition 1.6) and Lemma 1.10. ∎

3.2 Asymptotic sample path behaviour

The purpose of this section is to investigate the almost-sure asymptotic behaviour of the last-passage-time process; here is the main result (see also [9]).

Theorem 3.4 Let $f : (0, \infty) \to (0, \infty)$ be a continuous strictly increasing function with $\lim_{t \to \infty} f(t)/t = 0$ and $\liminf_{t \to \infty} f(t)/f(2t) > 0$. Then, with probability one,

$$\liminf_{t \to \infty} g_t/f(t) = 0 \ or \ \infty$$

according as the integral

$$\int_{[1,\infty)} U(f(t)) \, \Pi(dt) \tag{3.4}$$

diverges or converges.

When we specialize Theorem 3.4 to the case when $\mathcal{R}$ is the zero set of a one-dimensional Brownian motion, we get $\liminf_{t \to \infty} g_t/f(t) = 0$ or ∞ a.s. according as the integral $\int^{\infty} \sqrt{f(t)} t^{-3} \, dt$ diverges or converges. In particular,

$$\liminf_{t \to \infty} \frac{g_t \log^2 t}{t} = 0 \quad \text{and} \quad \lim_{t \to \infty} \frac{g_t \log^{2+\varepsilon} t}{t} = \infty \quad a.s.$$

for any $\varepsilon > 0$. This result goes back to Chung and Erdős [36], see also Hobson [73] and Hu and Shi [79] for recent developments in the same vein.

Checking Theorem 3.4 when the killing rate $\mathbf{k}$ is positive, is straightforward. Indeed, $\mathcal{R}$ is then bounded, and so is g_t a.s. On the other hand, the renewal measure is also bounded and the integral (3.4) always converges. So with no loss of generality, we may assume henceforth that $\mathbf{k} = 0$. The proof of Theorem 3.4 relies on two simple properties of subordinators. Informally, we have to compare the relative size of a subordinator and its jumps. Our first lemma reduces this comparison to that of certain integrals. Recall Proposition 1.3.

Lemma 3.5 For every Borel function $b : [0, \infty) \to [1, \infty)$, the events

$$\{\Delta_t > b(\sigma_{t-}) \text{ infinitely often as } t \to \infty\}$$

and

$$\left\{ \int_0^{\infty} \overline{\Pi} \circ b(\sigma_t) dt = \infty \right\}$$

coincide up to a set of probability zero.

Proof: This is a variant of the Lévy-Borel-Cantelli lemma. Specifically, the fact that the jump process Δ is a Poisson point process with characteristic measure Π entails that the compensated sum

$$\sum_{s \leq t} \mathbf{1}_{\{\Delta_s > b(\sigma_{s-})\}} - \int_0^t \overline{\Pi} \circ b(\sigma_s) ds \qquad (t \geq 0)$$

is a martingale. On the event

$$\{\Delta_t > b(\sigma_{t-}) \text{ infinitely often as } t \to \infty\} \cap \left\{\int_0^\infty \overline{\Pi} \circ b(\sigma_t)dt < \infty\right\},$$

this martingale converges to ∞; whereas on the event

$$\{\Delta_t \le b(\sigma_{t-}) \text{ for all sufficiently large } t\} \cap \left\{\int_0^\infty \overline{\Pi} \circ b(\sigma_t)dt = \infty\right\},$$

it converges to $-\infty$. As the jumps of this martingale are bounded by 1, both events have probability zero (see e.g. the corollary on page 484 in [144]). ∎

Motivated by the preceding lemma, we then establish an easy result on the convergence of integrals of a subordinator.

Lemma 3.6 *Let $h : [0, \infty) \to [0, \infty)$ be a decreasing function. The following assertions are equivalent.*

(i)
$$\int_0^\infty h(x)U(dx) < \infty$$

(ii)
$$\mathbb{P}\left(\int_0^\infty h(\sigma_t)dt < \infty\right) = 1$$

(iii)
$$\mathbb{P}\left(\int_0^\infty h(\sigma_t)dt < \infty\right) > 0$$

Proof: The derivations (i)⇒(ii)⇒(iii) are obvious. Suppose that (iii) holds and pick $\varepsilon > 0$ and $k > 0$ such that

$$\mathbb{P}\left(\int_0^\infty h(\sigma_t)dt < k\right) > \varepsilon.$$

Next, consider for every integer $n > 0$ the stopping time

$$T_n = \inf\left\{t : \int_0^t h(\sigma_s)ds \ge kn\right\},$$

and apply the Markov property (Proposition 1.1) at time T_n. We see that conditionally on $\{T_n < \infty\}$, the process $\sigma' = \sigma_{T_n+} - \sigma_{T_n}$ is a subordinator distributed as σ. Then, using the hypothesis that h decreases, we get

$$
\begin{aligned}
\mathbb{P}(T_{n+1} = \infty \mid T_n < \infty) &= \mathbb{P}\left(\int_{T_n}^\infty h(\sigma_t)dt < k \mid T_n < \infty\right) \\
&= \mathbb{P}\left(\int_0^\infty h(\sigma_t' + \sigma_{T_n})dt < k \mid T_n < \infty\right) \\
&\ge \mathbb{P}\left(\int_0^\infty h(\sigma_t')dt < k \mid T_n < \infty\right) \\
&= \mathbb{P}\left(\int_0^\infty h(\sigma_t)dt < k\right) > \varepsilon.
\end{aligned}
$$

This shows that $k^{-1} \int_0^\infty h(\sigma_t)dt$ is bounded from above by a geometric variable. As a consequence, it has finite expectation and (i) follows. ∎

We point out that when one specializes Lemma 3.6 to the case when σ is a stable subordinator with index $1/2$, one recovers a result of Donati-Martin, Rajeev and Yor (Theorem 6.2 in [46] and Theorem 1.3 in [131]) on the a.s. convergence of certain integrals involving the Brownian local time. Theorem 3.4 now follows readily from Lemmas 3.5 and 3.6.

Proof of Theorem 3.4: Write f^{-1} for the inverse function of f, so $f(\Delta_t) > \sigma_{t-}$ if and only if $\Delta_t > f^{-1}(\sigma_{t-})$. An immediate combination of Lemmas 3.6 and 3.5 shows that

$$\mathbb{P}\left(f(\Delta_t) > \sigma_{t-} \text{ infinitely often as } t \to \infty\right) = 0 \text{ or } 1$$

according as the integral $\int^\infty \overline{\Pi} \circ f^{-1}(x)dU(x)$ converges or diverges. By a change of variables and an integration by parts, the latter is equivalent to the integral (3.4) being finite or infinite. Next, recall that $g_t = \sigma(L_t-)$ for all $t \geq 0$ a.s. It follows that $f(\Delta_t) > \sigma_{t-}$ infinitely often as $t \to \infty$ if and only if $f(t - g_t) > g_t$ infinitely often. We deduce that a.s.,

$$\liminf_{t \to \infty} g_t/f(t - g_t) \geq 1 \text{ or } \leq 1$$

according as (3.4) converges or diverges.

First, assume that (3.4) diverges. By the subadditivity of the renewal function, the same holds when f is replaced by εf for an arbitrary $\varepsilon \in (0,1)$. It follows that $\liminf_{t\to\infty} g_t/f(t - g_t) = 0$ a.s., and because f increases, we conclude that $\liminf_{t\to\infty} g_t/f(t) = 0$ a.s.

Finally, assume that (3.4) converges. By the same argument based on the subadditivity of the renewal function as above, we have that $\lim_{t\to\infty} g_t/f(t - g_t) = \infty$ a.s. It is then straightforward to derive from the assumptions $\lim_{t\to\infty} f(t)/t = 0$ and $\liminf_{t\to\infty} f(t)/f(2t) > 0$ that $\lim_{t\to\infty} g_t/f(t) = \infty$ a.s. (simply distinguish the cases $g_t \leq t/2$ and $g_t > t/2$). ∎

We now conclude this chapter with an interesting application of the techniques developed so far to the case when the regenerative set is given in the form $\mathcal{R} = \overline{\{t \geq 0 : X_t = r\}}$, where X is some Markov process started from a regular point r. Theorem 3.2 provides a necessary and sufficient condition for g_t/t to converge in probability; and it is natural to ask whether the convergence then holds almost surely. To this end, the equivalence

$$\lim_{t \to \infty} g_t/t = 0 \quad \text{a.s.} \quad \Longleftrightarrow \quad r \text{ is a transient state}$$

is obvious (if r is a recurrent state, then $g_t = t$ infinitely often). The problem of the convergence towards 1 is less obvious. Its solution is essentially a variation of a result of Kesten on the asymptotic behaviour of the largest step of increasing random walks.

Proposition 3.7 (Kesten [97]) *The following assertions are equivalent:*

(i) $\lim_{t\to\infty} g_t/t = 1$ *a.s.*

(ii) $\mathbb{P}\left(\liminf_{t\to\infty} g_t/t > 0\right) > 0.$

(iii) *The Markov process X is positive recurrent.*

Proof: (i) $\Leftrightarrow$ (ii) It is immediate to see that (i) holds if and only if for every $\varepsilon > 0$, $\Delta_t \le \varepsilon \sigma_{t-}$ for all sufficiently large t, a.s. By Lemmas 3.5 and 3.6, we deduce that

$$\text{(i)} \iff \int^\infty \overline{\Pi}(\varepsilon t)dU(t) < \infty \quad \text{for every } \varepsilon > 0.$$

Similarly, (ii) holds if and only if the event $\{\Delta_t \le k\sigma_{t-}$ for all sufficiently large $t\}$ has positive probability for some $k < \infty$. Again by Lemmas 3.5 and 3.6, we deduce that

$$\text{(ii)} \iff \int^\infty \overline{\Pi}(kt)dU(t) < \infty \quad \text{for some } k < \infty.$$

Because the renewal function is subadditive, an integration by parts now shows that (i) and (ii) are equivalent.

(i) $\Leftrightarrow$ (iii) Let us exclude the degenerate case when σ is a pure drift, and recall from Proposition 1.4 that then $U(t) \asymp t/I(t)$ as $t \to \infty$ where I stands for the integrated tail of the Lévy measure. On the other hand, we know from the preceding argument that

$$\text{(i)} \iff \int^\infty \overline{\Pi}(t)dU(t) < \infty \iff \int^\infty \frac{t\Pi(dt)}{I(t)} < \infty,$$

where the second equivalence follows from an integration by parts.

Recall from (2.3) that X is positive recurrent if and only if $\mathbb{E}(\sigma_1) < \infty$, that is if and only if $I(\infty) = \int_0^\infty \overline{\Pi}(t)dt = \int_{(0,\infty)} t\Pi(dt) < \infty$. Because I is an increasing function, it is plain that (i) holds in this case.

We next suppose that (i) holds. It is immediately checked that the mapping $t \to t/I(t)$ increases, and the convergence of the preceding integral thus forces $t\overline{\Pi}(t) = o(I(t))$. An integration by parts shows that

$$\int^\infty \overline{\Pi}(t)\left(\frac{1}{I(t)} - \frac{t\overline{\Pi}(t)}{I(t)^2}\right)dt < \infty,$$

and hence we must have $\int^\infty \overline{\Pi}(t)I^{-1}(t)dt < \infty$. The latter is clearly equivalent to $I(\infty) < \infty$, that is to (iii). ∎

In the positive recurrent case, an application of Lemmas 3.5 and 3.6 and the renewal theorem (Proposition 1.6) shows that the sample path behaviour of the last passage time process is specified as follows: For every increasing function $f : (0,\infty) \to (0,\infty)$

$$\mathbb{P}\left(t - g_t > f(t) \text{ infinitely often as } t \to \infty\right) = 0 \text{ or } 1$$

according as the integral $\int^\infty \overline{\Pi} \circ f(t)dt$ converges or diverges.

Chapter 4

Rates of growth of local time

We present the remarkable law of the iterated logarithm for the local time due to
Fristedt and Pruitt, and also investigate the modulus of continuity of the local time
on a path. The independence and stationarity of the increments of a subordinator are
the key to the proper application of the Borel-Cantelli lemma.

4.1 Law of the iterated logarithm

The main result of this section is the following version of the law of the iterated
logarithm for local times.

Theorem 4.1 (Fristedt and Pruitt [61] [1]) *There exists a positive and finite constant*
c_Φ *such that*

$$\limsup_{t \to 0+} \frac{L_t \Phi \left(t^{-1} \log \log \Phi(t^{-1}) \right)}{\log \log \Phi(t^{-1})} = c_\Phi \qquad a.s.$$

The exact value of c_Φ does not seem to be known explicitly in general. When Φ is
regularly varying with index $\alpha \in [0, 1]$ at ∞, then $c_\Phi = c_\alpha$, where

$$c_\alpha = \alpha^{-\alpha}(1 - \alpha)^{-(1-\alpha)}, \tag{4.1}$$

with the convention $0^{-0} = 1$; see Barlow, Perkins and Taylor [5], or [8]. The sharpest
result related to Theorem 4.1 is in Pruitt [130].

There is also a version of Theorem 4.1 for large times, which follows from a simple
variation of the arguments for small times. Specifically, suppose that the killing rate
is $\mathbf{k} = 0$. Then there exists $c'_\Phi \in (0, \infty)$ such that

$$\limsup_{t \to \infty} \frac{L_t \Phi \left(t^{-1} \log | \log \Phi(t^{-1})| \right)}{\log | \log \Phi(t^{-1})|} = c'_\Phi \qquad a.s. \tag{4.2}$$

---

[1]Theorem 4.1 is slightly more explicit than the result stated in [61]. Specifically, the normalizing
function there is the inverse function of $t \to \varphi \left(t^{-1} \log \log \varphi(t^{-1}) \right)^{-1} \log \log \varphi(t^{-1})$, where φ denotes
the inverse function of Φ. However, after some tedious calculation, one can check that the normalizing
function in [61] and that in Theorem 4.1 are of the same order, and therefore the two statements
agree.

When L is the local time at a regular point for some recurrent Markov process, the ergodic theorem asserts that if A is a positive additive functional associated with a measure μ with finite mass, then $A_t \sim \mu(E)L_t$ as $t \to \infty$, a.s. A law of the iterated logarithm for A thus follows from (4.2). Further developments in the direction of a second order law, were made recently by Csáki *et al.* [39], Marcus and Rosen [115, 116], Bertoin [8], Khoshnevisan [98]...

The proof of Theorem 4.1 relies on two technical lemmas. We write

$$f(t) = \frac{\log \log \Phi(t^{-1})}{\Phi\left(t^{-1} \log \log \Phi(t^{-1})\right)}, \qquad t \text{ small enough,}$$

and denote the inverse function of Φ by φ.

Lemma 4.2 *For every integer $n \geq 2$, put*

$$t_n = \frac{\log n}{\varphi(e^n \log n)} \quad, \quad a_n = f(t_n).$$

(i) *The sequence $(t_n : n \geq 2)$ decreases, and we have $a_n \sim e^{-n}$.*

(ii) *The series $\Sigma \mathbb{P}\left(L_{t_n} > 3a_n\right)$ converges*

Proof: (i) The first assertion follows readily from the fact that φ is convex and increasing. On the one hand, since Φ increases, we have for $n \geq 3$

$$\Phi(t_n^{-1}) = \Phi(\varphi(e^n \log n)/\log n) \leq \Phi(\varphi(e^n \log n)) = e^n \log n.$$

On the other hand, since Φ is concave, we have for $n \geq 3$

$$\Phi(t_n^{-1}) = \Phi(\varphi(e^n \log n)/\log n) \geq \Phi(\varphi(e^n \log n))/\log n = e^n.$$

This entails

$$\log \log \Phi(t_n^{-1}) \sim \log n \tag{4.3}$$

and then

$$t_n^{-1} \log \log \Phi(t_n^{-1}) \sim \varphi(e^n \log n).$$

Note that if $\alpha_n \sim \beta_n$, then $\Phi(\alpha_n) \sim \Phi(\beta_n)$ (because Φ is concave and increasing). We deduce that

$$\Phi\left(t_n^{-1} \log \log \Phi(t_n^{-1})\right) \sim e^n \log n, \tag{4.4}$$

and our assertion follows from (4.3).

(ii) The probability of the event $\{L_{t_n} > 3a_n\} = \{\sigma_{3a_n} < t_n\}$ is bounded from above by

$$\exp\{\lambda t_n\}\mathbb{E}\left(\exp\{-\lambda \sigma_{3a_n}\}\right) = \exp\{\lambda t_n - 3a_n \Phi(\lambda)\}$$

for every $\lambda \geq 0$. We choose $\lambda = \varphi(e^n \log n)$; so $\Phi(\lambda) = e^n \log n$ and $\lambda t_n = \log n$. Our statement follows now from (i). ∎

Lemma 4.3 *For every integer $n \geq 2$, put*

$$s_n = \frac{2 \log n}{\varphi(2 \exp\{n^2\} \log n)} \quad, \quad b_n = f(s_n).$$

(i) *We have $b_n \sim \exp\{-n^2\}$.*

(ii) *The series $\Sigma \mathbb{P}\left(\sigma(b_n/3) < 2s_n/3\right)$ diverges*

Proof: (i) Just note that $s_n = t_{n^2}$ and apply Lemma 4.2(i).

(ii) For every b, s and $\lambda \geq 0$, we have

$$\mathbb{P}(\sigma_b \geq s) \leq \left(1 - e^{-\lambda s}\right)^{-1} \mathbb{E}\left(1 - \exp\{-\lambda \sigma_b\}\right),$$

which entails

$$\mathbb{P}(\sigma_b < s) \geq \frac{e^{-b\Phi(\lambda)} - e^{-\lambda s}}{1 - e^{-\lambda s}}. \tag{4.5}$$

Apply this to $b = b_n/3$, $s = 2s_n/3$ and $\lambda = \varphi(2\exp\{n^2\}\log n)$, and observe that then $\Phi(\lambda) = 2\exp\{n^2\}\log n$, $\lambda s = \frac{4}{3}\log n$ and $b\Phi(\lambda) \sim \frac{2}{3}\log n$ (by (i)). In particular $e^{-b\Phi(\lambda)} \geq n^{-3/4}$ for every sufficiently large n; we thus obtain

$$2\mathbb{P}\left(\sigma(b_n/3) < 2s_n/3\right) \geq \frac{n^{-3/4} - n^{-4/3}}{1 - n^{-4/3}},$$

and our claim follows. ∎

We are now able to establish the law of the iterated logarithm, using a standard method based on the Borel-Cantelli lemma.

Proof of Theorem 4.1: 1. To prove the upper-bound, we use the notation of Lemma 4.2. Take any $t \in [t_{n+1}, t_n]$, so, provided that n is large enough

$$f(t) \geq \frac{\log\log \Phi(t_n^{-1})}{\Phi(t_{n+1}^{-1} \log\log \Phi(t_{n+1}^{-1}))}$$

(because Φ increases). By (4.3), the numerator is equivalent to $\log n$, and, by (4.4), the denumerator to $e^{n+1}\log(n+1)$. By Lemma 4.2, we thus have

$$\limsup_{t \to 0+} f(t_n)/f(t) \leq e.$$

On the other hand, an application of the Borel-Cantelli to Lemma 4.2 shows that

$$\limsup_{n \to \infty} L_{t_n}/f(t_n) \leq 3 \qquad a.s.$$

and we deduce that

$$\limsup_{t \to 0+} \frac{L_t}{f(t)} \leq \left(\limsup_{n \to \infty} \frac{L_{t_n}}{f(t_n)}\right)\left(\limsup_{t \to 0+} \frac{f(t_n)}{f(t)}\right) \leq 3e \qquad a.s.$$

2. To prove the lower-bound, we use the notation of Lemma 4.3 and observe that the sequence $(b_n, n \geq 2)$ decreases ultimately (by Lemma 4.3(i)). First, by Lemma 4.3(ii), we have

$$\sum \mathbb{P}\left(\sigma(b_n/3) - \sigma(b_{n+1}/3) < 2s_n/3\right) \geq \sum \mathbb{P}\left(\sigma(b_n/3) < 2s_n/3\right) = \infty;$$

so by the Borel-Cantelli lemma for independent events,

$$\liminf_{n \to \infty} \frac{\sigma(b_n/3) - \sigma(b_{n+1}/3)}{s_n} \leq \frac{2}{3}.$$

If we admit for a while that

$$\limsup_{n\to\infty} \frac{\sigma(b_{n+1}/3)}{s_n} \le \frac{1}{4},\tag{4.6}$$

we can conclude that

$$\liminf_{n\to\infty} \frac{\sigma(b_n/3)}{s_n} < \frac{11}{12}.$$

This implies that the set $\{s : \sigma(f(s)/3) < s\}$ is unbounded a.s. Plainly, the same then holds for $\{s : L_s > f(s)/3\}$, and as a consequence:

$$\limsup_{t\to 0+} L_t/f(t) \ge 1/3 \qquad a.s.\tag{4.7}$$

Now we establish (4.6). The obvious inequality (which holds for any $\lambda > 0$)

$$\mathbb{P}\left(\sigma(b_{n+1}/3) > s_n/4\right) \le (1 - \exp\{-\lambda s_n/4\})^{-1}\, \mathbb{E}\left(1 - \exp\{-\lambda\sigma(b_{n+1}/3)\}\right)$$

entails for the choice

$$\lambda = \varphi(2\exp\{n^2\}\log n) = \frac{2\log n}{s_n}$$

that

$$\mathbb{P}\left(\sigma(b_{n+1}/3) > s_n/4\right) \le \frac{2b_{n+1}\exp\{n^2\}\log n}{3\left(1 - \exp\{-\frac{1}{2}\log n\}\right)}.$$

By Lemma 4.3(i), the numerator is bounded from above for every sufficiently large n by

$$3\exp\{n^2 - (n+1)^2\}\log n \le e^{-n}$$

and the denumerator is bounded away from 0. We deduce that the series

$$\sum \mathbb{P}\left(\sigma(b_{n+1}/3) > s_n/4\right)$$

converges, and the Borel-Cantelli lemma entails (4.6). The proof of (4.7) is now complete.

3. The two preceding parts show that

$$\limsup_{t\to 0+} L_t/f(t) \in [1/3, 3e] \qquad a.s.$$

By the Blumenthal zero-one law, it must be a constant number c_Φ, a.s. ∎

To conclude this section, we mention that the independence and homogeneity of the increments of the inverse local time are also very useful in investigating the class of lower functions for the local time. We now state without proof the main result in that field, which has been proven independently by Fristedt and Skorohod. See [57], [67], or Theorem III.9 in [11], where the result is given in terms of the rate of growth of the subordinator.

Proposition 4.4 (i) *When* $d > 0$, *one has* $\lim_{t\to 0+} L_t/t = 1/d$ *a.s.*

(ii) *When* $d = 0$ *and* $f : [0,\infty) \to [0,\infty)$ *is an increasing function such that* $t \to f(t)/t$ *decreases, one has*

$$\liminf_{t\to 0+} L_t/f(t) = 0 \quad a.s. \quad \Longleftrightarrow \quad \int_{0+} f(x)\Pi(dx) = \infty.$$

Moreover, if these assertions fail, then $\lim_{t\to 0+} L_t/f(t) = \infty$ *a.s.*

4.2 Modulus of continuity

Once a law of the iterated logarithm has been established for a continuous process, it is natural to look for information on its modulus of continuity. Again we have a general result that holds for any local time of a Markov process.

Theorem 4.5 *For every $T > 0$, we have a.s.*

$$\limsup_{t \to 0+} \left\{ \sup_{0 \le \tau \le T} \frac{(L_{\tau+t} - L_\tau) \, \Phi\left(t^{-1} \log \Phi(t^{-1})\right)}{\log \Phi(t^{-1})} \right\} \le 12$$

and

$$\liminf_{t \to 0+} \left\{ \sup_{0 \le \tau \le T} \frac{(L_{\tau+t} - L_\tau) \, \Phi\left(t^{-1} \log \Phi(t^{-1})\right)}{\log \Phi(t^{-1})} \right\} \ge 1/6$$

Theorem 4.5 has been obtained in a less explicit form by Fristedt and Pruitt [62], following an earlier work of Hawkes [69] in the stable case. The bounds 1/6 and 12 are clearly not optimal, and a much more precise result is available under the condition that Φ is regularly varying with index $\alpha \in [0, 1]$ at ∞: In that case, one has a.s.

$$\lim_{t \to 0+} \left\{ \sup_{0 \le \tau \le T} \frac{(L_{\tau+t} - L_\tau) \, \Phi\left(t^{-1} \log \Phi(t^{-1})\right)}{\log \Phi(t^{-1})} \right\} = c_\Phi,$$

where c_Φ is the constant that appears in Theorem 4.1; see e.g. [8]. Whether or not this identity holds in any case is an open problem.

To start with, we write

$$g(t) = \frac{\log \Phi(t^{-1})}{\Phi\left(t^{-1} \log \Phi(t^{-1})\right)}, \qquad t \text{ small enough,}$$

and recall that φ stands for the inverse function of Φ. We then introduce for every integer $n \ge 2$:

$$t(n) = \frac{n}{\varphi(ne^n)} \quad , \quad a(n) = g(t(n)).$$

Lemma 4.6 (i) *The sequence $(t(n) : n \ge 2)$ decreases. Moreover we have:*

$$\log \Phi(t(n)^{-1}) \sim n \quad , \quad \Phi(t(n)^{-1} \log \Phi(t(n)^{-1})) \sim ne^n \quad , \quad a_n \sim e^{-n}.$$

(ii) *For n large enough and any $t \in [t(n+1), t(n)]$, we have*

$$a(n)/3 \le g(t) \le 3a(n+1).$$

Proof: (i) follows from an argument similar to that in Lemma 4.2.

(ii) Since Φ increases, we have

$$g(t) \ge \frac{\log \Phi(t(n)^{-1})}{\Phi(t(n+1)^{-1} \log \Phi(t(n+1)^{-1}))}.$$

We know from (i) that the numerator is equivalent to n, and the denumerator to $(n+1)e^{n+1}$. Using (i) again, we deduce that for n large enough, $g(t) \ge a(n)/3$. The proof of the second inequality is similar. ∎

Next, we establish the following upper bound.

Lemma 4.7 *We have for every $\rho > 0$*

$$\limsup_{t \to 0+} \left\{ \sup_{0 \le \tau \le \sigma_\rho} (L_{\tau+t} - L_\tau)/g(t) \right\} \le 12, \qquad a.s.$$

Proof: Consider for every $n \in \mathbb{N}$ and every integer $j = 0, 1, \cdots, [\rho/a(n)]$ the event

$$A_{jn} = \{ \sigma_{(j+3)a(n)} - \sigma_{ja(n)} \le t(n) \}.$$

By the Markov property of σ, we have for every $\lambda > 0$

$$
\begin{aligned}
\mathbb{P}(A_{jn}) = \mathbb{P}\left(\sigma_{3a(n)} \le t(n) \right) &\le \exp\{\lambda t(n)\} \mathbb{E}\left(\exp\left\{ -\lambda \sigma_{3a(n)} \right\} \right) \\
&= \exp\left\{ \lambda t(n) - 3a(n)\Phi(\lambda) \right\}.
\end{aligned}
$$

The choice $\lambda = \varphi(ne^n)$ together with Lemma 4.6(i) yield

$$\mathbb{P}(A_{jn}) \le \exp\{ n - 3ne^n a(n) \} = o(e^{-2n});$$

so that (using again Lemma 4.6(i)) $\mathbb{P}\left(\bigcup_j A_{jn} \right) = o(e^{-n})$. Hence $\sum_n \mathbb{P}\left(\bigcup_j A_{jn} \right) < \infty$. We conclude that $\sigma_{(j+3)a(n)} - \sigma_{ja(n)} > t(n)$ for all large enough n and all integers $j \le [\rho/a(n)]$, a.s.

We now work on the event that

$$\limsup_{t \to 0+} \left\{ \sup_{0 \le \tau \le \sigma_\rho} (L_{\tau+t} - L_\tau)/g(t) \right\} > 12.$$

Then, for some arbitrarily large n, we can find $t \in [t(n+1), t(n)]$ and $\tau \in [0, \sigma_\rho]$ such that $L_{\tau+t} - L_\tau > 12g(t)$. On the other hand, we have $(j-1)a(n) \le L_\tau \le ja(n)$ for some integer $j \le [\rho/a(n)]$. By Lemma 4.6(ii), this implies

$$L_{\tau+t} > (j-1)a(n) + 12g(t) \ge (j-1)a(n) + 4a(n) = (j+3)a(n);$$

and therefore we then have both

$$\sigma_{ja(n)} \ge \tau \quad \text{and} \quad \sigma_{(j+3)a(n)} < \tau + t.$$

In conclusion, we must have $\sigma_{(j+3)a(n)} - \sigma_{ja(n)} < t \le t(n)$; and we know that the probability of the latter event goes to zero as $n \to \infty$. ∎

The first part of Theorem 4.5 derives from Lemma 4.7 by an immediate argument of monotonicity. Similarly, the second part is a consequence of the following lemma.

Lemma 4.8 *We have for every $\eta > 0$:*

$$\liminf_{t \to 0+} \left\{ \sup_{0 \le \tau \le \sigma_\eta} (L_{\tau+t} - L_\tau)/g(t) \right\} \ge 1/2, \qquad a.s.$$

Proof: We keep the notation of Lemma 4.6. Consider for every $n \in \mathbb{N}$ and every integer $j = 0, 1, \cdots, [\eta/a(n)]$ the event

$$B_{jn} = \{\sigma_{(j+1)a(n)/2} - \sigma_{ja(n)/2} \geq t(n)\}.$$

By the independence and stationarity of the increments of σ, we have

$$\mathbb{P}\left(\bigcap_j B_{jn}\right) = \mathbb{P}(B_{0,n})^{[\eta/a(n)]} \leq \exp\left\{-\frac{\eta}{a(n)}\left(1 - \mathbb{P}(B_{0n})\right)\right\}.$$

To estimate the right-hand side, we apply (4.5) with $b = a(n)/2$, $s = t(n)$ and $\lambda = \varphi(ne^n)$, so $\Phi(\lambda) = ne^n$. Using Lemma 4.6(i), we get

$$1 - \mathbb{P}(B_{0n}) = \mathbb{P}(\sigma_{a(n)/2} < t(n)) \leq \frac{\exp\{-2n/3\} - \exp\{-n\}}{1 - \exp\{-n\}} \sim \exp\{-2n/3\}.$$

Applying Lemma 4.6(i) again, we deduce that

$$\mathbb{P}\left(\bigcap_j B_{jn}\right) = O\left(\exp\left\{-\eta\exp\{n/2\}\right\}\right).$$

and the right-hand side induces a summable series.

Applying the Borel-Cantelli lemma, this entails that a.s., for every sufficiently large integer n, we are able to pick an integer $j \in \{0, 1, \cdots, [\eta/a(n)]\}$ such that

$$\sigma_{(j+1)a(n)/2} - \sigma_{ja(n)/2} < t(n).$$

Writing $\tau(n) = \sigma_{ja(n)/2}$, we thus have $L_{\tau(n)} = ja(n)/2$ and $L_{\tau(n)+t(n)} > (j+1)a(n)/2$. This forces

$$L_{\tau(n)+t(n)} - L_{\tau(n)} > a(n)/2 = g(t(n))/2.$$

As a consequence, for every $t \in [t(n+1), t(n)]$, Lemma 4.6(ii) and an obvious argument of monotonicity yield

$$L_{\tau(n+1)+t} - L_{\tau(n+1)} > g(t(n+1))/2 \geq g(t)/6;$$

which establishes the lemma. ∎

The law of the iterated logarithm specifies the rate of growth of the local time at the origin of times. By the regenerative property and the additivity of the local time, we see that for any stopping time T which takes its values in the subset of points in $\mathcal{R}$ which are not isolated on their right, the rate of growth of L at time T is the same as at the origin. Theorem 4.5 can be combined with a condensation argument due to Orey and Taylor [124] to investigate the maximal rate of growth on a path. More precisely, it is immediate from the first part of Theorem 4.5 that

$$\limsup_{t \to 0+} \frac{(L_{\tau+t} - L_\tau)\,\Phi\left(t^{-1}\log\Phi(t^{-1})\right)}{\log\Phi(t^{-1})} < 12, \qquad \text{for all } \tau \geq 0,$$

and the second part, combined with the condensation argument (cf. [124] for details), yields that a.s.

$$\limsup_{t \to 0+} \frac{(L_{\tau+t} - L_\tau)\, \Phi\left(t^{-1} \log \Phi(t^{-1})\right)}{\log \Phi(t^{-1})} \geq 1/2, \qquad \text{for some } \tau \geq 0.$$

An instant τ for which the preceding lower bound holds, is referred to as a *rapid point* for the local time, in the terminology of Kahane [90]. Adapting arguments of Orey and Taylor [124] for Brownian motion, Laurence Marsalle [118] has obtained interesting results about the Hausdorff dimension of the set of fast points when Φ is regularly varying at ∞.

It is also natural to investigate the minimal rate of growth of the local time at instants $\tau \in \mathcal{R}$ which are not isolated on their right $\mathcal{R}$ (otherwise the rate of growth is plainly zero). To this end, Marsalle [118] (extending earlier results of Fristedt [59] in the stable case) has shown recently that under some rather mild conditions on the Laplace exponent Φ, the minimal rate of growth has the same order as $1/\Phi(1/t)$. Specifically, one has a.s.

$$\limsup_{t \to 0+} (L_{\tau+t} - L_\tau)\, \Phi(1/t) > 0 \qquad \text{for every } \tau \in \mathcal{R} \text{ not isolated on its right.,}$$

and

$$\limsup_{t \to 0+} (L_{\tau+t} - L_\tau)\, \Phi(1/t) < \infty \qquad \text{for some } \tau \geq 0.$$

An instant τ which fulfils the preceding conditions is referred to as a *slow point*.

Finally, we mention that functional (i.e. *à la* Strassen) laws of the iterated logarithm for certain local times have been obtained by Marcus and Rosen [115], Csáki *et al.* [40] and Gantert and Zeitouni [65].

Chapter 5

Geometric properties of regenerative sets

This chapter is concerned with two geometric aspects of regenerative sets. We first discuss fractal dimensions and then consider the intersection with a given Borel set. The intersection of two independent regenerative sets receives special attention.

5.1 Fractal dimensions

5.1.1 Box-counting dimension

The box-counting dimension is perhaps the simplest notion amongst the variety of fractal dimensions in use; see Falconer [52]. For every non-empty bounded subset $F \subseteq [0, \infty)$, let $N_\varepsilon(F)$ be the smallest number of intervals of length (at most) $\varepsilon > 0$ which can cover F. The *lower* and *upper box-counting dimensions* of F are defined as

$$\underline{\dim}_B(F) = \liminf_{\varepsilon \to 0+} \frac{\log N_\varepsilon(F)}{\log 1/\varepsilon} \quad , \quad \overline{\dim}_B(F) = \limsup_{\varepsilon \to 0+} \frac{\log N_\varepsilon(F)}{\log 1/\varepsilon} ,$$

respectively. When these two quantities are equal, their common value is referred to as the box dimension (or also the Minkowski dimension) of F.

Following Blumenthal and Getoor [22], we next introduce the so-called lower and upper indices of the Laplace exponent Φ

$$\underline{\mathrm{ind}}\,(\Phi) = \sup\left\{\rho > 0 : \lim_{\lambda \to \infty} \Phi(\lambda)\lambda^{-\rho} = \infty\right\} = \liminf_{\lambda \to \infty} \frac{\log \Phi(\lambda)}{\log \lambda} ,$$

$$\overline{\mathrm{ind}}\,(\Phi) = \inf\left\{\rho > 0 : \lim_{\lambda \to \infty} \Phi(\lambda)\lambda^{-\rho} = 0\right\} = \limsup_{\lambda \to \infty} \frac{\log \Phi(\lambda)}{\log \lambda} ,$$

with the usual convention $\sup \emptyset = 0$. For instance, in the stable case $\Phi(\lambda) = \lambda^\alpha$, the lower and upper indices both equal α; and for a Gamma process, both the lower and upper indices are zero. Making use of Proposition 1.4, it is easy to exhibit a Laplace exponent such that $\underline{\mathrm{ind}}\,(\Phi) = a$ and $\overline{\mathrm{ind}}\,(\Phi) = b$ for arbitrary $0 \leq a \leq b \leq 1$.

Theorem 5.1 *We have a.s. for every $t > 0$*

$$\underline{\dim}_B(\mathcal{R} \cap [0, t]) = \underline{\mathrm{ind}}\,(\Phi) \quad and \quad \overline{\dim}_B(\mathcal{R} \cap [0, t]) = \overline{\mathrm{ind}}\,(\Phi).$$

Proof: The argument for the upper dimension is essentially a variation of that for the lower dimension, and we shall merely consider the latter. As we are concerned with a local path property of subordinators, there is no loss of generality in assuming that the killing rate is $k = 0$. Fix $\varepsilon > 0$ and introduce by induction the following sequence of finite stopping times: $T(0, \varepsilon) = 0$ and

$$T(n + 1, \varepsilon) = \inf\{t > T(n, \varepsilon) : \sigma_t - \sigma_{T(n,\varepsilon)} > \varepsilon\}, \qquad n = 0, 1, \cdots$$

Because the points $\sigma_{T(0,\varepsilon)}, \sigma_{T(1,\varepsilon)}, \cdots$ are at distance at least ε from each others, we see that for every fixed $t > 0$, if $T(n, \varepsilon) \leq t$, then the minimal number of intervals of length ε that is needed to cover $\mathcal{R} \cap [0, t]$ cannot be less than $n + 1$. On the other hand, it is clear from the construction that the intervals $[\sigma_{T(n,\varepsilon)}, \sigma_{T(n,\varepsilon)} + \varepsilon]$ have length ε and do cover $\mathcal{R}$. We conclude that

$$N_\varepsilon(\mathcal{R} \cap [0, t]) = \mathrm{Card}\left\{n \in \mathbb{N} : \sigma_{T(n,\varepsilon)} \leq t\right\}. \tag{5.1}$$

Next, introduce an independent exponential time τ with parameter 1. The Markov property of σ applied at time $T(n, \varepsilon)$ and the lack of memory of the exponential law entail that

$$\begin{aligned}
\mathbb{P}\left(\sigma_{T(n+1,\varepsilon)} \leq \tau \mid \sigma_{T(n,\varepsilon)} \leq \tau\right) &= \mathbb{P}\left(\sigma_{T(n+1,\varepsilon)} - \sigma_{T(n,\varepsilon)} \leq \tau - \sigma_{T(n,\varepsilon)} \mid \sigma_{T(n,\varepsilon)} \leq \tau\right) \\
&= \mathbb{P}\left(\sigma_{T(1,\varepsilon)} \leq \tau\right).
\end{aligned}$$

In other words, the random variable in (5.1) has a geometric distribution with parameter $\mathbb{P}\left(\sigma_{T(1,\varepsilon)} \leq \tau\right) = \mathbb{P}(g_\tau \geq \varepsilon)$, i.e.

$$\mathbb{P}\left(N_\varepsilon(\mathcal{R} \cap [0, \tau]) > n\right) = (1 - \mathbb{P}(g_\tau < \varepsilon))^n. \tag{5.2}$$

In order to estimate the left-hand side, recall from Lemma 1.11 that the Laplace transform of g_τ is $\Phi(1)/\Phi(1+\cdot)$. It follows from the same argument based on the Tauberian theorem of de Haan and Stadtmüller that we used in the proof of Proposition 1.4, that

$$\mathbb{P}(g_\tau < \varepsilon) \asymp 1/\Phi(1/\varepsilon), \qquad (\varepsilon \to 0+). \tag{5.3}$$

Pick first $\rho > \underline{\mathrm{ind}}\,(\Phi)$, so (by (5.3)) there is a sequence of positive real numbers $\varepsilon_n \downarrow 0$ with $\lim_{n \to \infty} \varepsilon_n^{-\rho} \mathbb{P}(g_\tau < \varepsilon_n) = \infty$. It now follows from (5.2) that

$$\lim_{n \to \infty} \mathbb{P}\left(N_{\varepsilon_n}(\mathcal{R} \cap [0, \tau]) > \varepsilon_n^{-\rho}\right) = 0,$$

and this forces (by Fatou's lemma)

$$\liminf_{\varepsilon \to 0+} \frac{\log N_\varepsilon(\mathcal{R} \cap [0, \tau])}{\log 1/\varepsilon} \leq \rho \qquad \text{a.s.}$$

We have thus proven the upper bound $\underline{\dim}_B(\mathcal{R} \cap [0, t]) \leq \underline{\mathrm{ind}}\,(\Phi)$ a.s.

To establish the converse lower bound, we may suppose that $\underline{\mathrm{ind}}\,(\Phi) > 0$ since otherwise there is nothing to prove. Then pick $0 < \rho < \underline{\mathrm{ind}}\,(\Phi)$ and note that the series $\sum 2^{n\rho}/\Phi(2^n)$ converges. We deduce from (5.2) and (5.3) that

$$\sum_{n=0}^{\infty} \mathbb{P}\left(N_{2^{-n}}\left(\mathcal{R} \cap [0,\tau]\right) \leq 2^{n\rho}\right) < \infty$$

so by the Borel-Cantelli lemma and an immediate argument of monotonicity

$$\liminf_{\varepsilon \to 0+} \frac{\log N_\varepsilon\left(\mathcal{R} \cap [0,\tau]\right)}{\log 1/\varepsilon} \geq \rho \quad \text{a.s.}$$

This shows that $\underline{\dim}_B(\mathcal{R} \cap [0,t]) \geq \underline{\mathrm{ind}}\,(\Phi)$ a.s. ∎

5.1.2   Hausdorff and packing dimensions

Lower and upper box-counting dimensions are attractively simple notions which are rather easy to work with in practice. However they are not always relevant in discussing fractal dimension, due to the following fact (see Proposition 3.4 in [52]): The closure $\overline{F}$ of a set F has the same lower and upper box-counting dimensions as F. In particular, a countable dense subset of $[0,1]$ has box-dimension 1, which is a rather disappointing feature.

This motivated the definition of *modified box-counting dimensions* (see Falconer [52], section 3.3):

$$\underline{\dim}_{MB}(F) = \inf\left\{\sup_i \underline{\dim}_B(F_i) : F \subseteq \bigcup_{i=1}^{\infty} F_i\right\},$$

$$\overline{\dim}_{MB}(F) = \inf\left\{\sup_i \overline{\dim}_B(F_i) : F \subseteq \bigcup_{i=1}^{\infty} F_i\right\}.$$

It is clear that in general

$$\underline{\dim}_{MB}(F) \leq \underline{\dim}_B(F) \quad \text{and} \quad \overline{\dim}_{MB}(F) \leq \overline{\dim}_B(F),$$

and these inequalities can be strict. Nonetheless, the box dimension and its modified version always agree for regenerative sets.

Lemma 5.2 *We have a.s. for every $t > 0$*

$$\underline{\dim}_B(\mathcal{R} \cap [0,t]) = \underline{\dim}_{MB}(\mathcal{R} \cap [0,t]) \quad and \quad \overline{\dim}_B(\mathcal{R} \cap [0,t]) = \overline{\dim}_{MB}(\mathcal{R} \cap [0,t]).$$

Proof: The random set $\mathcal{R} \cap [0,t]$ is compact and an immediate application of the Markov property shows that

$$\underline{\dim}_B(\mathcal{R} \cap [0,t] \cap V) = \underline{\dim}_B(\mathcal{R} \cap [0,t]) \quad , \quad \overline{\dim}_B(\mathcal{R} \cap [0,t] \cap V) = \overline{\dim}_B(\mathcal{R} \cap [0,t])$$

for all open sets V that intersect $\mathcal{R} \cap [0,t]$. Our claim follows from Proposition 3.6 in Falconer [52]. ∎

Taylor and Tricot [148] introduced the so-called *packing* dimension dim P, which in fact coincides with the upper modified box-counting dimension $\overline{\dim}_{MB}$; see Proposition 3.8 in [52]. Combining Lemma 5.2 and Theorem 5.1 thus identifies the packing dimension of a regenerative set with the upper index of its Laplace exponent, which is a special case of a general result of Taylor [147] on the packing dimension of the image of a Lévy process. We refer to Fristedt and Taylor [63] for further results on the packing measure of the range of a subordinator.

We next turn our attention to the so-called Hausdorff dimension; let us first briefly introduce this notion and refer to Rogers [133] for a complete account. Fix $\rho > 0$. For every subset $F \subseteq [0, \infty)$ and every $\varepsilon > 0$, denote by $\mathcal{C}(\varepsilon)$ the set of all the coverings $C = \{I_i, i \in \mathcal{I}\}$ of E with intervals I_i of length $|I_i| < \varepsilon$ (here $\mathcal{I}$ stands for a generic at most countable set of indices). Then introduce

$$m_\varepsilon^\rho(F) = \inf_{C \in \mathcal{C}(\varepsilon)} \sum_{i \in I} |I_i|^\rho.$$

Plainly $m_\varepsilon^\rho(F)$ increases as ε decreases to $0+$, and the limit is denoted by

$$m^\rho(F) = \lim_{\varepsilon > 0} \inf_{C \in \mathcal{C}(\varepsilon)} \sum_{i \in I} |I_i|^\rho \in [0, \infty].$$

It can be shown that the mapping $F \to m^\rho(F)$ defines a measure on Borel sets, called the ρ-dimensional Hausdorff measure. It should be clear that when F is fixed, the mapping $\rho \to m^\rho(F)$ decreases. Moreover, it is easy to see that if $m^\rho(F) = 0$ then $m^{\rho'}(F) = 0$ for every $\rho' > \rho$; and if $m^\rho(F) > 0$ then $m^{\rho'}(F) = \infty$ for every $\rho' < \rho$. The critical value

$$\dim_H(F) = \sup\{\rho > 0 : m^\rho(F) < \infty\} = \inf\{\rho > 0 : m^\rho(F) = 0\},$$

is called the Hausdorff dimension of F. We now identify the Hausdorff dimension of $\mathcal{R}$ with the lower index of its Laplace exponent.

Corollary 5.3 (Horowitz [73]) *We have for every $t > 0$* $\dim_H(\mathcal{R} \cap [0, t]) = \underline{\mathrm{ind}}(\Phi)$ *a.s.*

Proof: The upper bound follows from Theorem 5.1, Lemma 5.2 and the obvious fact that

$$\dim_H(F) \le \overline{\dim}_B(F)$$

for all bounded sets.

To prove the lower bound, we may suppose that $\underline{\mathrm{ind}}(\Phi) > 0$ since otherwise there is nothing to prove. The argument is based on the fact that the local time is a.s. Hölder-continuous with exponent ρ on every compact time interval, for every $\rho < \underline{\mathrm{ind}}(\Phi)$. To establish the latter assertion, note first by an application of the Markov property of σ at L_t that for every $p > 0$ and $s, t \ge 0$:

$$\mathbb{E}\left((L_{t+s} - L_t)^p\right) \le \mathbb{E}(L_s^p).$$

It follows that

$$\mathbb{E}\left((L_{t+s} - L_t)^p\right) \le p \int_0^\infty x^{p-1} \mathbb{P}(L_s > x) dx = p \int_0^\infty x^{p-1} \mathbb{P}(\sigma_x \le s) dx.$$

Using the obvious inequality

$$\mathbb{P}(\sigma_x \leq s) \leq e\,\mathbb{E}(\exp\{-s^{-1}\sigma_x\}) = \exp\{1 - x\Phi(s^{-1})\},$$

we get

$$\mathbb{E}\left((L_{t+s} - L_t)^p\right) \leq e\Gamma(p+1)\Phi(s^{-1})^{-p}.$$

The Hölder-continuity now derives from Kolmogorov's criterion and the very definition of the lower index.

Next, consider a covering of $\mathcal{R} \cap [0, t]$ by finitely many intervals $[a_0, b_0], \cdots, [a_n, b_n]$, where $a_0 \leq b_0 \leq \cdots \leq a_n \leq b_n$ (there is no loss of generality in focussing on finite coverages, because $\mathcal{R} \cap [0, t]$ is compact). Observe that $L_{b_{i-1}} = L_{a_i}$ for $i = 1, \cdots, n$. Since L is a.s. Hölder continuous with exponent ρ on $[0, 1]$, we deduce that

$$\sum_{i=0}^{n}(b_i - a_i)^\rho \geq K \sum_{i=0}^{n}(L_{b_i} - L_{a_i}) = KL_{b_n} \geq KL_t > 0 \quad \text{a.s.}$$

where $K > 0$ is a certain random variable. This shows that the ρ-Hausdorff measure of $\mathcal{R} \cap [0, t]$ is positive a.s., so its Hausdorff dimension is at least ρ. ∎

To summarize the main results of this section, there are two natural fractal dimensions -which may coincide- associated with a regenerative set. The lower dimension agrees both with the Hausdorff dimension and the lower (modified) box-counting dimension; it is given by the lower index of the Laplace exponent. The upper dimension agrees both with the packing dimension and the upper (modified) box-counting dimension; it is given by the upper index of the Laplace exponent.

There exist many further results in the literature about Hausdorff dimension and subordinators; see section III.5 in [11] and [58] and references therein. To this end, we also recall that Fristedt and Pruitt [61] have been able to specify the exact Hausdorff measure of the range; which provides a remarkable refinement of the result of Horowitz. In a different direction, the multifractal structure of the occupation measure of a stable subordinator has been recently considered by Hu and Taylor [78].

5.2 Intersections with a regenerative set

5.2.1 Equilibrium measure and capacity

We are concerned with the probability that a regenerative set $\mathcal{R}$ intersects a given (deterministic) Borel set B. As $\mathcal{R}$ only differs from $\{\sigma_t : t > 0\}$, the set of points that are visited by the subordinator σ, by at most countably many points, it is readily seen that

$$\mathbb{P}(\mathcal{R} \cap B \neq \emptyset) = \mathbb{P}(\sigma_t \in B \text{ for some } t > 0).$$

This connection enables us to investigate the left-hand-side using the classical potential theory for Markov processes; see Chapter VI in Blumenthal and Getoor [23], Berg and Forst [6], and the references therein. To this end, it will be convenient to use the notation $\mathbb{P}^x$ for the law of the subordinator started from $x \in \mathbb{R}$, viz. the distribution of $x + \sigma$ under $\mathbb{P} = \mathbb{P}^0$.

For the sake of simplicity, we will assume that the renewal measure is absolutely continuous and that there is a version of the renewal density that is continuous on $(0, \infty)$. As a matter of fact, the results of this section hold more generally under the sole assumption of absolute continuity for the renewal measure; the continuity hypothesis for the renewal density just enables us to circumvent some technical difficulties inherent to the general case. The probability that a bounded Borel set B is hit by σ can be expressed in terms of renewal densities and the so-called equilibrium measure of B as follows (cf. Theorem VI(2.8) in [23]).

Proposition 5.4 *Suppose that U is absolutely continuous with a continuous density on $(0, \infty)$, and write $u(t)$ for the version of $U(dt)/dt$ such that $u \equiv 0$ on $(-\infty, 0]$ and u is continuous on $(0, \infty)$. Let $B \subseteq (-\infty, \infty)$ be a bounded Borel set. There is a Radon measure μ_B, called the equilibrium measure of B, with $\mathrm{Supp}\mu_B \subseteq \overline{B}$, and such that for every $x \in (-\infty, \infty)$*

$$\mathbb{P}^x\left(\sigma_t \in B \text{ for some } t > 0\right) = \int_{(-\infty, \infty)} u(y - x)\mu_B(dy).$$

Proof: The argument is a variation of that of Chung (cf. Chapter 5 in [35]). Fix x and introduce the last-passage time in B,

$$\gamma = \sup\{t > 0 : \sigma_t \in B\},$$

and note that $\sigma_{\gamma-} \in \overline{B}$ whenever $0 < \gamma < \infty$. Then consider for every $\varepsilon > 0$ and every bounded continuous function $f : \mathbb{R} \to [0, \infty)$ the quantity

$$I(\varepsilon) = \varepsilon^{-1} \mathbb{E}^x \left(\int_0^\infty f(\sigma_t) \mathbf{1}_{\{\gamma \in (t, t+\varepsilon)\}} dt \right).$$

The continuity of f and the identity

$$I(\varepsilon) = \mathbb{E}^x \left(\varepsilon^{-1} \int_{(\gamma-\varepsilon)+}^\gamma f(\sigma_t) dt \right)$$

make clear that

$$\lim_{\varepsilon \to 0+} I(\varepsilon) = \mathbb{E}^x \left(f(\sigma_{\gamma-}) \right), 0 < \gamma < \infty. \tag{5.4}$$

On the other hand, an application of the Markov property shows that

$$I(\varepsilon) = \mathbb{E}^x \left(\int_0^\infty f(\sigma_t) \varepsilon^{-1} \psi_\varepsilon(\sigma_t) dt \right) = \int_{-\infty}^\infty f(y) u(y - x) \varepsilon^{-1} \psi_\varepsilon(y) dy, \tag{5.5}$$

with $\psi_\varepsilon(y) = \mathbb{P}^y(0 < \gamma < \varepsilon)$.

It is readily seen from the resolvent equation (cf. [11] on page 23) that u is positive on $(0, \infty)$. First take the function f in the form

$$f(y) = \begin{cases} g(y)/u(y - x) & \text{if } y > x \\ 0 & \text{otherwise} \end{cases}$$

where g is a continuous function. As x is arbitrary, we see from (5.4) and (5.5) that the measure $\varepsilon^{-1}\psi_\varepsilon(y)dy$ converges weakly towards some Radon measure, say μ_B. We then deduce that

$$\mathbb{P}^x\left(\sigma_{\gamma-} \in dy, 0 < \gamma < \infty\right) = u(y-x)\mu_B(dy)$$

(recall that u is continuous except at 0 and that $u(0) = 0$). In particular μ_B has support in $\overline{B}$ and

$$\mathbb{P}^x\left(\sigma_t \in B \text{ for some } t > 0\right) = \mathbb{P}^x\left(0 < \gamma < \infty\right) = \int_{-\infty}^\infty u(y-x)\mu_B(dy),$$

which establishes our claim. ∎

The total mass of the equilibrium measure is called the capacity of B, and is denoted by

$$\mathrm{Cap}(B) = \mu_B(\mathbb{R}) = \mu_B(\overline{B}).$$

The set B is called polar if it has zero capacity, i.e. its equilibrium measure is trivial. We see from Proposition 5.4 that B is polar if and only if for every starting point $x \in \mathbb{R}$, the subordinator σ never visits B at any positive instant. The capacity can also be expressed as

$$\mathrm{Cap}(B) = \sup\left\{\mu(\mathbb{R}) : \mu(\mathbb{R} - B) = 0 \text{ and } \int_{\mathbb{R}} u(x-y)\mu(dy) \leq 1\right\},$$

see Blumenthal and Getoor [23] on page 286. As an immediate consequence, one obtains the following characterization of Borel sets $B \subseteq (0,\infty)$ that do not intersect a regenerative set $\mathcal{R}$:

$$\mathbb{P}(B \cap \mathcal{R} = \emptyset) = 1 \iff \sup_{x \in \mathbb{R}} U\mu(x) = \infty \quad \forall \mu \text{ probability measure with } \mu(B) = 1,$$

$$\tag{5.6}$$

where $U\mu(x) = \int u(y-x)\mu(dy)$.

5.2.2 Dimension criteria

The preceding characterization of polar sets is not always easy to apply, as it requires precise information on the renewal density. Our purpose in this section is to present more handy criteria in terms of the Hausdorff dimension (recall section 2.3). We refer to Hawkes [70] for further results connecting the polarity of sets and Hausdorff measures.

In order to present a simple test for non-intersection, we need first to estimate the probability that $\mathcal{R}$ intersects a given interval.

Lemma 5.5 *The following bounds hold for every $0 < a < b$*

$$\frac{U(b) - U(a)}{U(b-a)} \leq \mathbb{P}(\mathcal{R} \cap [a,b] \neq \emptyset) \leq \frac{U(2b-a) - U(a)}{U(b-a)}.$$

Proof: Applying the Markov property at $D_a = \inf\{x > a : x \in \mathcal{R}\} = \sigma_{L_a}$, we get

$$U(b) - U(a) = \mathbb{E}\left(\int_{L_a}^{\infty} \mathbf{1}_{\{\sigma_t \in (a,b]\}} dt\right) = \int_{[a,b]} \mathbb{P}(\sigma_{L_a} \in dx)\mathbb{E}\left(\int_0^{\infty} \mathbf{1}_{\{\sigma_t \in (a-x,b-x]\}} dt\right)$$

$$= \int_{[a,b]} \mathbb{P}(D_a \in dx)U(b-x)$$

$$\leq \mathbb{P}(D_a \leq b)U(b-a).$$

Since the events $\{D_a \leq b\}$ and $\{\mathcal{R} \cap (a,b] \neq \emptyset\}$ coincide, the lower bound is proven.

A similar argument yields the upper-bound. More precisely

$$U(2b - a) - U(a) = \int_{[a,2b-a]} \mathbb{P}(D_a \in dx)U(2b - a - x)$$

$$\geq \int_{[a,b]} \mathbb{P}(D_a \in dx)U(2b - a - x) \geq \mathbb{P}(D_a \leq b)U(b-a).$$

This entails

$$\mathbb{P}(\mathcal{R} \cap (a,b] \neq \emptyset) \leq \frac{U(2b-a) - U(a)}{U(b-a)},$$

and since the renewal function is continuous, our claim follows. ∎

Proposition 5.6 (Orey [123]) *Suppose that the renewal measure has a locally bounded density u on $(0,\infty)$. Let $B \subseteq (0,\infty)$ with $\dim_H(B) < 1 - \overline{\mathrm{ind}}(\Phi)$. Then $\mathcal{R} \cap B = \emptyset$ a.s.*

Proof: As $\dim_H(B) < 1 - \overline{\mathrm{ind}}(\Phi)$, there is $\rho < 1 - \overline{\mathrm{ind}}(\Phi)$ such that the ρ-dimensional Hausdorff measure of B is zero. This means that for every $\varepsilon > 0$, one can cover B with a family of intervals $([a_i, b_i] : i \in \mathcal{I})$ such that

$$\sum_{i \in I} |b_i - a_i|^{\rho} \leq \varepsilon. \tag{5.7}$$

We then invoke Lemma 5.5 to get

$$\mathbb{P}(\mathcal{R} \cap B \neq \emptyset) \leq \sum_I \mathbb{P}(\mathcal{R} \cap [a_i, b_i] \neq \emptyset) \leq \sum_I \frac{U(2b_i - a_i) - U(a_i)}{U(b_i - a_i)}.$$

With no loss of generality, we may (and will) suppose that for some $c > 1$, $1/c \leq a_i < b_i \leq c$ for every i. As U is Lipschitz-continuous on $[1/c, 2c]$, the right-hand side in the ultimate displayed equation is less than or equal to

$$M \sum_I \frac{b_i - a_i}{U(b_i - a_i)}$$

for some finite constant number M.

By Proposition 1.4, we know that there is a constant number $k > 0$ such that $1/U(t) \leq k\Phi(1/t)$. The very definition of the upper index entails that $\Phi(1/t) = o(t^{\rho-1})$. We conclude that

$$\mathbb{P}(\mathcal{R} \cap B \neq \emptyset) \leq C \sum_{i \in I} |b_i - a_i|^{\rho},$$

and by (5.7), the right-hand side can be made as small as we wish. ∎

We then give a test for intersection with positive probability.

Proposition 5.7 (Hawkes [70]) *Suppose that the renewal measure has a decreasing density u on $(0, \infty)$ with respect to the Lebesgue measure. Let $B \subseteq (0, \infty)$ with $\dim_H(B) > 1 - \underline{\mathrm{ind}}(\Phi)$. Then $\mathbb{P}(\mathcal{R} \cap B \neq \emptyset) > 0$.*

Proposition 5.7 follows from (5.6) and the following variation of Frostman's lemma.

Lemma 5.8 *Under the hypotheses of Proposition 5.7, there is a probability measure μ with compact support $K \subseteq B$ such that $\mu * u$ is a bounded function.*

Proof: Pick ρ strictly between $1 - \underline{\mathrm{ind}}(\Phi)$ and $\dim_H(B)$. According to Frostman's lemma (see e.g. Theorem 4.13 in [52] and its proof), there is a probability measure μ with compact support $K \subseteq B$ such that

$$\sup_{x \geq 0} \int_{[0,\infty)} |y - x|^{-\rho} \mu(dy) < \infty.$$

Applying Proposition 1.4 and the hypothesis that the renewal density u decreases, we get

$$u(t) \leq \frac{U(t)}{t} \leq \frac{c}{t\Phi(1/t)}.$$

On the other hand, we know from the very definition of the lower index that $\Phi(1/t)$ is bounded from below by $t^{\rho-1}$ for all small enough $t > 0$. In conclusion $u(t) = O(t^{-\rho})$ and our claim follows. ∎

We point out that, since the Laplace transform of the renewal measure is $1/\Phi$, the renewal density exists and is decreasing if and only if $\lambda/\Phi(\lambda)$ is the Laplace exponent of some subordinator (this is seen by an integration by parts), and then Propositions 5.6 and 5.7 are relevant. For instance, recall that the zero set of a d-dimensional Bessel process ($0 < d < 2$) can be viewed as the range of a stable subordinator with index $1 - d/2$. We deduce that a d-dimensional Bessel process never vanishes a.s. on a time-set $B \subseteq (0, \infty)$ with Hausdorff dimension strictly less than $d/2$, whereas it vanishes with positive probability on a time-set with Hausdorff dimension strictly greater than $d/2$.

5.2.3 Intersection of independent regenerative sets

We finally consider the intersection of two independent regenerative sets, say $\mathcal{R}^{(1)}$ and $\mathcal{R}^{(2)}$. It should be clear that the closed random set $\mathcal{R} = \mathcal{R}^{(1)} \cap \mathcal{R}^{(2)}$ inherits the regenerative property, and our main concern is then to characterize its distribution.

The case when both $\mathcal{R}^{(1)}$ and $\mathcal{R}^{(2)}$ are heavy is straightforward. Specifically, write $d^{(1)}$ and $d^{(2)}$ for the positive drift coefficients of $\mathcal{R}^{(1)}$ and $\mathcal{R}^{(2)}$, respectively, and recall that the renewal densities $u^{(1)}$ and $u^{(2)}$ are continuous and positive on $[0, \infty)$ (cf. Proposition 1.9). Because $\mathcal{R}^{(1)}$ and $\mathcal{R}^{(2)}$ are independent, we have for every $x \geq 0$

$$\mathbb{P}(x \in \mathcal{R}) = \mathbb{P}\left(x \in \mathcal{R}^{(1)}\right) \mathbb{P}\left(x \in \mathcal{R}^{(2)}\right) = \mathrm{d}^{(1)}\mathrm{d}^{(2)}u^{(1)}(x)u^{(2)}(x).$$

The right-hand side is a continuous everywhere positive function of x; we conclude by an application of Proposition 1.9 that $\mathcal{R}$ is a heavy regenerative set whose renewal density is proportional to $u^{(1)}u^{(2)}$. We present below a more general result.

Proposition 5.9 (Hawkes [71]) *Let $\mathcal{R}^{(1)}$ and $\mathcal{R}^{(2)}$ be two independent regenerative sets and $\mathcal{R} = \mathcal{R}^{(1)} \cap \mathcal{R}^{(2)}$. Suppose that $\mathcal{R}^{(1)}$ and $\mathcal{R}^{(2)}$ both possess renewal densities $u^{(1)}$ and $u^{(2)}$ which are continuous and positive on $(0, \infty)$, and that $\mathcal{R}$ does not reduce to $\{0\}$ a.s. Then $\mathcal{R}$ has a renewal density given by $u = c u^{(1)} u^{(2)}$, where $c > 0$ the constant of normalization.*

Proof: The idea of the proof is the same as for Proposition 5.4. We assume first that $\mathcal{R}^{(1)}$ is bounded, and hence so is $\mathcal{R}$. Introduce the last passage times

$$\gamma^{(1)} = \sup\{t > 0 : \sigma_t^{(1)} \in \mathcal{R}^{(2)}\} \quad , \quad \gamma^{(2)} = \sup\{t > 0 : \sigma_t^{(2)} \in \mathcal{R}^{(1)}\}$$

which are positive and finite by assumption. Note also that the largest point of $\mathcal{R}$ can be expressed as the common value $g_\infty = \sigma^{(1)}\left(\gamma^{(1)}-\right) = \sigma^{(2)}\left(\gamma^{(2)}-\right)$. Take a bounded continuous function $f : [0, \infty) \times [0, \infty) \to [0, \infty)$, and consider for every $\varepsilon > 0$ the quantity

$$I(\varepsilon) = \mathbb{E}\left(\varepsilon^{-2} \int_0^\infty ds \int_0^\infty dt\, f\left(\sigma_s^{(1)}, \sigma_t^{(2)}\right) \mathbf{1}_{\{\gamma^{(1)} \in (s, s+\varepsilon), \gamma^{(2)} \in (t, t+\varepsilon)\}}\right).$$

On the one hand, we can write $I(\varepsilon)$ as

$$\mathbb{E}\left(\varepsilon^{-2} \int_{\gamma^{(1)}-\varepsilon}^{\gamma^{(1)}} ds \int_{\gamma^{(2)}-\varepsilon}^{\gamma^{(2)}} dt\, f\left(\sigma_s^{(1)}, \sigma_t^{(2)}\right)\right)$$

and then apply the theorem of dominated convergence to get

$$\lim_{\varepsilon \to 0+} I(\varepsilon) = \mathbb{E}\left(f(g_\infty, g_\infty)\right). \tag{5.8}$$

On the other hand, taking conditional expectation (i.e. an optional projection) yields

$$I(\varepsilon) = \mathbb{E}\left(\varepsilon^{-2} \int_0^\infty ds \int_0^\infty dt\, f\left(\sigma_s^{(1)}, \sigma_t^{(2)}\right) Y_{s,t}\right),$$

with $Y_{s,t} = \mathbb{P}\left(\gamma^{(1)} \in (s, s+\varepsilon), \gamma^{(2)} \in (t, t+\varepsilon) \mid \mathcal{F}_s^{(1)} \otimes \mathcal{F}_t^{(2)}\right)$. An application of the Markov property shows that

$$Y_{s,t} = \psi_\varepsilon(\sigma_s^{(1)} - \sigma_t^{(2)})$$

where $\psi_\varepsilon(y)$ denotes the probability that the random sets

$$\{v > 0 : \sigma_v^{(1)} + y \in \mathcal{R}^{(2)}\} \quad \text{and} \quad \{v \geq 0 : \sigma_v^{(2)} - y \in \mathcal{R}^{(1)}\}$$

are both non-empty and contained into $(0, \varepsilon)$. We thus have

$$
\begin{aligned}
I(\varepsilon) &= \mathbb{E}\left(\varepsilon^{-2} \int_0^\infty ds \int_0^\infty dt f\left(\sigma_s^{(1)}, \sigma_t^{(2)}\right) \psi_\varepsilon(\sigma_s^{(1)} - \sigma_t^{(2)})\right) \\
&= \int_0^\infty \int_0^\infty f(y, z) u^{(1)}(y) u^{(2)}(z) \varepsilon^{-2} \psi_\varepsilon(y - z) dy\,dz.
\end{aligned} \tag{5.9}
$$

Next, take the function f in the form

$$
f(y, z) = \frac{h(y - z)}{u^{(1)}(y) u^{(2)}(z)} \varphi(y) \varphi(z)
$$

where $h : (-\infty, \infty) \to [0, \infty)$ is a continuous bounded function and φ a continuous function with compact support included into $(0, \infty)$. We deduce from (5.8) and (5.9) that the measure $\varepsilon^{-2} \psi_\varepsilon(x) dx$ converges weakly as $\varepsilon \to 0+$ towards $c\delta_0$ for some $c > 0$. Finally take f in the form $f(y, z) = f(z)$ to get

$$
\mathbb{P}(g_\infty \in dt) = c u^{(1)}(t) u^{(2)}(t) dt.
$$

The comparison with Lemma 1.10 entails that the renewal measure $U(dt)$ of $\mathcal{R}$ is absolutely continuous with a density proportional to $u^{(1)} u^{(2)}$.

Proposition 5.9 is thus proven when $\mathcal{R}^{(1)}$ is bounded. The case when $\mathcal{R}^{(1)}$ is unbounded follows by approximation, introducing a small killing rate in $\sigma^{(1)}$. ∎

To apply Proposition 5.9, it is crucial to know whether $\mathcal{R}^{(1)} \cap \mathcal{R}^{(2)} = \{0\}$ a.s. Because a renewal measure is a Radon measure on $[0, \infty)$, Proposition 5.9 entails that if $\mathcal{R}^{(1)}$ and $\mathcal{R}^{(2)}$ both possess renewal densities $u^{(1)}$ and $u^{(2)}$ which are continuous and positive on $(0, \infty)$, then

$$
\int_{0+} u^{(1)}(x) u^{(2)}(x) dx = \infty \implies \mathcal{R}^{(1)} \cap \mathcal{R}^{(2)} = \{0\} \quad \text{a.s.}
$$

By a recent result in [16], the necessary and sufficient condition for $\mathcal{R}^{(1)} \cap \mathcal{R}^{(2)} = \{0\}$ a.s. is that the convolution $u^{(1)} \star u^{(2)}$ is unbounded. I know no examples in which $u^{(1)} \star u^{(2)}$ is unbounded and $\int_{0+} u^{(1)}(x) u^{(2)}(x) dx < \infty$. See also Evans [51], Rogers [135] and Fitzsimmons and Salisbury [56] for results in that direction.

The problem of characterizing the distribution of the intersection of two independent regenerative sets in the general case seems still open. We refer to [16] for the most recent results, and to Hawkes [71], Fitzsimmons et al. [54], Fristedt [60] and Molchanov [120] other works this topic. See also [14] for another geometric problem on regenerative sets involving the notion of embedding, which is connected to the preceding.

Chapter 6

Burgers equation with Brownian initial velocity

This chapter is adapted from [15]; its purpose is to point out an interesting connection between the inviscid Burgers equation with Brownian initial velocity and certain subordinators. Applications to statistical properties of the solution are discussed.

6.1 Burgers equation and the Hopf-Cole solution

Burgers equation with viscosity parameter $\varepsilon > 0$

$$\partial_t u + \partial_x \left(u^2/2 \right) = \varepsilon \partial_{xx}^2 u \tag{6.1}$$

has been introduced by Burgers as a model of hydrodynamic turbulence, where the solution $u_\varepsilon(x, t)$ is meant to describe the velocity of a fluid particle located at x at time t. Although it is now known that this is not a good model for turbulence, it still is widely used in physical problems as a simplified version of more elaborate models (e.g. the Navier-Stokes equation). A most important feature of (6.1) is that it is one of the very few non-linear equations that can be solved explicitly. Specifically, Hopf [75] and Cole [37] observed that applying the transformation $\gamma = 2\varepsilon \log g$ to the potential function γ given by $\partial_x \gamma = -u_\varepsilon$, yields the heat equation $\partial_t g = \varepsilon \partial_{xx}^2 g$. This enables one to determine g and hence u_ε.

The asymptotic behaviour of the solution u_ε of (6.1) as ε tends to 0 is an interesting question. Roughly, u_ε converges to a certain function $u_0 = u$, which provides a (weak) solution of the inviscid limit equation

$$\partial_t u + \partial_x \left(u^2/2 \right) = 0. \tag{6.2}$$

More precisely, u can be expressed implicitly in terms of the initial velocity $u(\cdot, 0)$ as follows (cf. Hopf [75], and also [142] and [140] for a brief account). Under simple conditions such as $u(\cdot, 0) = 0$ on $(-\infty, 0)$ and $\liminf_{x \to \infty} u(x, 0)/x \geq 0$, the function

$$s \to \int_0^s (tu(r, 0) + r - x) dr \tag{6.3}$$

tends to ∞ as $s \to \infty$, for every $x \geq 0$ and $t > 0$. We then denote by $a(x, t)$ the largest location of the overall minimum of (6.3). The mapping $x \to a(x, t)$ is right-continuous increasing; it is known as the *inverse Lagrangian function*. The Hopf-Cole solution to (6.2) is given by

$$u(x, t) = \frac{x - a(x, t)}{t}. \tag{6.4}$$

6.2 Brownian initial velocity

Sinai [140] and She *et al.* [142] have considered the inviscid Burgers equation when the initial velocity is given by a Brownian motion; see also Carraro-Duchon [33] where (6.2) is understood in some weak statistical sense. More precisely

$$u(\cdot, 0) = 0 \text{ on } (-\infty, 0], \text{ and } (u(x, 0), x \geq 0) \quad \text{is a Brownian motion} \tag{6.5}$$

is enforced from now on. Our main purpose is to point out that for each fixed $t > 0$, the inverse Lagrangian function is then a subordinator; here is the precise statement.

Theorem 6.1 *For each fixed $t > 0$, the process $(a(x, t) : x \geq 0)$ is a subordinator started from $a(0, t)$. Its Laplace exponent Φ is given by*

$$\Phi(q) = \frac{\sqrt{2t^2 q + 1} - 1}{t^2}.$$

In other words, $(a(x, t) - a(0, t) : x \geq 0)$ has the same distribution as the first passage process of a Brownian motion with variance t^2 and unit drift.

One can prove that the random variable $a(0, t)$ has a gamma distribution, which completes the description of the law of the inverse Lagrangian function. As this is not relevant to the applications we have in mind, we omit the proof and refer to [15] for an argument (see also Lachal [105] for the law of further variables related to $a(0, t)$).

Theorem 6.1 has several interesting consequences; we now briefly present a few, and refer to [18] for some further applications connected to the multifractal spectrum of the solution (see also Jaffard [86]).

The discontinuities of the Eulerian velocity u are a major object of interest. Call $x > 0$ an Eulerian regular point if u is continuous at x, and an Eulerian shock point otherwise. In the latter case the amplitude of the jump $u(x, t) - u(x-, t)$ is necessarily negative (see (6.4) and Theorem 6.1); from the viewpoint of hydrodynamic turbulence, it corresponds to the velocity of the fluid particle absorbed into the shock. For each fixed $t > 0$, let us write

$$\Delta(t) = (a(x, t) - a(x-, t), x \geq 0)$$

for the process of the jumps of the inverse Lagrangian function taken at time t, and recall from (6.4) that $u(x, t) - u(x-, t) = -\frac{1}{t} \Delta_x(t)$.

Proposition 1.3 and the Lévy-Khintchine formula

$$\sqrt{2q + 1} - 1 = \frac{1}{\sqrt{2\pi}} \int_0^\infty \left(1 - e^{-qy}\right) y^{-3/2} \exp\left\{-y/2\right\} dy$$

yield the following statistical description of the shocks.

Corollary 6.2 *For each fixed $t > 0$, $\Delta(t)$ is a Poisson point process valued in $(0, \infty)$ with characteristic measure*

$$\frac{1}{t\sqrt{2\pi y^3}} \exp\left\{-\frac{y}{2t^2}\right\} dy \qquad (y > 0).$$

Next, we turn our attention to the fractal properties of the so-called *Lagrangian regular points*, that are the points $y \geq 0$ for which there exists some $x \geq 0$ such that the function (6.3) reaches its overall minimum at $y = a(x, t)$ and nowhere else. A moment of reflection shows that the set $\mathcal{R}_c$ of Lagrangian regular points can be viewed as the range of the inverse Lagrangian function on its continuity set, i.e.

$$\mathcal{R}_c = \{a(x, t) : x \geq 0 \text{ regular Eulerian point }\}.$$

As $\mathcal{R}_c$ only differs from the range of $a(\cdot, t)$ by at most countably many points, we thus obtain as an immediate application of section 5.1 the following.

Corollary 6.3 *The Hausdorff dimension and the packing dimension of $\mathcal{R}_c$ both equal $1/2$ a.s.*

That the Hausdorff dimension of $\mathcal{R}_c$ is $1/2$ was the main result of Sinai [140]; see also Aspandiiarov and Le Gall [1].

Finally, we mention that Theorems 4.1 and 4.5 respectively yield the law of the iterated logarithm and the modulus of continuity of the *Lagrangian function* $a \to x(a, t)$, that is the inverse of the function $x \to a(x, t)$; the precise statements are left to the reader. The relevance of the Lagrangian function in hydrodynamic turbulence stems from the fact that it can be viewed as the position at time t of the fluid particle started from the location a. This can be seen from the identity $\partial_t x(a, t) = u(x(a, t), t)$ that follows easily from (6.4) and (6.2).

6.3   Proof of the theorem

Let Ω denote the set of càdlàg paths $\omega : [0, \infty) \to \mathbb{R} \cup \{\infty\}$ such that $\lim_{s \to \infty} \omega(s) = \infty$; we write $X_s : \omega \to \omega(s)$ for the canonical projections. Consider also the shift operators $(\theta_s : s \geq 0)$ and the killing operators $(\mathbf{k}_s : s \geq 0)$

$$X_r \circ \theta_s = X_{r+s} \quad , \quad X_r \circ \mathbf{k}_s = \begin{cases} X_r & \text{if } r < s \\ \infty & \text{otherwise} \end{cases}$$

For every $x \in \mathbb{R}$, let $\mathbb{P}^x$ stand for the law of the Brownian motion with variance t^2 and unit drift started at x, which is viewed as a probability measure on Ω. We next introduce the indefinite integral of X

$$I_s = \int_0^s X_r dr, \qquad s \geq 0,$$

its past-minimum function

$$m_s = \min_{0 \leq r \leq s} I_r, \qquad s \geq 0,$$

and the largest location of the overall minimum of I

$$a = \max\{s \geq 0 : I_s = m_\infty\}.$$

Plainly, a is not a stopping time. Nonetheless, there is a Markov type property at a which is a special case of the so-called the Markov property at last passage times, and this provides the key to the proof of Theorem 6.1.

Lemma 6.4 *For every $x \geq 0$, the processes $X \circ \mathsf{k}_a$ and $X \circ \theta_a$ are independent under $\mathbb{P}^{-x}$, and the law of $X \circ \theta_a$ does not depend on x.*

Proof: The proof is based on the fact that, loosely speaking, splitting the path of a Markov process at its last passage time at a given point produces two independent processes; and more precisely, the law of the part after the last passage time does not depend of the initial distribution of the Markov process. We refer to [44] on pages 299-300 and the related references quoted therein for a precise and much more general statement.

Consider the integral process reflected at its past minimum, $I - m$. The additive property of the integral $I_{s+r} = I_s + I_r \circ \theta_s$ and the strong Markov property of Brownian motion readily entail that the pair $(X, I - m)$ is a strong Markov process; see the proof of Proposition VI.1 in [11] for a closely related argument. On the other hand, it should be clear that for every $x \geq 0$, we have $a < \infty$ and $X_a = 0$, $\mathbb{P}^{-x}$-a.s. In particular a can be viewed as the last passage time of $(X, I - m)$ at $(0,0)$, and it now follows from the aforementioned Markov property at last-passage times that the processes $(X, I - m) \circ \mathsf{k}_a$ and $(X, I - m) \circ \theta_a$ are independent and that the law of the latter does not depend on x. This establishes our claim. ∎

We are now able to prove Theorem 6.1.

Proof: Fix $x \geq 0$ and $t > 0$. We know from (6.5) that $(tu(s,0) + s - x : s \geq 0)$ is a Brownian motion with variance t^2 and unit drift started at $-x$; it has the law of $X = (X_s : s \geq 0)$ under $\mathbb{P}^{-x}$. In this framework, we can make the following identifications: The function (6.3) coincides with the integral $s \to I_s$, and the inverse Lagrangian function evaluated at x is simply $a(x,t) = a$. Moreover, it is readily seen that for every $0 \leq z \leq x$, $a(z,t)$ only depends on the killed path $X \circ \mathsf{k}_a$.

Write $X' = X \circ \theta_a$, $I'_s = \int_0^s X'_r dr$, and for $y \geq 0$, $a'(y,t)$ for the largest location of the overall minimum of $s \to I'_s - ys$. We then observe the identity

$$a(x + y, t) - a = a'(y, t). \tag{6.6}$$

More precisely, $a(x+y,t)$ is the largest location of the overall minimum of $s \to I_s - sy$. This location is bounded from below by $a(x,t) = a$, so that $a(x+y,t) - a$ is the largest location of the overall minimum of $s \to I_{a+s} - (a+s)y$. Because $I_{a+s} = I_a + I'_s$, (6.6) follows.

According to Lemma 6.4, X' and $X \circ \mathsf{k}_a$ are independent. We deduce from (6.6) that the increment $a(x+y,t) - a(x,t)$ is independent of $(a(z,t) : 0 \leq z \leq x)$. Because the law of X' does not depend on x, the same holds for $a'(y,t) = a(x+y,t) - a(x,t)$.

We have thus proven the independence and homogeneity of the increments of the inverse Lagrangian function.

Next, introduce $T = \min\{s \geq 0 : X_s = 0\}$, the first hitting time of 0 by X. By the strong Markov property, $\tilde{X} = X \circ \theta_T$ is independent of $X \circ k_T$ and has the law $\mathbb{P}^0$. The very same argument as above shows that

$$a(x, t) = T + \tilde{a}(0, t) \tag{6.7}$$

where $\tilde{a}(0, t)$ stands for the largest location of the minimum of $s \to \tilde{I}_s = \int_0^s \tilde{X}_r dr$. Because $\tilde{a}(0, t)$ is independent of T and has the same law as $a(0, t)$, the decompositions $a(x, t) = (a(x, t) - a(0, t)) + a(0, t)$ and (6.7), and the independence of the increments property show that T and $a(x, t) - a(0, t)$ have the same law. In other words, the process $(a(x, t) - a(0, t) : x \geq 0)$ has the same one-dimensional distributions as the first passage process $(T_x : x \geq 0)$ of a Brownian motion with variance t^2 and unit drift started at zero. Because both have independent and homogeneous increments, we conclude that these two processes have the same law.

Finally, the assertion that the Laplace exponent of the first passage process of a Brownian motion with variance t^2 and unit drift is given by $\Phi(q) = t^{-2}\left(\sqrt{2t^2q + 1} - 1\right)$ is well-known; see e.g. Formula 2.0.1 on page 223 in Borodin and Salmimen [26]. ∎

Chapter 7

Random covering

We consider the closed subset $\mathcal{R}$ of the nonnegative half-line left uncovered by a family of random open intervals formed from a Poisson point process. This set is regenerative; one can express its Laplace exponent in terms of the characteristic measure of the Poisson point process. This enables us to determine the cases when $\mathcal{R}$ is degenerate, or bounded, or light, and also to specify its fractal dimensions. The approach relies on the correspondence between regenerative sets and subordinators.

7.1   Setting

Consider a Poisson point process $\ell = (\ell_t, t \geq 0)$ taking values in the positive half-line $(0, \infty)$; let μ denote its characteristic measure. This means that if $(\mathcal{M}_t)_{t \geq 0}$ stands for the completed natural filtration generated by ℓ, then for every Borel set $B \subseteq [0, \infty)$, the counting process $\text{Card}\{0 \leq s \leq t : \ell_s \in B\}$, $t \geq 0$, is an $(\mathcal{M}_t)$-Poisson process with intensity $\mu(B)$. Recall that this implies that to disjoint Borel sets correspond independent Poisson processes.

We associate to each $t \geq 0$ the open interval $I_t = (t, t + \ell_t)$ (of course, there are only a countable numbers of times when $\ell_t \in (0, \infty)$, so there are countably many non-empty intervals). We then consider the closed set of points in $[0, \infty)$ which are left uncovered by these random intervals:

$$\mathcal{R} = [0, \infty) - \bigcup_{t \geq 0} I_t \, .$$

For short, we will refer to $\mathcal{R}$ as the uncovered set in the sequel. If $\mu((\varepsilon, \infty)) = \infty$ for some $\varepsilon > 0$, then the set $\{t : \ell_t > \varepsilon\}$ is everywhere dense a.s., and it follows that $\mathcal{R} = \{0\}$ a.s. This trivial case is henceforth excluded, and we denote by $\overline{\mu}(x) = \mu((x, \infty))$, $x > 0$, the tail of μ.

The problem of finding a necessary and sufficient condition for $\mathcal{R}$ to reduce to $\{0\}$, was raised by Mandelbrot [113] and solved by Shepp [143]. Previously, Dvoretzky asked a closely related question on covering the circle with random arcs; see chapter 11 in Kahane [91] for further references on this topic. To tackle this question, we will follow a method due to Fitzsimmons, Fristedt and Shepp [55], which also enables us to settle many other natural questions about $\mathcal{R}$. The approach relies on the following intuitively obvious observation:

Lemma 7.1 *If 0 is not isolated in $\mathcal{R}$ a.s., then $\mathcal{R}$ is a perfect regenerative set.*

Proof: We first verify that the uncovered set is progressively measurable. Take any $0 < s < t$ and note that

$$[s,t] \subseteq \bigcup_{v \geq 0} I_v \iff [s,t] \subseteq \bigcup_{0 \leq v \leq t,\, \ell_v > 1/n} I_v \quad \text{for some large enough } n\,.$$

Indeed, the interval I_v does not intersect $[s,t]$ for $v \geq t$; and from any cover of $[s,t]$ by a family of open intervals, we can extract a cover by a finite sub-family. Next, fix an integer n. The Poisson point process ℓ restricted to $(1/n, \infty)$ is discrete (since $\overline{\mu}(1/n) < \infty$); and it can be easily deduced that the event

$$\{[s,t] \text{ is covered by } (I_v : \ell_v > 1/n \text{ and } 0 \leq v \leq t)\}$$

is $\mathcal{M}_t$-measurable. Hence, the event $\{[s,t] \text{ is covered by } (I_v, v \geq 0)\}$ is also $\mathcal{M}_t$-measurable. Writing $G_t = g_{t+} = \sup\{u \leq t : u \in \mathcal{R}\}$, the equivalence

$$G_t < s \iff [s,t] \subseteq \bigcup_{v \geq 0} I_v$$

shows that the right-continuous process $(G_t : t \geq 0)$ is adapted, and thus optional. It follows that $\mathcal{R} = \{t : t - G_t = 0\}$ is progressively measurable.

We next check that $\mathcal{R}$ has no isolated points a.s. For any fixed $t > 0$, it is easily seen that $D_{t-} = \inf\{s \geq t : s \in \mathcal{R}\}$ is an announceable stopping time [1]. It is well known that a Poisson point process does not jump at an announceable stopping time, so the shifted point process $\ell' = \left(\ell_{D_{t-}+s}, s \geq 0\right)$ is again a Poisson point process with intensity μ. Since the collection of intervals $(I_v : 0 \leq v < D_{t-})$ do not cover D_{t-}, they do not cover any $s > D_{t-}$ either. In other words, $s > D_{t-}$ is covered by the intervals $(I_v : v \geq 0)$ if and only if $s - D_{t-}$ is covered by $((v, v + \ell'_v) : v \geq 0)$. We know by assumption that 0 is not isolated in $\mathcal{R}$ a.s., and this implies that D_{t-} is not isolated in $\mathcal{R}$ either. Any positive instant in $\mathcal{R}$ which is isolated on its left can be expressed in the form D_{t-} for some rational number $t > 0$. We conclude that $\mathcal{R}$ has no isolated points a.s.

Finally, we establish the regenerative property. Let T be an arbitrary $(\mathcal{M}_t)$-stopping time, which is a right-accumulation point of $\mathcal{R}$ a.s. on $\{T < \infty\}$. Then T is not a jump time of ℓ, for if it were, then I_T would be a right-neighbourhood of T. As a consequence, conditionally on $\{T < \infty\}$, the shifted point process $\ell' = (\ell_{T+t}, t \geq 0)$ is independent of $\mathcal{M}_T$ and is again a Poisson point process with intensity μ. By the same argument as in the preceding paragraph, an instant $s > T$ is covered by the intervals $(I_t : t \geq 0)$, if and only if $s - T$ is covered by $((t, t + \ell'_t) : t \geq 0)$. This shows that $\mathcal{R}$ is regenerative. ∎

---

[1]Specifically, consider the process $X_u = \sup\{s + \ell_s - u, 0 \leq s < u\}$, $u \geq 0$; note that X is adapted with càdlàg paths and no negative jumps. In this setting D_{t-} coincides with the limit of the increasing sequence of stopping times $\inf\{s \geq q : X_s \leq 1/n\}$, $n = 1, 2, \cdots$.

7.2 The Laplace exponent of the uncovered set

Lemma 7.1 enables us to identify the uncovered set as the range of some subordinator σ, whenever 0 is not isolated in $\mathcal{R}$. This will allow us to derive information on $\mathcal{R}$ from known results of subordinators, if we are able to characterize σ in terms of the characteristic measure μ of the Poisson point process. This motivates the main result of this section, which provides an explicit formula for the Laplace exponent Φ of σ. Recall that $\bar{\mu}$ denotes the tail of μ.

Theorem 7.2 (Fitzsimmons, Fristedt and Shepp [55]) *If*

$$\int_0^1 \exp\left\{ \int_t^1 \bar{\mu}(s)ds \right\} dt = \infty,$$

then $\mathcal{R} = \{0\}$ *a.s. Otherwise,* $\mathcal{R}$ *is a perfect regenerative set, and the Laplace exponent of the corresponding subordinator is given by*

$$\frac{1}{\Phi(\lambda)} = c \int_0^\infty e^{-\lambda t} \exp\left\{ \int_t^1 \bar{\mu}(s)ds \right\} dt, \qquad \lambda > 0,$$

where $c > 0$ *is the constant of normalization (recall that* $\Phi(1) = 1$*).*

Using the fact that the Laplace transform of the renewal measure is $1/\Phi$, one can rephrase the statement as follows: When the uncovered set is not trivial, the renewal measure is absolutely continuous with density

$$u(t) = c \exp\left\{ \int_t^1 \bar{\mu}(s)ds \right\}, \qquad t \geq 0. \tag{7.1}$$

For instance, when the tail of the characteristic measure is $\bar{\mu}(x) = \beta x^{-1}$ for some $\beta > 0$, then $\exp\left\{ \int_t^1 \bar{\mu}(s)ds \right\} = t^{-\beta}$. We get from Theorem 7.2 that $\mathcal{R}$ reduces to $\{0\}$ a.s. if $\beta \geq 1$, and otherwise $\Phi(\lambda) = \lambda^{1-\beta}$, that is $\mathcal{R}$ is the range of a stable subordinator of index $1 - \beta$.

Proof: We will prove the theorem first in the simple case when the Poisson point process is discrete, and then deduce the general case by approximation. So we first suppose that $\bar{\mu}(0+) < \infty$; in particular the integral in Theorem 7.2 converges. Then ℓ is a discrete Poisson point process and $\mathcal{R}$ plainly contains some right-neighbourhood of the origin. *A fortiori* 0 is not isolated in $\mathcal{R}$ a.s., and by Lemma 7.1, $\mathcal{R}$ is a heavy regenerative set.

A fixed time $t > 0$ is uncovered if and only if $\ell_s \leq t - s$ for every $s < t$; which entails that

$$\mathbb{P}(t \in \mathcal{R}) = \exp\left\{ -\int_0^t \bar{\mu}(t-s)ds \right\} > 0.$$

It then follows from Proposition 1.9(ii) that the renewal density of $\mathcal{R}$ at t is proportional to $\exp\{-\int_0^t \bar{\mu}(t-s)ds\}$, which is the same as (7.1), and this proves the theorem in the discrete case.

We then deduce the general case when $\bar{\mu}(0+) = \infty$ by approximation. For every integer $n > 0$, let $\ell^{(n)} = (\ell_t : t \geq 0$ and $\ell_t > 1/n)$ denote the discrete Poisson point process restricted to $(1/n, \infty)$, and $\mathcal{R}^{(n)}$ the corresponding uncovered set. We know that the Laplace exponent associated with $\mathcal{R}^{(n)}$ is given by

$$\frac{1}{\Phi^{(n)}(\lambda)} = c_n \int_0^\infty e^{-\lambda t} \exp\left\{\int_t^1 \overline{\mu}(s \vee n^{-1})ds\right\} dt \,.$$

For every $s > 0$, $\overline{\mu}(s \vee n^{-1})$ increases to $\overline{\mu}(s)$ as $n \to \infty$. It follows that the probability measure on $[0, \infty)$,

$$c_n e^{-t} \exp\left\{\int_t^1 \overline{\mu}(s \vee n^{-1})ds\right\} dt$$

converges in the weak sense towards

$$c_\infty e^{-t} \exp\left\{\int_t^1 \overline{\mu}(s)ds\right\} dt$$

(where c_∞ is the normalizing constant) if $\int_0^1 \exp\left\{\int_t^1 \overline{\mu}(s)ds\right\} dt < \infty$, and towards the Dirac point mass at 0 otherwise. Considering Laplace transforms, we deduce that for every $\lambda > 0$, $\lim_{n \to \infty} \Phi^{(n)}(\lambda) = \Phi^{(\infty)}(\lambda)$, where

$$\frac{1}{\Phi^{(\infty)}(\lambda)} = \begin{cases} 1 & \text{if } \int_0^1 \exp\left\{\int_t^1 \overline{\mu}(s)ds\right\} dt = \infty \,, \\ c_\infty \int_0^\infty e^{-\lambda t} \exp\left\{\int_t^1 \overline{\mu}(s)ds\right\} dt & \text{otherwise.} \end{cases} \tag{7.2}$$

On the other hand, $\left(\mathcal{R}^{(n)} : n \in \mathbb{N}\right)$ is a decreasing sequence of random closed sets and $\mathcal{R} = \bigcap \mathcal{R}^{(n)}$. As a consequence, we have

$$G_t^{(n)} = \sup\{s \leq t : s \in \mathcal{R}^{(n)}\} \longrightarrow \sup\{s \leq t : s \in \mathcal{R}\} = G_t \qquad (\text{as } n \to \infty).$$

We deduce from Lemma 1.11 that for every $\lambda > 0$

$$\int_0^\infty e^{-t} \mathbb{E}\left(\exp\{-\lambda G_t\}\right) dt = \frac{1}{\Phi^{(\infty)}(\lambda+1)}. \tag{7.3}$$

Suppose first that $\int_0^1 \exp\left\{\int_t^1 \overline{\mu}(s)ds\right\} dt < \infty$. We see from (7.2) that $\Phi^{(\infty)}(\lambda)$ goes to ∞ as $\lambda \to \infty$. Together with (7.3), this forces $\mathbb{P}(G_t = 0) = 0$ for almost every $t \geq 0$; which means that 0 is not isolated in $\mathcal{R}$. We then know from Lemma 7.1 that $\mathcal{R}$ is regenerative; comparing (7.3) and Lemma 1.11 shows that its Laplace exponent must be $\Phi = \Phi^{(\infty)}$.

Finally, suppose that $\int_0^1 \exp\left\{\int_t^1 \overline{\mu}(s)ds\right\} dt = \infty$, so $\lim_{n \to \infty} \Phi^{(n)}(\lambda) = 1$ for every $\lambda > 0$. We deduce from (7.3) that $\mathbb{P}(G_t = 0) = 1$ for almost every $t \geq 0$, that is $\mathcal{R} = \{0\}$ a.s. ∎

7.3 Some properties of the uncovered set

We suppose throughout this subsection that

$$\int_0^1 \exp\left\{\int_t^1 \overline{\mu}(s)ds\right\} dt < \infty \,,$$

that is that $\mathcal{R}$ is not degenerate to the single point $\{0\}$, a.s. We immediately get the following features.

Corollary 7.3 $\mathcal{R}$ *is heavy or light according as the integral $\int_0^1 \overline{\mu}(t)dt$ converges or diverges.*

Proof: We know from Proposition 1.8 that a regenerative set is heavy or light according as the drift coefficient d is zero or positive. On the other hand, recall that

$$d = \lim_{\lambda \to \infty} \lambda^{-1} \Phi(\lambda).$$

According to Theorem 7.2, we have by an integration by parts

$$\frac{\lambda}{\Phi(\lambda)} = c \int_0^\infty \left(1 - e^{-\lambda t}\right) \overline{\mu}(t) \exp\left\{\int_t^1 \overline{\mu}(s)ds\right\} dt, \qquad \lambda > 0;$$

and we deduce by monotone convergence that

$$\frac{1}{d} = c \int_0^\infty \overline{\mu}(t) \exp\left\{\int_t^1 \overline{\mu}(s)ds\right\} dt = c \exp\left\{\int_0^1 \overline{\mu}(s)ds\right\} - c \exp\left\{-\int_1^\infty \overline{\mu}(s)ds\right\}.$$

We conclude that d $= 0$ iff $\int_0^1 \overline{\mu}(s)ds = \infty$.

Alternatively, one may also deduce the result from Proposition 1.9 and the easy fact that the probability that the point 1 is left uncovered equals $\exp\left\{-\int_0^1 \overline{\mu}(s)ds\right\}$. ∎

Corollary 7.4 *If $\int_1^\infty \exp\left\{-\int_1^t \overline{\mu}(s)ds\right\} dt = \infty$, then $\mathcal{R}$ is unbounded a.s. Otherwise, $\mathcal{R}$ is bounded a.s. and the distribution of the largest uncovered point*

$$g_\infty = \sup\{s \ge 0 : s \in \mathcal{R}\}$$

is given by

$$\mathbb{P}(g_\infty \in dt) = k^{-1} \exp\left\{\int_t^1 \overline{\mu}(s)ds\right\} dt, \qquad \text{with } k = \int_0^\infty \exp\left\{\int_t^1 \overline{\mu}(s)ds\right\} dt.$$

Proof: According to (2.2), the probability that $\mathcal{R}$ is bounded equals 0 or 1 according as the killing rate k $= \Phi(0)$ is zero or positive. It follows immediately from Theorem 7.2 that

$$k = 0 \iff \int_1^\infty \exp\left\{-\int_1^t \overline{\mu}(s)ds\right\} = \infty.$$

When $\mathcal{R} \ne \{0\}$ is bounded a.s., the formula for the distribution of g_∞ follows from Lemma 1.11 and the expression (7.1) for the density of the renewal measure. ∎

Motivated by the limit theorem 3.2 for the process of the last passage times in $\mathcal{R}$, we next investigate the asymptotic behaviour of the Laplace exponent Φ.

Corollary 7.5 *For every $\alpha \in (0, 1]$, the following assertions are equivalent:*

(i) $\overline{\mu}(s) \sim (1 - \alpha)s^{-1}$ *as $s \to \infty$ (for $\alpha = 1$, this means that $\overline{\mu}(s) = o(s^{-1})$).*

(ii) Φ *is regularly varying at 0+ with index α.*

Proof: Recall from Proposition 1.5 that Φ is regularly varying at $0+$ with index α if and only if the renewal function U is regularly varying at ∞ with index α. We know from (7.1) that U has a decreasing derivative u, so the monotone density theorem applies and (ii) holds if and only if u is regularly varying at ∞ with index $\alpha - 1$ (cf. [20] on page 39).

Using again (7.1), we have

$$t^{1-\alpha}u(t) = c\exp\left\{\int_1^t \left((1-\alpha)s^{-1} - \bar{\mu}(s)\right)ds\right\},$$

and it is then plain from the theorem of representation of slowly varying functions (cf. [20] on page 12) that (i) implies that u is regularly varying at ∞ with index $\alpha - 1$. Conversely, suppose that u is regularly varying at ∞ with index $\alpha - 1$, so that by the theorem of representation of slowly varying functions

$$\int_1^t \left((1-\alpha)s^{-1} - \bar{\mu}(s)\right)ds = c(t) + \int_1^t \varepsilon(s)s^{-1}ds,$$

where $\lim_{t\to\infty} c(t) \in \mathbb{R}$ and $\lim_{t\to\infty} \varepsilon(t) = 0$. It then follows readily from the monotonicity of $\bar{\mu}$ that this representation is possible only if (i) holds. ∎

We next turn our attention to the fractal dimensions of the uncovered set, which are given by the lower and upper indices of the Laplace exponent, see Theorem 5.1.

Corollary 7.6 *The lower and upper indices are given by*

$$\underline{\operatorname{ind}}(\Phi) = \sup\left\{\rho : \lim_{t\to 0+} t^{1-\rho}\exp\left\{\int_t^1 \bar{\mu}(s)ds\right\} = 0\right\} = 1 - \limsup_{t\to 0+}\left(\frac{\int_t^1 \bar{\mu}(s)ds}{\log 1/t}\right),$$

$$\overline{\operatorname{ind}}(\Phi) = \inf\left\{\rho : \lim_{t\to 0+} t^{1-\rho}\exp\left\{\int_t^1 \bar{\mu}(s)ds\right\} = \infty\right\} = 1 - \liminf_{t\to 0+}\left(\frac{\int_t^1 \bar{\mu}(s)ds}{\log 1/t}\right).$$

Proof: For the sake of conciseness, we focus on the lower index. We get from the formula for Φ in Theorem 7.2

$$\underline{\operatorname{ind}}(\Phi) = \sup\left\{\rho : \lim_{\lambda\to\infty} \lambda^\rho \int_0^\infty e^{-\lambda t}\exp\left\{\int_t^1 \bar{\mu}(s)ds\right\}dt = 0\right\}$$

$$= \sup\left\{\rho : \lim_{\lambda\to\infty} \lambda^{\rho-1} \int_0^\infty e^{-t}\exp\left\{\int_{t/\lambda}^1 \bar{\mu}(s)ds\right\}dt = 0\right\}.$$

Using the immediate inequality

$$\int_0^\infty e^{-t}\exp\left\{\int_{t/\lambda}^1 \bar{\mu}(s)ds\right\}dt \geq e^{-1}\exp\left\{\int_{1/\lambda}^1 \bar{\mu}(s)ds\right\},$$

we deduce

$$\underline{\operatorname{ind}}(\Phi) \leq \sup\left\{\rho : \lim_{t\to 0+} t^{1-\rho}\exp\left\{\int_t^1 \bar{\mu}(s)ds\right\} = 0\right\}.$$

To prove the converse inequality, we may suppose that there is $\rho > 0$ such that

$$\lim_{t\to 0+} t^{1-\rho}\exp\left\{\int_t^1 \bar{\mu}(s)ds\right\} = 0$$

(otherwise there is nothing to prove). Recall that the renewal measure has density u given by (7.1), so that $u(t) = o(t^{\rho-1})$ and then $U(t) = o(t^{\rho})$ as $t \to 0+$, for every $\varepsilon > 0$. Applying Proposition 1.4, this entails $\lim_{\lambda \to \infty} \lambda^{-\rho} \Phi(\lambda) = \infty$, and thus $\underline{\mathrm{ind}}\,(\Phi) \geq \rho$. ∎

The identification of the uncovered set in terms of a subordinator σ enables us to invoke results of section 3.2 to decide whether a given Borel set $B \subseteq (0, \infty)$ is completely covered by the random intervals. Typically, recall Propositions 5.6 and 5.7 which are relevant as the renewal density u is a decreasing function (by (7.1)). If the Hausdorff dimension of B is greater that $1 - \underline{\mathrm{ind}}\,(\Phi)$, then the probability that B is not completely covered is positive. On the other hand, if the Hausdorff dimension of B is less that $1 - \overline{\mathrm{ind}}\,(\Phi)$, then B is completely covered a.s. Of course, (5.6) provides a complete (but not quite explicit) characterization of sets which are completely covered by the random intervals.

Finally, let us mention an interesting problem raised by Pat Fitzsimmons (private communication). It is easily seen that the uncovered set $\mathcal{R}$ is an infinitely divisible regenerative set, in the sense that for every integer n, $\mathcal{R}$ can be expressed as the intersection of n-independent regenerative sets with the same distribution. Conversely, can any (perfect) infinitely divisible regenerative set be viewed of as a set left uncovered by random intervals sampled from a Poisson point process? Kendall [93] gave a positive answer in the heavy case. The light case seems to be still open.

Chapter 8

Lévy processes

Real-valued Lévy processes give rise to two interesting families of regenerative sets: the set of times when a fixed point is visited, and the set of times when a new supremum is reached. Some applications are given in the special case when the Lévy process has no positive jumps. Some applications of Bochner's subordination to Lévy processes are also discussed.

8.1 Local time at a fixed point

Throughout this chapter, $(X_t : t \geq 0)$ will denote a real-valued Lévy process, i.e. X has independent and homogeneous increments and càdlàg paths. For instance the difference of two independent strict subordinators is a Lévy process. For every $x \in \mathbb{R}$, write $\mathbf{P}^x$ for the distribution of the process $X + x$; it is well-known that $X = (\Omega, \mathcal{M}, \mathcal{M}_t, X_t, \theta_t, \mathbf{P}^x)$ is a Feller process (see e.g. [11], Chapter I). The purpose of this section is to study the regularity of a fixed point r, and then to determine the distribution of its local time. To this end, we need information on the resolvent operator V^q.

To start with, recall that the characteristic function of X_t can be expressed in the form

$$\mathbf{E}^0\left(e^{i\lambda X_t}\right) = e^{-t\Psi(\lambda)}, \qquad \lambda \in \mathbb{R}, t \geq 0,$$

where $\Psi : \mathbb{R} \to \mathbb{C}$. One calls Ψ the characteristic exponent of X; it can be expressed via the Lévy-Khintchine's formula (which is more general than that which we discussed in Section 1.2 in the special case of subordinator):

$$\Psi(\lambda) = ia\lambda + \frac{1}{2}b\lambda^2 + \int_{\mathbb{R}}(1 - e^{i\lambda x} + i\lambda x\mathbf{1}_{\{|x|<1\}})\Lambda(dx), \qquad (8.1)$$

where $a \in \mathbb{R}$, $b \geq 0$ is called the Gaussian coefficient, and Λ a measure on $\mathbb{R} - \{0\}$ with $\int(1 \wedge |x|^2)\Lambda(dx) < \infty$ called the Lévy measure. It follows that for every Lebesgue-integrable function f and $q > 0$, we have

$$\int_{-\infty}^{\infty} e^{i\lambda x} V^q f(x) dx = \int_{-\infty}^{\infty} e^{i\lambda x} \left(\int_0^{\infty} \mathbf{E}^x \left(f(X_t) \right) e^{-qt} dt \right) dx$$

$$= \int_0^{\infty} e^{-qt} \left(\int_{-\infty}^{\infty} e^{i\lambda x} \mathbf{E}^0 \left(f(X_t + x) \right) dx \right) dt$$

$$= \int_0^{\infty} e^{-qt} \left(\int_{-\infty}^{\infty} e^{i\lambda y} f(y) \mathbf{E}^0 \left(e^{-i\lambda X_t} \right) dy \right) dt$$

$$= \left(\int_0^{\infty} e^{-qt} \exp\{-t\Psi(-\lambda)\} dt \right) \left(\int_{-\infty}^{\infty} e^{i\lambda y} f(y) dy \right)$$

$$= \frac{1}{q + \Psi(-\lambda)} \left(\int_{-\infty}^{\infty} e^{i\lambda y} f(y) dy \right).$$

In other words, if $\mathcal{F}(g)$ stands for the Fourier transform of an integrable function g, then

$$\mathcal{F}(V^q f)(\lambda) = \frac{\mathcal{F}(f)(\lambda)}{q + \Psi(-\lambda)}. \tag{8.2}$$

We are now able to prove the following basic result which goes back to Orey [123].

Proposition 8.1 [1] *Suppose that the characteristic exponent Ψ satisfies*

$$\int_{-\infty}^{\infty} |q + \Psi(\lambda)|^{-1} d\lambda < \infty$$

for some (and then all) $q > 0$. Then every point $r \in \mathbb{R}$ is regular for itself and the Laplace exponent Φ of the inverse local time is given by

$$\frac{1}{\Phi(q)} = c \int_{-\infty}^{\infty} \frac{d\lambda}{q + \Psi(\lambda)}, \qquad q > 0,$$

where $c > 0$ is the constant of normalization.

Proof: The function

$$v^q(x) = \frac{1}{2\pi} \int_{-\infty}^{\infty} \frac{e^{-ix\lambda}}{q + \Psi(\lambda)} d\lambda, \qquad x \in \mathbb{R},$$

is continuous and its Fourier transform is $\lambda \to 1/(q + \Psi(\lambda))$. By Fourier inversion, we deduce from (8.2) that

$$V^q f(x) = \int_{-\infty}^{\infty} f(y) v^q(y - x) dy.$$

---

[1]We mention for completeness that Bretagnolle [30] has established a sharper and much more difficult result: a necessary and sufficient condition for 0 to be regular for itself is

$$\int_{-\infty}^{\infty} \Re\left(\frac{1}{1 + \Psi(\lambda)} \right) d\lambda < \infty \quad \text{and} \quad X \text{ has unbounded variation.}$$

In other words, the q-resolvent operator of X has a continuous density kernel $v^q(x,y) = v^q(y-x)$ with respect to the Lebesgue measure. Plainly $\widehat{X} = -X$ is also a Lévy process and the very same calculations show that its q-resolvent operator is given by

$$\widehat{V}^q f(x) = \int_{-\infty}^{\infty} f(y) v^q(x - y) dy .$$

Hence, X and $\widehat{X}$ are in duality with respect to the Lebesgue measure, and the condition (iii) of Proposition 2.2 is fulfilled. This yields our statement. ∎

It is easily seen that when local times exist, they can be expressed as occupation densities, in the sense that the local time at level $r \in \mathbb{R}$ is given by $L(r, \cdot) = \lim_{\varepsilon \to 0+} (2\varepsilon)^{-1} \int_0^{\cdot} \mathbf{1}_{\{|X_t - r| \le \varepsilon\}} dt$. See section V.1 in [11] for details. A major problem in this field is to decide whether the mapping $(r,t) \to L(r,t)$ has a continuous version. This has been solved in a remarkable paper by Barlow [4], see also [3] and Marcus and Rosen [114] in the symmetric case.

Proposition 8.1 provides a simple expression for the Laplace exponent Φ of the inverse local time, which is explicit in terms of the characteristic exponent Ψ. This enables one to directly apply the general results proven in the preceding chapters; here is an example. Suppose for simplicity that X is symmetric and that the condition of Proposition 8.1 is fulfilled. We should like to express the condition

$$\Phi \text{ is regularly varying with index } \rho \in (0,1) \qquad (\text{at } 0+, \text{ resp. at } \infty) \qquad (8.3)$$

in terms of Ψ. This question is motivated for instance by the Dynkin-Lamperti theorem 3.2. Alternatively, (8.3) has an important rôle in the law of the iterated logarithm for local times (which has been considered in particular by Marcus and Rosen [115, 116]). The assumption of symmetry ensures that the characteristic exponent Ψ is an even real-valued function. We write $\Psi^\uparrow$ for the so-called increasing rearrangement of Ψ, viz.

$$\Psi^\uparrow(x) = m(\lambda \in \mathbb{R} : \Psi(\lambda) \le x) \qquad (x \ge 0)$$

where m refers to the Lebesgue measure. By Proposition 8.1, we have

$$\frac{1}{c\Phi(q)} = \int_{[0,\infty)} \frac{1}{q+x} d\Psi^\uparrow(x) = \int_{[0,\infty)} \left(\int_0^\infty e^{-(q+x)t} dt \right) d\Psi^\uparrow(x)$$
$$= \int_0^\infty e^{-qt} \mathcal{L}\Psi^\uparrow(t) dt$$

where $\mathcal{L}\Psi^\uparrow(t) = \int_{[0,\infty)} e^{-tx} d\Psi^\uparrow(x)$ is the Laplace transform of the measure with distribution function $\Psi^\uparrow$. Because Φ is regularly varying with index ρ if and only if $1/\Phi$ is regularly varying with index $-\rho$, we deduce from a tauberian theorem that (8.3) holds if and only if the indefinite integral of $\mathcal{L}\Psi^\uparrow$, $\int_0^t \mathcal{L}\Psi^\uparrow(t) dt$, is regularly varying with index ρ (at ∞, resp. at $0+$). Plainly, the indefinite integral of $\mathcal{L}\Psi^\uparrow$ has a decreasing derivative, so by the monotone density theorem, the latter is equivalent to $\mathcal{L}\Psi^\uparrow$ varying regularly with index $\rho - 1$ (at ∞, resp. at $0+$). We then again invoke a tauberian theorem to conclude that

$$(8.3) \qquad \Longleftrightarrow \qquad \Psi^\uparrow \text{ varies regularly with index } 1 - \rho. \qquad (\text{at } 0+, \text{ resp. at } \infty).$$

More precisely, the preceding argument shows that when (8.3) holds, then

$$\Phi(q) \sim c'q/\Psi^{\dagger}(q) \qquad \text{(at 0+, resp. at } \infty)$$

for some positive finite constant number c' which can be expressed explicitly in terms of our data.

8.2 Local time at the supremum

We next turn our attention the supremum process $S. = \sup\{X_s : 0 \leq s \leq \cdot\}$. It is easy to check that the so-called reflected process $S - X$ is a Feller process; see Proposition VI.1 in [11]. The closed zero set of the reflected process

$$\mathcal{R} = \overline{\{t \geq 0 : X_t = S_t\}}$$

coincides with the set of times when the Lévy process reaches a new supremum. It is known as the ladder time set. There is a simple criterion due to Rogozin [133] to decide whether 0 is regular for itself with respect to the reflected process:

$$\mathcal{R} \text{ is perfect} \quad \Longleftrightarrow \quad \int_{0+} t^{-1}\mathbf{P}^0(X_t \geq 0)dt = \infty.$$

See also [13] for an equivalent condition in terms of the Lévy measure of X.

We henceforth suppose that $\mathcal{R}$ is perfect; the Laplace exponent of the ladder time set can be expressed as follows:

$$\Phi(q) = \exp\left\{\int_0^\infty \left(e^{-t} - e^{-qt}\right) t^{-1}\mathbf{P}^0(X_t \geq 0)dt\right\}, \qquad q \geq 0. \qquad (8.4)$$

Formula (8.4) is a special case of a result of Fristedt (see e.g. Corollary VI.10 in [11] and the comments thereafter), and goes back to Spitzer in discrete time. The main drawback of (8.4) is that it involves the probabilities $\mathbf{P}^0(X_t \geq 0)$ which are usually not known explicitly. For instance, Bingham [19] has raised the question of determining the class of Laplace exponents which can arise in connection with ladder time sets. This interesting problem seems to be still open. [2]

As an example of an application of (8.4) motivated by Chapter 4, we consider the question of whether the Laplace exponent of a ladder time set has the asymptotic behaviour that is required in the Dynkin-Lamperti Theorem 3.2.

Proposition 8.2 *For each fixed $\alpha \in [0,1]$, Φ is regularly varying with index α at 0+ (respectively, at ∞) if and only if*

$$\lim \frac{1}{t}\int_0^t \mathbf{P}^0(X_s \geq 0)ds = \alpha \qquad \text{as } t \to \infty \text{ (respectively, as } t \to 0+). \qquad (8.5)$$

---

[2]By an application of the Frullani integral to (8.4), one sees that the function $q \to q/\Phi(q)$ must be the Laplace exponent of a subordinator; cf. the proof of Theorem 8.3 below. In particular ladder time processes form a strict sub-class of subordinators.

Proof: We know from Theorem 3.2 that Φ is regularly varying with index α at $0+$ (respectively, at ∞) if and only if $\lim q\Phi'(q)/\Phi(q) = \alpha$ as $q \to 0+$ (respectively, as $q \to \infty$). According to (8.4), the logarithmic derivative of Φ is given by

$$\frac{\Phi'(q)}{\Phi(q)} = \int_0^\infty e^{-qt}\mathbf{P}^0(X_t \geq 0)dt.$$

By a Tauberian theorem, the right-hand side is equivalent to α/q if and only if (8.5) holds. ∎

One refers to (8.5) as Spitzer's condition; it has a crucial rôle in developing fluctuation theory for Lévy processes, in particular in connection with estimates for the distribution of first passage times and for the asymptotic behaviour of the time spent by the Lévy process in the positive semi-axis. See Chapter VI in [11]. It is natural to compare (8.5) with the apparently stronger condition

$$\lim \mathbf{P}^0(X_t \geq 0) = \alpha \qquad \text{as } t \to \infty \text{ (respectively, as } t \to 0+). \tag{8.6}$$

We will refer to (8.6) as Doney's condition, for Doney [47] has recently proven that the discrete time versions of (8.5) and (8.6) are equivalent, settling a question that has puzzled probabilists for a long time. We present here the analogous result in continuous time.

Theorem 8.3 *The conditions of Spitzer and Doney are equivalent.*

Proof: We shall only prove the theorem for $0 < \alpha < 1$ and $t \to 0+$, and we refer to [17] for the complete argument. The implication (8.6) $\Rightarrow$ (8.5) is obvious, so we assume that (8.5) holds. Notice that the case when X is a compound Poisson process with a possible drift is then ruled out; this ensures that $\mathbf{P}^0(X_t = 0) = 0$ for all $t > 0$, and as a consequence, the mapping $t \to \mathbf{P}^0(X_t \geq 0)$ is continuous on $(0, \infty)$.

Introduce the Laplace exponent $\hat{\Phi}$ of the dual ladder time set which corresponds to the Lévy process $\hat{X} = -X$. This means

$$\begin{aligned}
\hat{\Phi}(q) &= \exp\left\{\int_0^\infty \left(e^{-t} - e^{-qt}\right)t^{-1}\mathbf{P}^0(X_t < 0)dt\right\} \\
&= \exp\left\{\int_0^\infty \left(e^{-t} - e^{-qt}\right)t^{-1}\left(1 - \mathbf{P}^0(X_t \geq 0)\right)dt\right\} \\
&= q/\Phi(q),
\end{aligned}$$

where the last equality follows from the Frullani integral. As a consequence, (8.4) yields

$$\int_0^\infty e^{-qt}\mathbf{P}^0(X_t \geq 0)dt = \Phi'(q)/\Phi(q) = \Phi'(q)\hat{\Phi}(q)/q. \tag{8.7}$$

We know from Proposition 8.2 that Φ is regularly varying at ∞ with index α, and also that $\hat{\Phi}$ is regularly varying at ∞ with index $1 - \alpha$. Because Φ and $\hat{\Phi}$ are Laplace exponents of subordinators with zero drift, we obtain from the Lévy-Khintchine formula that

$$\Phi'(q) = \int_{(0,\infty)} e^{-qx}x\,d\left(-\overline{\Pi}(x)\right) \quad , \quad \hat{\Phi}(q)/q = \int_0^\infty e^{-qx}\overline{\hat{\Pi}}(x)dx,$$

where $\overline{\overline{\Pi}}$ (respectively, $\widehat{\overline{\overline{\Pi}}}$) is the tail of the Lévy measure of the ladder time process of X (respectively, of $\widehat{X}$). We now get from (8.7)

$$\mathbf{P}^0(X_t \geq 0) = \int_{(0,t)} \widehat{\overline{\overline{\Pi}}}(t-s)sd\left(-\overline{\overline{\Pi}}(s)\right) \qquad \text{for a.e. } t > 0. \qquad (8.8)$$

By a change of variables, the right-hand side can be re-written as

$$t\int_{(0,1)} \widehat{\overline{\overline{\Pi}}}(t(1-u))ud\left(-\overline{\overline{\Pi}}(tu)\right) = \int_{(0,1)} \frac{\widehat{\overline{\overline{\Pi}}}(t(1-u))}{\widehat{\Phi}(1/t)}ud\left(-\frac{\overline{\overline{\Pi}}(tu)}{\Phi(1/t)}\right).$$

Now, apply the second part of Proposition 1.5. For every fixed $\varepsilon \in (0,1)$, we have uniformly on $u \in [\varepsilon, 1-\varepsilon]$ as $t \to 0+$:

$$\frac{\overline{\overline{\Pi}}(tu)}{\Phi(1/t)} \to \frac{u^{-\alpha}}{\Gamma(1-\alpha)} \quad , \quad \frac{\widehat{\overline{\overline{\Pi}}}(t(1-u))}{\widehat{\Phi}(1/t)} \to \frac{(1-u)^{\alpha-1}}{\Gamma(\alpha)}.$$

Recall $\mathbf{P}^0(X_t \geq 0)$ depends continuously on $t > 0$. We deduce from (8.8) that

$$\liminf_{t\to 0+} \mathbf{P}^0(X_t \geq 0) \geq \frac{\alpha}{\Gamma(\alpha)\Gamma(1-\alpha)} \int_\varepsilon^{1-\varepsilon} (1-u)^{\alpha-1}u^{-\alpha}du,$$

and as ε can be picked arbitrarily small, $\liminf_{t\to 0+} \mathbf{P}^0(X_t \geq 0) \geq \alpha$. The same argument for the dual process gives $\liminf_{t\to 0+} \mathbf{P}^0(X_t < 0) \geq 1-\alpha$, which establishes (8.5). $\blacksquare$

If, as usual, we denote by σ the inverse local time at 0 of the reflected process $S-X$, then it is easy to check from the stationarity and independence of the increments of X that the time-changed process $H = X \circ \sigma = S \circ \sigma$ is again a subordinator. One calls H the ladder height process; it has several interesting applications in fluctuation theory for Lévy processes. We refer to sections 4 and 5 of chapter VI in [11] for more on this topic.

8.3 The spectrally negative case

Throughout this section, we suppose that the real-valued Lévy process X has no positive jumps, one sometimes says that X is spectrally negative. The degenerate case when either X is the negative of a subordinator or a deterministic drift has no interest and will be implicitly excluded in the sequel. We refer to Chapter VII in [11] for a detailed account of the theory of such processes.

The absence of positive jumps enables to use the same argument as that in section 1.1 to show that the first passage process of X

$$\sigma_t = \inf\{s \geq 0 : X_s > t\} \qquad (t \geq 0)$$

is a subordinator. The inverse of σ coincides with the (continuous) supremum process S of X, so S serves as a local time on the set of times when X reaches a new supremum, that when $S = X$. In other words, S is proportional to the local time at 0 of the reflected process $S - X$.

As usual, we denote the Laplace exponent of σ by Φ. Note that if T stands for an independent exponential time, say with parameter $q > 0$, then

$$\mathbf{P}^0 \left(S_T > x \right) = \mathbf{P}^0 \left(\sigma_x < T \right) = \mathbf{E}^0 \left(\exp\{-q\sigma_x\} \right) = e^{-x\Phi(q)}$$

for every $x > 0$, so that S_T has an exponential distribution with parameter $\Phi(q)$. By taking q sufficiently large, we see that for every fixed $t > 0$, S_t has a finite exponential moment of any order. As a consequence, though X may take values of both signs, its exponential moments are finite. This enables us to study X using the Laplace transform instead of the Fourier transform. More precisely, the characteristic exponent can be continued analytically on the lower half-plane $\{z \in \mathbb{C} : \Im(z) < 0\}$. We then put $\psi(\lambda) = \Psi(-i\lambda)$ for $\lambda > 0$, so that

$$\mathbf{E}^0(\exp\{\lambda X_t\}) = \exp\{t\psi(\lambda)\}, \qquad \lambda \geq 0.$$

Invoking Hölder's inequality, we see that the mapping $\psi : [0, \infty) \to (-\infty, \infty)$ is strictly convex. On the other hand, we also deduce from the monotone convergence theorem that $\lim_{\lambda \to \infty} \psi(\lambda) = \infty$.

We are now able to specify the Laplace exponent Φ.

Proposition 8.4 *We have* $\Phi \circ \psi(\lambda) = \lambda$ *for every* $\lambda \geq 0$ *such that* $\psi(\lambda) > 0$.

Proof: It follows from the independence and stationarity of the increments that the process

$$\exp\{\lambda X_s - \psi(\lambda)s\}, \qquad s \geq 0$$

is a nonnegative martingale. As X cannot jump above the level t, we must have $X_{\sigma_t} = t$ on $\{\sigma_t < \infty\}$. On the other hand, the assumption that $\psi(\lambda) > 0$ ensures that the martingale converges a.s. to 0 as $s \to \infty$ on the event $\{\sigma_t = \infty\}$. An application of the optional sampling theorem at the stopping time σ_t yields

$$\mathbf{E}^0 \left(\exp\{\lambda t - \psi(\lambda)\sigma_t\}, \sigma_t < \infty \right) = 1.$$

Recall the convention $e^{-\infty} = 0$; the preceding identity can be re-written as

$$\exp\{-\lambda t\} = \mathbf{E}^0 \left(\exp\{-\psi(\lambda)\sigma_t\} \right) = \exp\{-t\Phi(\psi(\lambda))\},$$

which establishes our claim. ∎

In comparison with (8.4), Proposition 8.4 provides an explicit expression for the Laplace exponent Φ directly in terms of our data (namely, ψ) which is much easier to deal with. For instance, it is immediately seen that Φ is regularly varying with index $\rho \in [0, 1]$ if and only if ψ is regularly varying with index $1/\rho$ (which forces in fact ρ to be greater than or equal to $1/2$). In the same vein, the lower and upper indices of Φ are given by

$$\underline{\mathrm{ind}}\,(\Phi) = \sup\left\{ \rho > 0 : \lim_{\lambda \to \infty} \psi(\lambda)\lambda^{-1/\rho} = 0 \right\}$$

$$\overline{\mathrm{ind}}\,(\Phi) = \inf\left\{ \rho > 0 : \lim_{\lambda \to \infty} \psi(\lambda)\lambda^{-1/\rho} = \infty \right\}.$$

As another example of application, we derive the following extension of Khintchine's law of the iterated logarithm (see also [10] for further results in the same vein).

Corollary 8.5 *There is a positive constant c such that*

$$\limsup_{t \to 0+} \frac{X_t \Phi(t^{-1} \log |\log t|)}{\log |\log t|} = c \qquad a.s.$$

Proof: Consider the functions

$$f(t) = \frac{\Phi\left(t^{-1} \log \log \Phi(t^{-1})\right)}{\log \log \Phi(t^{-1})} \quad \text{and} \quad \tilde{f}(t) = \frac{\Phi\left(t^{-1} \log \log \Phi(t^{-1})\right)}{\log \log \Phi(t^{-1} \log \log \Phi(t^{-1}))}.$$

The function $s \to s / \log \log s$ is monotone increasing on some neighbourhood of ∞ and the function $t \to \Phi(t^{-1} \log \log \Phi(t^{-1}))$ decreases. We deduce that the compound function $\tilde{f}$ decreases on some neighbourhood of 0. Moreover, it is easily seen that

$$\log \log \Phi(t^{-1} \log \log \Phi(t^{-1})) \sim \log \log \Phi(t^{-1})$$

(cf. the proof of Lemma 4.2), so that $f(t) \sim \tilde{f}(t)$ as $t \to 0+$.

Because the supremum process S is proportional to the local time at 0 of $S - X$, we deduce from Theorem 4.1 that $\limsup_{t \to 0+} S_t \tilde{f}(t) = c$ a.s. for some positive constant c. By an obvious argument of monotonicity, we may replace S by X in the preceding identity. So all that we need now is to check that

$$\tilde{f}(t) \sim \frac{\Phi(t^{-1} \log |\log t|)}{\log |\log t|} \qquad (t \to 0+).$$

On the one hand, it is easily seen from the Lévy-Khintchine formula for ψ that $\limsup_{\lambda \to \infty} \lambda^{-2} \psi(\lambda) < \infty$, which in turn implies that $\liminf_{\lambda \to \infty} \lambda^{-1/2} \Phi(\lambda) > 0$. On the other hand, recall that Φ is concave, so that $\limsup_{\lambda \to \infty} \lambda^{-1} \Phi(\lambda) < \infty$. We deduce that

$$\log \log \Phi(t^{-1}) \sim \log |\log t| \qquad \text{as } t \to 0+,$$

and then, since Φ is concave and increasing, that

$$\Phi\left(t^{-1} \log \log \Phi(t^{-1})\right) \sim \Phi(t^{-1} \log |\log t|) \qquad \text{as } t \to 0+.$$

Our claim follows. ∎

We refer to Jaffard [86] and the references therein for further results on the regularity of the paths of Levy processes, in particular precise information on their multifractal structure.

8.4 Bochner's subordination for Lévy processes

Bochner [25] introduced the concept of subordination (after which subordinators were named) of Markov processes as follows. Let $M = (\Omega, \mathcal{M}, \mathcal{M}_t, M_t, \theta_t, \mathbf{P}^x)$ be some time-homogeneous Markov process and $\sigma = (\sigma_t : t \geq 0)$ a subordinator that is independent of M. The process $\tilde{M} = \left(\tilde{M}_t = M_{\sigma_t} : t \geq 0\right)$ obtained from M by time-substitution based on σ (with the convention that $M_\infty = \Upsilon$ where Υ is a cemetery point for M) is referred to as the subordinate process of M with directing process σ. It is easily seen that the homogeneous Markov property is preserved by this time-substitution, in the sense that the process $\tilde{M} = \left(\Omega, \mathcal{M}, \tilde{\mathcal{M}}_t, \tilde{M}_t, \tilde{\theta}_t, \mathbf{P}^x\right)$ is again Markovian, where $\tilde{\mathcal{M}}_t = \mathcal{M}_{\sigma_t}$ and $\tilde{\theta}_t = \theta_{\sigma_t}$. More precisely, the semigroup $\left(\tilde{Q}_t : t \geq 0\right)$ of $\tilde{M}$ is given in terms of the semigroup $(Q_t : t \geq 0)$ of M and the distribution of σ by

$$\tilde{Q}_t(x, dy) = \int_{[0,\infty)} Q_s(x, dy) \mathbb{P}(\sigma_s \in dt). \tag{8.9}$$

We refer to Feller [53], Bouleau [27] and Hirsch [72] for more on this topic. See also Bakry [2], Jacob and Schilling [84, 85], Meyer [119] and the references therein for applications in analysis (in particular to the Riesz transform and the Paley-Wiener theory); and Bouleau and Lépingle [28] for applications to simulation methods.

We now consider the special case when the Markov process is a Lévy process, i.e. $M = X$. In order to avoid problems related to killing, we will also suppose that σ is a strict subordinator. From an analytic viewpoint, this means that the semigroup $(Q_t : t \geq 0)$ is a Markovian convolution semigroup, namely

$$Q_t f(x) = \int_{\mathbb{R}} f(x + y) \mathbf{P}^0(X_t \in dy)$$

for every Borel bounded function f. It follows from (8.9) that $\left(\tilde{Q}_t : t \geq 0\right)$ is also a Markovian convolution semigroup, i.e. the subordinate process $\tilde{X}$ is again a Lévy process.

Because the law of a Lévy process is specified by the characteristic exponent Ψ, it is natural to search for an expression of the characteristic exponent $\tilde{\Psi}$ of the subordinate Lévy process $\tilde{X}$. To this end, observe first that Ψ maps $\mathbb{R}$ into $\mathbb{C}_+ = \{z \in \mathbb{C} : \Re z \geq 0\}$, and second (from the Lévy-Khintchine formula) that the Laplace exponent Φ of a subordinator can be continued analytically on $\mathbb{C}_+$; we will still denote by Φ this extension. It should be clear that

$$\mathbb{E}\left(e^{-z\sigma_t}\right) = e^{-t\Phi(z)} \qquad \text{for any } z \in \mathbb{C}_+.$$

As X and σ are independent, we then get

$$\mathbf{E}^0\left(\exp\{i\lambda X_{\sigma_t}\}\right) = \mathbb{E}\left(\exp\{-\Psi(\lambda)\sigma_t\}\right) = \exp\{-t\Phi(\Psi(\lambda))\},$$

which proves the following statement:

Proposition 8.6 (Bochner [25]) *Let X be a Lévy process with characteristic exponent Ψ and σ an independent subordinator with Laplace exponent Φ. Then the subordinate process $\tilde{X} = X \circ \sigma$ is a Lévy process with characteristic exponent*

$$\tilde{\Psi} = \Phi \circ \Psi.$$

We refer to the second chapter of Chateau [34] for a study of the so-called subordination process, in which the subordinator σ is viewed as a parameter.

We now quote without proof a result of Huff [80], who has been able to make explicit the Lévy-Khintchine formula (8.1) for the subordinate process $\tilde{X}$. In the obvious notation, we have

$$\tilde{a} = \mathrm{d}a + \int_{(0,\infty)} \mathbf{E}^0\left(X_t, |X_t| < 1\right) \Pi(dt) \quad , \quad \tilde{b} = \mathrm{d}b,$$

$$\tilde{\Lambda}(dx) = \mathrm{d}\Lambda(dx) + \int_{(0,\infty)} \mathbf{P}^0\left(X_t \in dx\right) \Pi(dt).$$

Here is a classical example of Proposition 8.6 due to Spitzer [145]. Suppose that (X, Y) is a planar Brownian motion and let σ be the first-passage process of Y (see Section 1.1). Thus, the characteristic exponent of X is $\Psi(\lambda) = \frac{1}{2}\lambda^2$ for $\lambda \in \mathbb{R}$ and the Laplace exponent of σ is $\Phi(q) = \sqrt{2q}$ for $q \geq 0$. The characteristic exponent of the subordinate process $\tilde{X} = X \circ \sigma$ is thus $\tilde{\Psi}(\lambda) = |\lambda|$, i.e. $\tilde{X}$ is a standard symmetric Cauchy process. In the more general case when σ is a stable subordinator of index $\alpha \in (0,1)$ independent of X, then $\tilde{X}$ is a symmetric stable process with index 2α. See Molchanov and Ostrovski [121] and also Le Gall [107, 108] for connections with the so-called cone points of planar Brownian motion.

We now end this chapter with an application of the subordination technique to the so-called iterated Brownian motion. Consider $B^+ = (B^+(t), t \geq 0)$, $B^- = (B^-(t), t \geq 0)$ and $B = (B_t, t \geq 0)$ three independent linear Brownian motions started from 0. The process $Y = (Y_t, t \geq 0)$ given by

$$Y_t = \begin{cases} B^+(B_t) & \text{if } B_t \geq 0 \\ B^-(-B_t) & \text{if } B_t < 0 \end{cases}$$

is called an iterated Brownian motion. Its study has been motivated by certain limit theorems and a connection with partial differential equations involving the square of the Laplacian, and has been undertaken by numerous authors (cf. Khoshnevisan and Lewis [99] for a list of references). Our purpose here is to investigate the supremum process of Y,

$$\overline{Y}_t = \sup\{Y_s : 0 \leq s \leq t\} \qquad (t \geq 0)$$

via Bochner's subordination. To this end, we consider the supremum processes S^+, S^-, S and I, of B^+, B^-, B and $-B$, respectively. Observing that

$$S^+(S_t) = \sup\{Y_s : 0 \leq s \leq t \text{ and } B_t \geq 0\},$$

and a similar relation for $S^-(I_t)$, we see that the study of $\overline{Y}$ reduces to that of the compound processes $S^+ \circ S$ and $S^- \circ I$, via the identity

$$\overline{Y} = \left(S^+ \circ S\right) \vee \left(S^- \circ I\right). \tag{8.10}$$

Next, we introduce the right-continuous inverse of S, $\sigma_\cdot = \inf\{s : S_s > \cdot\}$, and recall that σ is a stable subordinator with index $1/2$, more precisely with Laplace exponent $\Phi(\lambda) = \sqrt{2\lambda}$. The inverse σ^+ of S^+ has the same law as σ and is independent of σ. By an immediate variation of Proposition 8.6 (involving Laplace transform instead of Fourier transform), $\tilde{\sigma} = \sigma \circ \sigma^+$ is a subordinator with Laplace exponent $\tilde{\Phi}(\lambda) = (8\lambda)^{1/4}$. Plainly $\sigma \circ \sigma^+$ is the right-continuous inverse of $S^+ \circ S$ and we conclude that the right-continuous inverse of the supremum of an iterated Brownian motion can be expressed as

$$\inf\{t : \overline{Y}_t > \cdot\} = \sigma^{(1)} \wedge \sigma^{(2)}$$

where $\sigma^{(1)}$ and $\sigma^{(2)}$ are both subordinators with Laplace exponent $\tilde{\Phi}$.

An application of the law of the iterated logarithm for the inverse of a stable subordinator (see Theorem 4.1) now gives

$$\limsup \frac{S^+ \circ S_t}{t^{1/4}(\log|\log t|)^{3/4}} = 2^{5/4} 3^{-3/4} \qquad a.s. \qquad (8.11)$$

both as $t \to 0+$ and $t \to \infty$. Using (8.10), one can replace $S^+ \circ S_t$ by $\overline{Y}_t$ (or even by Y_t) in (8.11), which establishes the law of the iterated logarithm for the iterated Brownian motion proven previously by Csáki et al. [38] and Deheuvels and Mason [41] for large times, and by Burdzy [31] for small times. We refer to [12] for further applications of this technique.

Chapter 9

Occupation times of a linear Brownian motion

We consider the occupation time process $A. = \int_0^{\cdot} f(B_s)ds$ where B is a linear Brownian motion and $f \geq 0$ a locally integrable function. The time-substitution based on the inverse of the local time of B at 0 turns A into a subordinator. This enables us to derive several interesting properties for the occupation time process and for linear diffusions.

9.1 Occupation times and subordinators

Let $B = (B_t, t \geq 0)$ be a one-dimensional Brownian motion started from 0. To agree with the usual normalization, we call the process

$$\ell_t = \lim_{\varepsilon \to 0+} \frac{1}{2\varepsilon} \int_0^t \mathbf{1}_{\{|B_s| < \varepsilon\}} ds, \qquad t \geq 0$$

Lévy's local time[1] of B at 0. Consider a locally integrable function $f : \mathbb{R} \to [0, \infty)$ and the corresponding occupation time process of B

$$A_t = \int_0^t f(B_s)ds, \qquad t \geq 0.$$

(More generally, we might have considered the additive functional associated with some Radon measure μ, see e.g. section X.2 in Revuz and Yor [132], but for the sake of simplicity, we will stick to the case when $\mu(dx) = f(x)dx$ is absolutely continuous.)

Let $\tau(t) = \inf\{s : \ell_s > t\}$ be the right-continuous inverse of ℓ. A routine argument based on the additivity, the fact that ℓ only increases on the zero-set of B and the strong Markov property, shows that the time-changed process

$$\sigma_t = A_{\tau(t)} = \int_0^{\tau(t)} f(B_s)ds, \qquad t \geq 0$$

is a subordinator.

---

[1]This means that the local time at 0 in the sense of section 2.2 is $L_t = 2^{-1/2}\ell_t$, in order to agree with (2.1).

Results on subordinators can be very useful in investigating occupation times. To this end we need information on the Laplace exponent Φ and the Lévy measure Π of σ; and this motivates the next section.

9.2 Lévy measure and Laplace exponent

9.2.1 Lévy measure via excursion theory

Our first purpose is to express the Lévy measure of σ in terms of Itô's excursion measure. The obvious hint for this is that, since the occupation time A is a continuous process, the jumps of the subordinator $\sigma = A \circ \tau$ correspond to the increments of A on the intervals of times when B has an excursion away from 0.

Recall the setting of section 3.2 and specialize it to the Brownian case. Let n be the measure of the excursions of B away from 0, that is the characteristic measure of the Poisson point process

$$e_t(s) = \begin{cases} B_{\tau(t-)+s} & \text{if } 0 \leq s < \tau(t) - \tau(t-) \\ 0 & \text{otherwise} \end{cases}$$

We denote the generic excursion by $\epsilon = (\epsilon(s) : s \geq 0)$ and its first return time to 0 by $\rho(\epsilon) = \inf \{s > 0 : \epsilon(s) = 0\}$.

Proposition 9.1 *The drift coefficient and the killing rate of σ are* $d = 0$ *and* $k = 0$, *respectively. The Lévy measure of σ coincides with the distribution of* $\int_0^{\rho(\epsilon)} f(\epsilon(s))ds$ *under n, i.e.*

$$\Pi(dx) = n \left(\int_0^{\rho(\epsilon)} f(\epsilon(s))ds \in dx \right).$$

Proof: We split the time interval $[0, \tau(1)]$ into excursion intervals. Since Brownian motion spends zero time at 0, we have

$$\int_0^{\tau(1)} f(B_s)ds = \sum_{0 \leq t \leq 1} \int_{\tau(t-)}^{\tau(t)} f(B_s)ds = \sum_{0 \leq t \leq 1} \int_0^{\tau(t)-\tau(t-)} f(B_{\tau(t-)+s})ds$$

$$= \sum_{0 \leq t \leq 1} \int_0^{\rho(e_t)} f(e_t(s))ds,$$

where $e = (e_t : t \geq 0)$ is the excursion process (see above). Applying the exponential formula for Poisson point processes (see e.g. Proposition 12 in section XII.1 in [132]), we get

$$\mathbf{E}^0 \left(\exp \left\{ -\lambda \sum_{0 \leq t \leq 1} \int_0^{\rho(e_t)} f(e_t(s))ds \right\} \right)$$

$$= \exp \left\{ -n \left(1 - \exp \left\{ -\lambda \int_0^{\rho(\epsilon)} f(\epsilon(s))ds \right\} \right) \right\}$$

$$= \exp \left\{ -\int_{(0,\infty)} (1 - e^{-\lambda x}) n \left(\int_0^{\rho(\epsilon)} f(\epsilon(s))ds \in dx \right) \right\}.$$

Comparison with the Lévy-Khintchine formula establishes the claim. ∎

Another useful observation which stems from excursion theory is the following independence property.

Corollary 9.2 *Let* $f_+, f_- : \mathbb{R} \to [0, \infty)$ *be two locally integrable functions with* $\text{Supp}(f_+) \subseteq [0, \infty)$ *and* $\text{Supp}(f_-) \subseteq (-\infty, 0]$, *respectively. Then the subordinators*

$$\sigma_t^+ = \int_0^{\tau(t)} f_+(B_s)ds \quad \text{and} \quad \sigma_t^+ = \int_0^{\tau(t)} f_-(B_s)ds$$

are independent. If moreover $f_-(x) = f_+(-x)$, *then* σ^+ *and* σ^- *have the same law.*

Proof: We know from the foregoing that σ^+ and σ^- are two subordinators in the same filtration, both with zero drift and zero killing rate. They are determined by their jump processes. Since jumps correspond to increments of the occupation times on an interval of excursion of B away from 0, σ^+ jumps only when the excursion process e takes values in the space of nonnegative paths, whereas σ^- jumps only when e takes values in the space of non-positive paths. In particular, σ^+ and σ^- never jump simultaneously. By a well-known property of Poisson point processes, their respective jump processes are independent. Because σ^+ and σ^- are both characterized by their jumps, they are independent.

Finally, the excursion measure is symmetric, that is n is invariant by the mapping $\epsilon \to -\epsilon$. It follows that σ^+ and σ^- have the same Lévy measure, and hence the same law, whenever $f_-(x) = f_+(-x)$. ∎

9.2.2 Laplace exponent via the Sturm-Liouville equation

The main result of this subsection characterizes the Laplace exponent Φ in terms of the solution of a Sturm-Liouville equation.

Proposition 9.3 *For every* $\lambda > 0$, *there exists a unique function* $y_\lambda : \mathbb{R} \to [0, 1]$ *such that:*

• y_λ *is a convex increasing function on* $(-\infty, 0)$, *and a convex decreasing function on* $(0, \infty)$.

• y_λ *solves the Sturm-Liouville equation* $y'' = 2\lambda y f$ *on both* $(-\infty, 0)$ *and* $(0, \infty)$, *and* $y_\lambda(0) = 1$.

The Laplace exponent of σ *is then given by*

$$\Phi(\lambda) = \frac{1}{2}\left(y_\lambda'(0-) - y_\lambda'(0+)\right).$$

Proof: We present a proof due to Jeulin and Yor [89], which is based on stochastic calculus. One can also establish the result by analytic arguments that rely on the Feynman-Kac formula and Proposition 2.2; see e.g. Itô and McKean [83], Jeanblanc *et al.* [88] and Pitman and Yor [128].

The existence and uniqueness of y_λ is a well-known result on the Sturm-Liouville equation; see for instance Dym and McKean [49]. By stochastic calculus (more precisely, by an application of the Itô-Tanaka formula), the process

$$M_t = y_\lambda(B_t) \exp\left\{\frac{1}{2}(y_\lambda'(0-) - y_\lambda'(0+))\ell_t - \lambda \int_0^t f(B_s)ds\right\}, \qquad t \geq 0,$$

is a local martingale. Because $M_s \leq \exp\left\{\frac{1}{2}(y_\lambda'(0-) - y_\lambda'(0+))t\right\}$ for every $s \leq \tau(t)$, we can apply Doob's optional sampling theorem for M at time $\tau(t)$. Since $\ell_{\tau(t)} = t$ and $B_{\tau(t)} = 0$, we get

$$\mathbf{E}^0\left(\exp\left\{\frac{1}{2}(y_\lambda'(0-) - y_\lambda'(0+))t - \lambda \int_0^{\tau(t)} f(B_s)ds\right\}\right) = 1,$$

that is

$$\exp\{-t\Phi(\lambda)\} = \mathbf{E}^0\left(\exp\left\{-\lambda \int_0^{\tau(t)} f(B_s)ds\right\}\right) = \exp\left\{-\frac{1}{2}(y_\lambda'(0-) - y_\lambda'(0+))t\right\}.$$

This completes the proof. ∎

The solutions of Sturm-Liouville equations are not explicitly known in general (see however the hand-book by Borodin and Salminen [26] for a number of explicit formulas in some important special cases). Nonetheless one can deduce handy bounds for the Laplace exponent Φ in terms of the function f that will be quite useful in the sequel.

Corollary 9.4 *Put $F(x) = \int_0^x f(t)dt$ ($x \in \mathbb{R}$) and*

$$G(t) = 2\int_0^t (F(x) - F(-x))\,dx, \quad t \geq 0,$$

so G in a convex increasing function on $[0, \infty)$. We write $H(s) = \inf\{t \geq 0 : G(t) > s\}$, $s \geq 0$, for its inverse. Then we have

(i)
$$\Phi(\lambda) \asymp \frac{1}{H(1/\lambda)}.$$

As a consequence, if U stands for the renewal measure of σ and I for the integrated tail of its Lévy measure (c.f. Lemma 1.4), then

(ii)
$$U(x) \asymp H(x) \quad and \quad I(x) \asymp \frac{x}{H(x)}.$$

Proof: (i) We first suppose that f vanishes on $(-\infty, 0)$ and start with the integral Sturm-Liouville equation:

$$y_\lambda(x) = 1 + xy_\lambda'(0+) + 2\lambda \int_0^x \left(\int_0^t y_\lambda(s)f(s)ds\right)dt, \qquad x \geq 0, \lambda > 0. \qquad (9.1)$$

Using the fact that $0 \leq y_\lambda \leq 1$, we deduce the inequality

$$-xy_\lambda'(0+) \leq 1 + 2\lambda \int_0^x \left(\int_0^t f(s)ds\right)dt = 1 + \lambda G(x).$$

Using this with $x = H(1/\lambda)$ gives $-y'_\lambda(0+)H(1/\lambda) \leq 2$.

To establish an lowerbound, we use the fact that y_λ decreases on $[0, \infty)$ in (9.1) to get

$$y_\lambda(x) - xy'_\lambda(0+) \geq 1 + 2\lambda \int_0^x \left(\int_0^t y_\lambda(x)f(s)ds \right) dt = 1 + \lambda y_\lambda(x)G(x).$$

Specifying this for $x = H(1/\lambda)$ gives $-y'_\lambda(0+)H(1/\lambda) \geq 1$.

We have thus established that

$$1 \leq -y'_\lambda(0+)H(1/\lambda) \leq 2, \qquad \lambda > 0$$

in the special case when f vanishes on $(-\infty, 0)$. By a symmetry argument, the bounds

$$1 \leq y'_\lambda(0-)H(1/\lambda) \leq 2, \qquad \lambda > 0$$

hold when f vanishes on $(0, \infty)$. It is immediate to deduce that

$$\frac{1}{H(1/\lambda)} \asymp y'_\lambda(0-) - y'_\lambda(0+)$$

in the general case; and our statement then derives from Proposition 9.3.

(ii) The estimate for the renewal measure now follows from Lemma 1.4. Since we know that the drift of σ is zero, the second estimate also follows from Lemma 1.4. ∎

A sharper estimate for Φ has been obtained in the form of a Tauberian type theorem by Kasahara[2], under the condition that the indefinite integral F of f is regularly varying. We quote the result for completeness and refer to Kotani and Watanabe [102] on page 240 for details of the proof. Thanks to Corollary 9.2, we may restrict our attention to the case when f vanishes on $(-\infty, 0)$.

Proposition 9.5 *Suppose that $f \equiv 0$ on $(-\infty, 0)$. Then Φ is regularly varying at $0+$ (respectively, at ∞) with index $\alpha \in (0, 1)$ if and only if F is regularly varying at ∞ (respectively, at $0+$) with index $(1/\alpha) - 1$. In that case,*

$$\Phi(\lambda) \sim (\alpha(1 - \alpha))^\alpha \frac{\Gamma(1 - \alpha)}{\Gamma(1 + \alpha)} \lambda^\alpha l(1/\lambda) \qquad (\lambda \to 0+),$$

where l is a slowly varying function at ∞ (respectively, at $0+$) such that an asymptotic inverse of $x \to xF(x)$ is $x \to x^\alpha l(x)$.

9.2.3 Spectral representation of the Laplace exponent

The so-called spectral theory of vibrating strings, which has been chiefly developed by M. G. Krein and his followers, is a most powerful tool for investigating the Sturm-Liouville boundary value problem that appears in Proposition 9.3. In this subsection, we will merely state the -tiny- portion of the theory that will be useful for the applications we have in mind; and refer to Dym and McKean [49] for a complete exposition.

---

[2]There is a typographical error in the definition of the constant D_α on p. 70 of [92]; see Kotani-Watanabe [102].

Proposition 9.6 (Krein) *Let y_λ be the function which appears in Proposition 9.3.*

(i) *There exists a unique measure ν on $[0,\infty)$ with $\int_{[0,\infty)}(1+\xi)^{-1}\nu(d\xi) < \infty$, such that for every $\lambda > 0$:*

$$\frac{2}{y_\lambda'(0-) - y_\lambda'(0+)} = \int_{[0,\infty)} \frac{\nu(d\xi)}{\lambda + \xi}.$$

(ii) *There exists a unique measure $\hat{\nu}$ on $[0,\infty)$ with $\int_{[0,\infty)}(1+\xi)^{-1}\hat{\nu}(d\xi) < \infty$, such that for every $\lambda > 0$:*

$$\frac{y_\lambda'(0-) - y_\lambda'(0+)}{2\lambda} = \int_{[0,\infty)} \frac{\hat{\nu}(d\xi)}{\lambda + \xi}.$$

When f vanishes on $(-\infty, 0)$, the measure $f(x)dx$ is sometimes called a *string* (in fact Krein's theory deals with a completely general family of measures). The measure $\frac{1}{2}\nu$ in Proposition 9.6(i) is then known as the *spectral measure* of the string, and the measure $2\hat{\nu}$ in (ii) coincides with the spectral measure of the so-called *dual* string $d\hat{F}(x)$, where $\hat{F}$ is the right continuous inverse of the distribution function $F(x) = \int_0^x f(x)dx$.

Krein's theory yields the following remarkable formulas for the Laplace exponent Φ of σ and the tail of its Lévy measure $\overline{\Pi}$, which seem to have been first observed by Knight [101] (see also Kotani and Watanabe [102] and Küchler [103]).

Corollary 9.7 *Suppose that $f \equiv 0$ on $(-\infty, 0)$. We have*

(i) *There exists a unique measure ν on $[0,\infty)$ with $\int_{[0,\infty)}(1+\xi)^{-1}\nu(d\xi) < \infty$ such that*

$$\frac{1}{\Phi(\lambda)} = \int_{[0,\infty)} \frac{\nu(d\xi)}{\lambda + \xi}, \qquad \lambda > 0.$$

As a consequence, the renewal measure $U(dx)$ of σ is absolutely continuous with density u given by

$$u(x) = \int_{[0,\infty)} e^{-x\xi}\nu(d\xi), \qquad x > 0.$$

(ii) *There exists a unique measure $\hat{\nu}$ on $[0,\infty)$ with $\int_{[0,\infty)}(1+\xi)^{-1}\hat{\nu}(d\xi) < \infty$ such that*

$$\overline{\Pi}(x) = \int_{[0,\infty)} e^{-x\xi}\hat{\nu}(d\xi), \qquad x > 0$$

Proof: (i) The first assertion follows immediately from Propositions 9.3 and 9.6. To get the second, just recall that the Laplace transform of the renewal measure is $1/\Phi$, so that by Fubini's theorem

$$\int_{[0,\infty)} e^{-\lambda x} U(dx) = \int_{[0,\infty)} \frac{\nu(d\xi)}{\lambda + \xi} = \int_0^\infty e^{-\lambda x} \left(\int_{[0,\infty)} e^{-x\xi}\nu(d\xi) \right) dx.$$

(ii) Recall that σ has zero drift. By an integration by parts in the Lévy-Khintchine formula, we get

$$\int_0^\infty e^{-\lambda x} \overline{\Pi}(x)dx = \frac{\Phi(\lambda)}{\lambda} = \frac{y_\lambda'(0-) - y_\lambda'(0+)}{2\lambda} \qquad \text{(by Proposition 9.3)}$$

$$= \int_{[0,\infty)} \frac{\hat{\nu}(d\xi)}{\lambda + \xi} \qquad \text{(by Proposition 9.6 (ii))}$$

$$= \int_0^\infty e^{-\lambda x} \left(\int_{[0,\infty)} e^{-\xi x}\hat{\nu}(d\xi) \right) dx \qquad \text{(by Fubini).}$$

As the tail of the Lévy measure is decreasing and the Laplace transform of the spectral measure continuous, this establishes our claim. ∎

In particular, the renewal measure and the Lévy measure both have completely monotone densities (Hawkes [71] observed that these two properties are equivalent for any subordinator). It seems there is no purely probabilistic proof for this remarkable feature.

It is immediately checked that $x \to \log u(x)$ is a decreasing convex function on $(0, \infty)$. In particular, the renewal density can also be expressed in the form

$$u(x) = c \exp\left\{\int_x^1 \overline{\mu}(t)dt\right\}$$

for some decreasing locally integrable function $\overline{\mu} : (0, \infty) \to \mathbb{R}$. In other words, $\overline{\mu}$ is the tail of some measure on $(0, \infty)$, and the comparison with Theorem 7.2 shows that the range of σ can be thought of as the set left uncovered by certain random intervals issued from a Poisson point process with characteristic measure μ. It would be quite interesting to have probabilistic evidence of this fact.

9.3 The zero set of a one-dimensional diffusion

The material developed in the preceding section can be applied to the study of the zero set of a regular linear diffusion in natural scale[3], using Feller's construction that we now recall.

For the sake of simplicity, we focus on the case when the speed measure is absolutely continuous, though this restriction is in fact superfluous. So let $f \geq 0$ be a locally integrable function such that the support of f is an interval which contains the origin. The occupation time process $A_t = \int_0^t f(B_s)ds$ increases exactly when the Brownian motion B visits $\mathrm{Supp}(f)$ and the time-changed process

$$X_t = B_{\alpha(t)}, \quad t \geq 0, \qquad \text{where } \alpha(t) = \inf\{s : A_s > t\},$$

is a continuous Markov process. One calls $X = (X_t, t \geq 0)$ the diffusion in $\mathrm{Supp}(f)$ with natural scale and speed measure $f(x)dx$. Its infinitesimal generator is $\mathcal{G}g = \frac{1}{2}g''/f$ with the Neumann reflecting condition at the boundary.

When one time-changes Lévy's local time ℓ of the Brownian motion by α, one obtains a continuous increasing process which increases exactly on the zero set of X. Using the approximation

$$\ell_{\alpha(t)} = \lim_{\varepsilon \to 0+} \frac{1}{2\varepsilon} \int_0^{\alpha(t)} 1_{\{|B_s| < \varepsilon\}} ds = \lim_{\varepsilon \to 0+} \frac{1}{2\varepsilon} \int_0^t 1_{\{|X_s| < \varepsilon\}} \frac{1}{f(X_s)} ds,$$

we see that $\ell_{\alpha(\cdot)}$ is an additive functional of the diffusion. Hence, the local time L of X at 0 must be $L_\cdot = c\ell_{\alpha(\cdot)}$ for some normalizing constant $c > 0$. We thus have

$$L^{-1}(t) = \inf\{s \geq 0 : L_s > t\} = \inf\{s \geq 0 : \ell_{\alpha(s)} > t/c\} = A_{\tau(t/c)}.$$

---

[3] Since we are only concerned with the zero set of the diffusion, this induces no loss of generality.

In other words, the inverse local time of the diffusion coincides with the subordinator σ up to a linear time-substitution.

As a first example of application, we present an explicit formula for the fractal dimensions of the zero set of the diffusion X in terms of its speed measure. Recall from Theorem 5.1 that the fractal dimensions are given by the lower and upper indices of the Laplace exponent.

Corollary 9.8 *The Hausdorff and packing dimensions of $\mathcal{R} = \{t \geq 0 : X_t = 0\}$ are given by*

$$\dim_H(\mathcal{R}) = \sup\left\{\rho \leq 1 : \lim_{x \to 0+} x^{1-1/\rho}\left(F(x) - F(-x)\right) = \infty\right\}$$

$$\dim_P(\mathcal{R}) = \inf\left\{\rho \leq 1 : \lim_{x \to 0+} x^{1-1/\rho}\left(F(x) - F(-x)\right) = 0\right\}$$

where $F(x) = \int_0^x f(t)dt$.

Proof: For the sake of conciseness, we shall only consider the Hausdorff dimension which coincides with the lower index

$$\underline{\mathrm{ind}}\,(\Phi) = \sup\left\{\rho \leq 1 : \lim_{\lambda \to \infty} \Phi(\lambda)\lambda^{-\rho} = \infty\right\}$$

(cf. chapter 3). We know from Corollary 9.4 that $\Phi(\lambda) \asymp 1/H(1/\lambda)$, where H is the inverse function of the indefinite integral $G(x) = 2\int_0^x \left(F(t) - F(-t)\right)dt$. It follows immediately that

$$\underline{\mathrm{ind}}\,(\Phi) = \sup\left\{\rho \leq 1 : \lim_{x \to 0+} G(x)x^{1/\rho} = \infty\right\}.$$

Finally, the obvious bound

$$x\left(F\left(x/2\right) - F\left(-x/2\right)\right) \leq G(x) \leq 2x\left(F(x) - F(-x)\right)$$

completes the proof. ∎

As a second illustration, we will use features on random covering to derive a result originally due to Tomisaki [149], which provides an explicit test to decide whether two independent diffusion processes ever visit a given point simultaneously. We first introduce some notation.

Let $X = (X_t : t \geq 0)$ and $Y = (Y_t : t \geq 0)$ be two independent regular diffusions in natural scale; for the sake of simplicity, we shall assume that both X and Y start from 0. Their speed measures are denoted by dF_X and dF_Y, respectively; we also write for $t \geq 0$

$$G_X(t) = 2\int_0^t \left(F_X(x) - F_X(-x)\right)dx \quad , \quad G_Y(t) = 2\int_0^t \left(F_Y(x) - F_Y(-x)\right)dx$$

and H_X and H_Y for the inverse functions of G_X and G_Y. Recall that H_X and H_Y are concave and increasing.

Corollary 9.9 (Tomisaki [149]) (i) *The probability of that $X_t = Y_t = $ for some $t > 0$ equals one if*

$$\int_0^1 H'_X(t)H'_Y(t)dt < \infty$$

and 0 otherwise.

(ii) *The probability of the event $\{X_t = Y_t = 0$ infinitely often as $t \to \infty\}$ equals one if*

$$\int_0^1 H'_X(t)H'_Y(t)dt < \infty \quad \text{and} \quad \int_1^\infty H'_X(t)H'_Y(t)dt = \infty$$

and 0 otherwise.

Proof: Let $\mathcal{R}_X$ and $\mathcal{R}_Y$ be the zero sets of X and Y, respectively. Denote by σ_X the inverse local times of X at 0. According to the observation made at the end of subsection 8.2.3, the range $\mathcal{R}_X$ of σ_X can be viewed as the set left uncovered by random intervals issued from a Poisson point process with characteristic measure μ_X. Idem for $\mathcal{R}_Y$ with a characteristic measure μ_Y. Because X and Y are independent, the intersection of their zero sets can thus be thought of as the closed subset of $[0, \infty)$ left uncovered by random intervals issued from a Poisson point process with characteristic measure $\mu_X + \mu_Y$.

(i) We apply Theorem 7.2. The probability that $\mathcal{R}_X \cap \mathcal{R}_Y$ reduces to $\{0\}$ is one if

$$\int_0^1 \exp\left\{\int_t^1 (\bar{\mu}_X(s) + \bar{\mu}_Y(s))\,ds\right\} dl = \infty \tag{9.2}$$

and zero otherwise. Writing u_X and u_Y for the renewal density of $\mathcal{R}_X$ and $\mathcal{R}_Y$ and applying (7.1), we see that (9.2) is equivalent to

$$\int_0^1 u_X(t)u_Y(t)dt = \infty.$$

Recall from Corollary 9.7 that u_X decreases, so the latter is also equivalent (in the obvious notation) to

$$\int_0^1 U_X(t)d(-u_Y(t)) = \infty.$$

Using then the estimate of Corollary 9.4(ii), we deduce that

$$(9.2) \iff \int_0^1 H_X(t)d(-u_Y(t)) = \infty.$$

Finally, integrate by parts and apply again Corollary 9.4(ii) to derive

$$(9.2) \iff \int_0^1 H'_X(t)H'_Y(t)\,dt = \infty.$$

(ii) The proof rests upon similar arguments and Corollary 7.4. ∎

In the literature, there exist many other examples of applications of the spectral representation of the Laplace exponent Φ. See in particular Bertoin [7], Kasahara [92], Kent [94, 95], Knight [101], Kotani and Watanabe [102], Küchler [103], Küchler and Salminen [104], Tomisaki [149], Watanabe [150, 151] and references therein.

References

[1] S. Aspandiiarov and J. F. Le Gall (1995). Some new classes of exceptional times of linear Brownian motion. *Ann. Probab.* **23**, 1605-1626.

[2] D. Bakry (1984). Etude probabiliste des transformées de Riesz et de l'espace H^1 sur les sphères. In: *Séminaire de Probabilités* XVIII, Lecture Notes in Maths. 1059 pp. 197-218. Springer, Berlin.

[3] M. T. Barlow (1985). Continuity of local times for Lévy processes. *Z. Wahrscheinlichkeitstheorie verw. Gebiete* **69**, 23-35.

[4] M. T. Barlow (1988). Necessary and sufficient conditions for the continuity of local time of Lévy processes. *Ann. Probab.* **16**, 1389-1427.

[5] M. T. Barlow, E. A. Perkins and S. J. Taylor (1986). Two uniform intrinsic constructions for the local time of a class of Lévy processes. *Illinois J. Math.* **30**, 19-65.

[6] C. Berg and G. Forst (1975). *Potential theory on locally compact Abelian groups.* Springer, Berlin.

[7] J. Bertoin (1989). Applications de la théorie spectrale des cordes vibrantes aux fonctionnelles additives principales d'un mouvement brownien réfléchi. *Ann. Inst. Henri Poincaré* **25**, 307-323.

[8] J. Bertoin (1995). Some applications of subordinators to local times of Markov processes. *Forum Math.* **7**, 629-644.

[9] J. Bertoin (1995). Sample path behaviour in connection with generalized arcsine laws. *Probab. Theory Relat. Fields* **103**, 317-327.

[10] J. Bertoin (1995). On the local rate of growth of Lévy processes with no positive jumps. *Stochastic Process. Appl.* **55**, 91-100.

[11] J. Bertoin (1996). *Lévy processes.* Cambridge University Press, Cambridge.

[12] J. Bertoin (1996). Iterated Brownian motion and stable($\frac{1}{4}$) subordinator. *Stat. Prob. Letters* **27**, 111-114.

[13] J. Bertoin (1997). Regularity of the half-line for Lévy processes. *Bull. Sci. Math.* **121**, 345-354.

[14] J. Bertoin (1997). Regenerative embedding of Markov sets. *Probab. Theory Relat. Fields* **108**, 559-571.

[15] J. Bertoin (1998). The inviscid Burgers equation with Brownian initial velocity. *Comm. Math. Phys.* **193**, 397-406.

[16] J. Bertoin (1999). Intersection of independent regenerative sets. To appear in *Probab. Theory Relat. Fields*.

[17] J. Bertoin and R. A. Doney (1997). Spitzer's condition for random walks and Lévy processes. *Ann. Inst. Henri Poincaré* **33**, 167-178.

[18] J. Bertoin and S. Jaffard (1997). Solutions multifractales de l'équation de Burgers. *Matapli* **52**, 19-28.

[19] N. H. Bingham (1975). Fluctuation theory in continuous time. *Adv. Appl. Prob.* **7**, 705-766.

[20] N. H. Bingham, C. M. Goldie and J. L. Teugels (1987). *Regular variation.* Cambridge University Press, Cambridge.

[21] R. M. Blumenthal (1992). *Excursions of Markov processes.* Birkhäuser, Boston.

[22] R. M. Blumenthal and R. K. Getoor (1961). Sample functions of stochastic processes with independent increments. *J. Math. Mech.* 10, 493-516.

[23] R. M. Blumenthal and R. K. Getoor (1968). *Markov processes and potential theory.* Academic Press, New-York.

[24] R. M. Blumenthal and R. K. Getoor (1970). Dual processes and potential theory. *Proc. 12th Biennal Seminar, Canad. Math. Congress*, 137-156.

[25] S. Bochner (1955). *Harmonic analysis and the theory of probability.* University of California Press, Berkeley.

[26] A. N. Borodin and P. Salminen (1996). *Handbook of Brownian motion - Facts and formulae.* Birkhäuser, Basel.

[27] N. Bouleau (1984). Quelques résultats sur la subordination au sens de Bochner. In: *Séminaire de Théorie du Potentiel* 7, Lecture Notes in Maths. 1061 pp. 54-81. Springer, Berlin.

[28] N. Bouleau and D. Lépingle (1994). *Numerical methods for stochastic processes.* Wiley, New York.

[29] L. Breiman (1968). A delicate law of the iterated logarithm for non-decreasing stable processes. *Ann. Math. Stat.* **39**, 1818-1824. [Correction id (1970). **41**, 1126.]

[30] J. Bretagnolle (1971). Résultats de Kesten sur les processus à accroissements indépendants. In: *Séminaire de Probabilités* V, Lecture Notes in Maths. 191 pp. 21-36. Springer, Berlin.

[31] K. Burdzy (1993). Some path properties of iterated Brownian motion. In: *Seminar on stochastic processes 1992*, pp. 67-87. Birkhäuser, Boston.

[32] J. M. Burgers (1974). *The nonlinear diffusion equation.* Dordrecht, Reidel.

[33] L. Carraro and J. Duchon (1998). Equation de Burgers avec conditions initiales à accroissements indépendants et homogènes. *Ann. Inst. Henri Poincaré: analyse non-linéaire* 15, 431-458.

[34] O. Chateau (1990). Quelques remarques sur les processus à accroissements indépendants et stationnaires, et la subordination au sens de Bochner. Thèse d'Université. Laboratoire de Probabilités de l'Université Pierre et Marie Curie.

[35] K. L. Chung (1982). *Lectures from Markov processes to Brownian motion.* Springer, Berlin.

[36] K. L. Chung and P. Erdős (1952). On the application of the Borel-Cantelli lemma. *Trans. Amer. Math. Soc.* 72, 179-186.

[37] J. D. Cole (1951). On a quasi-linear parabolic equation occuring in aerodynamics. *Quart. Appl. Math.* 9, 225-236.

[38] E. Csáki, M. Csörgő, A. Földes and P. Révész (1989). Brownian local time approximated by a Wiener sheet. *Ann. Probab.* 17, 516-537.

[39] E. Csáki, M. Csörgő, A. Földes and P. Révész (1992). Strong approximation of additive functionals. *J. Theoretic. Prob.* 5, 679-706.

[40] E. Csáki, P. Révész and J. Rosen (1997). Functional laws of the iterated logarithm for local times of recurrent random walks on $\mathbb{Z}^2$. Preprint.

[41] P. Deheuvels and D. M. Mason (1992) A functional LIL approach to pointwise Bahadur-Kiefer theorems. In: *Probability in Banach spaces 8* (eds R.M. Dudley, M.G. Hahn and J. Kuelbs) pp. 255-266. Birkhäuser, Boston.

[42] C. Dellacherie and P. A. Meyer (1975). *Probabilités et potentiel*, vol. I. Hermann, Paris.

[43] C. Dellacherie and P. A. Meyer (1980). *Probabilités et potentiel*, vol. II. Théorie des martingales. Hermann, Paris.

[44] C. Dellacherie and P. A. Meyer (1987). *Probabilités et potentiel*, vol. IV. Théorie du potentiel, processus de Markov. Hermann, Paris.

[45] C. Dellacherie, B. Maisonneuve and P. A. Meyer (1992). *Probabilités et potentiel*, vol. V. Processus de Markov, compléments de calcul stochastique. Hermann, Paris.

[46] C. Donati-Martin (1991). Transformation de Fourier et temps d'occupation browniens. *Probab. Theory Relat. Fields* 88, 137-166.

[47] R. A. Doney (1995). Spitzer's condition and ladder variables in random walk. *Probab. Theory Relat. Fields* 101, 577-580.

[48] R. A. Doney (1997). One-sided local large deviation and renewal theorems in the case of infinite mean. *Probab. Theory Relat. Fields* **107**, 451-465.

[49] H. Dym and H. P. McKean (1976). *Gaussian processes, function theory and the inverse spectral problem*. Academic Press.

[50] E. B. Dynkin (1961). Some limit theorems for sums of independent random variables with infinite mathematical expectation. In: *Selected Translations Math. Stat. Prob.* vol. 1, pp. 171-189. Inst. Math. Statistics Amer. Math. Soc.

[51] S. N. Evans (1987). Multiple points in the sample path of a Lévy process. *Probab. Theory Relat. Fields* **76**, 359-367.

[52] K. Falconer (1990). *Fractal Geometry. Mathematical foundations and applications*. Wiley, Chichester.

[53] W. E. Feller (1971). *An introduction to probability theory and its applications*, 2nd edn, vol. 2. Wiley, New-York.

[54] P. J. Fitzsimmons, B. E. Fristedt and B. Maisonneuve (1985). Intersections and limits of regenerative sets. *Z. Wahrscheinlichkeitstheorie verw. Gebiete* **70**, 157-173.

[55] P. J. Fitzsimmons, B. E. Fristedt and L. A. Shepp (1985). The set of real numbers left uncovered by random covering intervals. *Z. Wahrscheinlichkeitstheorie verw. Gebiete* **70**, 175-189.

[56] P. J. Fitzsimmons and T.S. Salisbury (1989). Capacity and energy for multiparameter Markov processes. *Ann. Inst. Henri Poincaré* **25**, 325-350.

[57] B. E. Fristedt (1967). Sample function behaviour of increasing processes with stationary independent increments. *Pac. J. Math.* **21**, 21-33.

[58] B. E. Fristedt (1974). Sample functions of stochastic processes with stationary, independent increments. In: *Advances in Probability 3*, pp. 241-396. Dekker, New-York.

[59] B. E. Fristedt (1979). Uniform local behavior of stable subordinators. *Ann. Probab.* **7**, 1003-1013.

[60] B. E. Fristedt (1996). Intersections and limits of regenerative sets. In: *Random Discrete Structures* (eds. D. Aldous and R. Pemantle) pp. 121-151. Springer, Berlin.

[61] B. E. Fristedt and W. E. Pruitt (1971). Lower functions for increasing random walks and subordinators. *Z. Wahrscheinlichkeitstheorie verw. Gebiete* **18**, 167-182.

[62] B. E. Fristedt and W. E. Pruitt (1972). Uniform lower functions for subordinators. *Z. Wahrscheinlichkeitstheorie verw. Gebiete* **24**, 63-70.

[63] B. E. Fristedt and S. J. Taylor (1983). Construction of local time for a Markov process. *Z. Wahrscheinlichkeitstheorie verw. Gebiete* **62**, 73-112.

[64] B. E. Fristedt and S. J. Taylor (1992). The packing measure of a general subordinator. *Probab. Theory Relat. Fields* **92**, 493-510.

[65] N. Gantert and O. Zeitouni (1998). Large and moderate deviations for the local time of a recurrent random walk on $\mathbb{Z}^2$. *Ann. Inst. Henri Poincaré* **34**, 687-704

[66] R. K. Getoor (1979). Excursions of a Markov process. *Ann. Probab.* **7**, 244-266.

[67] I. I. Gihman and A. V. Skorohod (1975). *The theory of stochastic processes II.* Springer, Berlin.

[68] P. E. Greenwood and J. W. Pitman (1980). Construction of local time and Poisson point processes from nested arrays. *J. London Math. Soc.* **22**, 182-192.

[69] J. Hawkes (1971). A lower Lipschitz condition for the stable subordinator. *Z. Wahrscheinlichkeitstheorie verw. Gebiete* **17**, 23-32.

[70] J. Hawkes (1975). On the potential theory of subordinators. *Z. Wahrscheinlichkeitstheorie verw. Gebiete* **33**, 113-132.

[71] J. Hawkes (1977). Intersection of Markov random sets. *Z. Wahrscheinlichkeitstheorie verw. Gebiete* **37**, 243-251.

[72] F. Hirsch (1984). Générateurs étendus et subordination au sens de Bochner In: *Séminaire de Théorie du Potentiel* 7, Lecture Notes in Maths. 1061 pp. 134-156. Springer, Berlin.

[73] D. G. Hobson (1994). Asymptotics for an arcsine type result. *Ann. Inst. Henri Poincaré* **30**, 235-243.

[74] J. Hoffmann-Jørgensen (1969). Markov sets. *Math. Scand.* 24, 145-166.

[75] E. Hopf (1950). The partial differential equation $u_t + uu_x = \mu u_{xx}$. *Comm. Pure Appl. Math.* **3**, 201-230.

[76] J. Horowitz (1968). The Hausdorff dimension of the sample path of a subordinator. *Israel J. Math.* **6**, 176-182.

[77] J. Horowitz (1972). Semilinear Markov processes, subordinators and renewal theory. *Z. Wahrscheinlichkeitstheorie verw. Gebiete* **24**, 167-193.

[78] X. Hu and S. J. Taylor (1997). The multifractal structure of stable occupation measure. *Stochastic Process. Appl.* **66**, 283-299.

[79] Y. Hu and Z. Shi (1997). Extreme lengths in Brownian and Bessel excursions. *Bernoulli* **3**, 387-402.

[80] B. Huff (1969). The strict subordination of a differential process. *Sankhya Sera.* **A 31**, 403-412.

[81] K. Itô (1942). On stochastic processes. I. (Infinitely divisible laws of probability). *Japan J. Math.* **18**, 261-301.

[82] K. Itô (1970). Poisson point processes attached to Markov processes. In: *Proc. 6th Berkeley Symp. Math. Stat. Prob.* **III**, 225-239.

[83] K. Itô and H. P. McKean (1965). *Diffusion processes and their sample paths.* Springer, Berlin.

[84] N. Jacob and R. L. Schilling (1996). Subordination in the sense of S. Bochner - An approach through pseudo differential operators. *Math. Nachr.* **178**, 199-231.

[85] N. Jacob and R. L. Schilling (1997). Some Dirichlet spaces obtained by subordinate reflected diffusions. Preprint.

[86] S. Jaffard: The multifractal nature of Lévy processes. Preprint.

[87] N. C. Jain and W. E. Pruitt (1987). Lower tail probabilities estimates for subordinators and nondecreasing random walks. *Ann. Probab.* **15**, 75-102.

[88] M. Jeanblanc, J. Pitman and M. Yor (1997). The Feynman-Kac formula and decomposition of Brownian paths. *Computat. Appl. Math.* **6.1**, 27-52.

[89] T. Jeulin and M. Yor (1981). Sur les distributions de certaines fonctionnelles du mouvement brownien. In: *Séminaire de Probabilités XV*, Lecture Notes in Math. 850, pp. 210-226. Springer, Berlin.

[90] J. P. Kahane (1985). *Some random series of functions.* 2nd edn. Cambridge University Press, Cambridge.

[91] J. P. Kahane (1990). Recouvrements aléatoires et théorie du potentiel. *Colloquium Mathematicum* LX/LXI, 387-411.

[92] Y. Kasahara (1975). Spectral theory of generalized second order differential operators and its applications to Markov processes. *Japan J. Math.* **1**, 67-84.

[93] D. G. Kendall (1968). Delphic semigroups, infinitely divisible regenerative phenomena, and the arithmetic of p-functions. *Z. Wahrscheinlichkeitstheorie verw. Gebiete* **9**, 163-195.

[94] J. T. Kent (1980). Eigenvalues expansions for diffusions hitting times. *Z. Wahrscheinlichkeitstheorie verw. Gebiete* **52**, 309-319.

[95] J. T. Kent (1982). The spectral decomposition of a diffusion hitting time. *Ann. Probab.* **10**, 207-219.

[96] H. Kesten (1969). Hitting probabilities of single points for processes with stationary independent increments. *Memoirs Amer. Math. Soc.* **93**.

[97] H. Kesten (1970). The limit points of random walk. *Ann. Math. Stat.* **41**, 1173-1205.

[98] D. Khoshnevisan (1997). The rate of convergence in the ratio ergodic theorem for Markov processes. Preprint.

[99] D. Khoshnevisan and T. M. Lewis (1997). Stochastic calculus for Brownian motion on a Brownian fracture. Preprint.

[100] J. F. C. Kingman (1972). *Regenerative phenomena.* Wiley, London.

[101] F. B. Knight (1981). Characterization of the Lévy measure of inverse local times of gap diffusions. In: *Seminar on Stochastic Processes 1981*, pp. 53-78, Birkhäuser.

[102] S. Kotani and S. Watanabe (1981). Krein's spectral theory of strings and general diffusion processes. In: *Functional Analysis in Markov Processes* (ed M. Fukushima), Proceeding Katata and Kyoto 1981, Lecture Notes in Math. 923 pp. 235-259, Springer.

[103] U. Küchler (1986). On sojourn times, excursions and spectral measures connected with quasi diffusions. *J. Math. Kyoto Univ.* **26**, 403-421.

[104] U. Küchler and P. Salminen (1989). On spectral measures of strings and excursions of quasi diffusions. In: *Séminaire de Probabilités XXIII*, Lecture Notes in Math. 1372 pp. 490-502, Springer.

[105] A. Lachal: Sur la distribution de certaines fonctionnelles de l'intégrale du mouvement brownien avec dérive parabolique et cubique. *Comm. Pure Appl. Math.* **XLIX-12**, 1299-1338.

[106] J. Lamperti (1962). An invariance principle in renewal theory. *Ann. Math. Stat.* **33**, 685-696.

[107] J. F. Le Gall (1987). Mouvement brownien, cônes et processus stables. *Probab. Theory Relat. Fields* **76**, 587-627.

[108] J. F. Le Gall (1992). Some properties of planar Brownian motion. In: *Ecole d'été de Probabilités de St-Flour XX*, Lecture Notes in Maths. 1527, pp. 111-235. Springer, Berlin.

[109] B. Maisonneuve (1971). Ensembles régénératifs, temps locaux et subordinateurs. In: *Séminaire de Probabilités V*, Lecture Notes in Math. 191, pp. 147-169. Springer, Berlin

[110] B. Maisonneuve (1974). Systèmes régénératifs. *Astérisque* **15**, Société Mathématique de France.

[111] B. Maisonneuve (1983). Ensembles régénératifs de la droite. *Z. Wahrscheinlichkeitstheorie verw. Gebiete* **63**, 501-510.

[112] B. Maisonneuve (1993). Processus de Markov: Naissance, retournement, régénération. In: *Ecole d'été de Probabilités de Saint-Flour XXI-1991*. Lecture Notes in Maths. 1541, Springer.

[113] B. B. Mandelbrot (1972). Renewal sets and random cutouts. *Z. Wahrscheinlichkeitstheorie verw. Gebiete* **22**, 145-157.

[114] M. B. Marcus and J. Rosen (1992). Sample path properties of the local times of strongly symmetric Markov processes via Gaussian processes. *Ann. Probab.* **20**, 1603-1684.

[115] M. B. Marcus and J. Rosen (1994). Laws of the iterated logarithm for the local times of symmetric Lévy processes and recurrent random walks. *Ann. Probab.* **22**, 626-658.

[116] M. B. Marcus and J. Rosen (1994). Laws of the iterated logarithm for the local times of recurrent random walks on Z^2 and of Lévy processes and recurrent random walks in the domain of attraction of Cauchy random variables. *Ann. Inst. Henri Poincaré* **30**, 467-499.

[117] L. Marsalle (1997). Applications des subordinateurs à l'étude de trois familles de temps exceptionnels. Thèse d'Université. Laboratoire de Probabilités de l'Université Pierre et Marie Curie.

[118] L. Marsalle (1999). Slow points and fast points of local times. To appear in *Ann. Probab.*

[119] P. A. Meyer (1984). Transformation de Riesz pour les lois gaussiennes. In: *Séminaire de Probabilités* XVIII, Lecture Notes in Maths. 1059 pp. 179-193. Springer, Berlin.

[120] I. S. Molchanov (1993). Intersection and shift functions of strong Markov random closed sets. *Prob. Math. Stats.* **14-2**, 265-279.

[121] I. S. Molchanov and E. Ostrovski (1969). Symmetric stable processes as traces of degenerate diffusion processes. *Th. Prob. Appl.* **14**, 128-131.

[122] J. Neveu (1961). Une généralisation des processus à accroissements positifs indépendants. *Abh. Math. Sem. Univ. Hamburg* **25**, 36-61.

[123] S. Orey (1967). Polar sets for processes with stationary independent increments. In: *Markov processes and potential theory*, pp. 117-126. Wiley, New-York.

[124] S. Orey and S. J. Taylor (1974). How often on a Brownian path does the law of the iterated logarithm fail? *Proc. London Math. Soc.* **28**, 174-192.

[125] J. W. Pitman (1986). Stationary excursions. *Séminaire de Probabilités XXI*, Lecture Notes in Maths. 1247 pp. 289-302. Springer, Berlin.

[126] J. W. Pitman and M. Yor (1992). Arc sine laws and interval partitions derived from a stable subordinator. *Proc. London Math. Soc.* **65**, 326-356.

[127] J. W. Pitman and M. Yor (1997). The two-parameter Poisson-Dirichlet distribution derived from a stable subordinator. *Ann. Probab.* **25**, 855-900.

[128] J. W. Pitman and M. Yor (1997). On the lengths of excursions of some Markov processes. *Séminaire de Probabilités XXXI*, Lecture Notes in Maths. 1655 pp. 272-286. Springer, Berlin.

[129] J. W. Pitman and M. Yor (1997). On the relative lengths of excursions derived from a stable subordinator. *Séminaire de Probabilités XXXI* Lecture Notes in Maths. 1655 pp. 287-305. Springer, Berlin.

[130] W. E. Pruitt (1991). An integral test for subordinators. In: *Random Walks, Brownian Motion and Iteracting Particle Systems: A Festschrift in honor of Frank Spitzer*, pp. 389-398. Birkhäuser, Boston.

[131] B. Rajeev and M. Yor (1995). Local times and almost sure convergence of semi-martingales. *Ann. Inst. Henri Poincaré* **31** 653-667.

[132] D. Revuz and M. Yor (1994). *Continuous martingales and Brownian motion*, 2nd edn. Springer, Berlin.

[133] C. A. Rogers (1970). *Hausdorff measure*. Cambridge University Press, Cambridge.

[134] L. C. G. Rogers (1983). Wiener-Hopf factorization of diffusions and Lévy processes. *Proc. London Math. Soc.* **47**, 177-191.

[135] L. C. G. Rogers (1989). Multiple points of Markov processes in a complete metric space. In: *Séminaire de Probabilités XXIII*. Lecture Note in Maths. 1372 pp. 186-197. Springer, Berlin.

[136] L. C. G. Rogers (1989). A guided tour through excursions. *Bull. London Math. Soc.* **21**, 305-341.

[137] L. C. G. Rogers and D. Williams (1987). *Diffusions, Markov processes, and martingales* vol. 2: Itô calculus. Wiley, New-York.

[138] L. C. G. Rogers and D. Williams (1994). *Diffusions, Markov processes, and martingales* vol. I: Foundations. (First edition by D. Williams, 1979) Wiley, New-York.

[139] B. A. Rogozin (1966). On the distribution of functionals related to boundary problems for processes with independent increments. *Th. Prob. Appl.* **11**, 580-591.

[140] Ya. Sinai (1992). Statistics of shocks in solution of inviscid Burgers equation. *Commun. Math. Phys.* **148**, 601-621.

[141] M. J. Sharpe (1989). *General theory of Markov processes*. Academic Press, New-York.

[142] Z. S. She, E. Aurell and U. Frisch (1992). The inviscid Burgers equation with initial data of Brownian type. *Commun. Math. Phys.* **148**, 623-641.

[143] L. A. Shepp (1972). Covering the line with random intervals. *Z. Wahrscheinlichkeitstheorie verw. Gebiete* **23**, 163-170.

[144] A. N. Shiryaev (1984). *Probability*. Springer-Verlag, New York.

[145] F. Spitzer (1958). Some theorems concerning two-dimensional Brownian motion. *Trans. Amer. Math. Soc.* **87**, 187-197.

[146] S. J. Taylor (1973). Sample path properties of processes with stationary independent increments. In: *Stochastic Analysis*, pp. 387-414. Wiley, London.

[147] S. J. Taylor (1986). The use of packing measure in the analysis of random sets. In: *Stochastic Processes and their Applications* (eds K. Itô and T. Hida), Proceedings Nagoya 1985, Lecture Notes in Maths 1203, pp. 214-222. Springer, Berlin.

[148] S. J. Taylor and C. Tricot (1985). Packing measure and its evaluation for a Brownian path. *Trans. Amer. Math. Soc.* **288**, 679-699.

[149] M. Tomisaki (1977). On the asymptotic behaviors of transition probability densities of one-dimensional diffusion processes. *Publ. RIMS Kyoto Univ.* **12**, 819-837.

[150] S. Watanabe (1975). On time-inversion of one-dimensional diffusion processes. *Z. Wahrscheinlichkeitstheorie verw. Gebiete* **31**, 115-124.

[151] S. Watanabe (1995). Generalized arcsine laws for one-dimensional diffusion processes and random walks. In: *Stochastic Analysis* (eds M. Cranston and M. Pinsky), Proceeding of Symposia in Pure Math. **57**, pp. 157-172. American Mathematical Society.

[152] M. Yor (1995). *Local times and excursions for Brownian motion: a concise introduction.* Lecciones en Matemáticas 1. Facultad de Ciencias, Universidad Central de Venezuela.

[153] M. Yor (1997). *Some aspects of Brownian motion. Part II: Some recent martingale problems.* Birkhäuser, Basel.

LECTURES ON GLAUBER DYNAMICS

FOR DISCRETE SPIN MODELS

Fabio MARTINELLI

Contents

Abstract.

These notes have been the subject of a course I gave in the summer 1997 for the school in probability theory in Saint-Flour. I review in a self-contained way the state of the art, sometimes providing new and simpler proofs of the most relevant results, of the theory of Glauber dynamics for classical lattice spin models of statistical mechanics. The material covers the dynamics in the one phase region, in the presence of boundary phase transitions, in the phase coexistence region for the two dimensional Ising model and in the so-called Griffiths phase for random systems.

1. Introduction

The aim of these notes is to present in a unified way progresses made in the last years in the theory of a special class of symmetric Markovian evolutions for lattice Gibbsian random fields, better known as Glauber dynamics or, in case of ± 1 variables, stochastic Ising models. Such evolutions, besides being interesting as models of non–equilibrium statistical mechanics, are important in many other different areas like images reconstruction, Monte Carlo Markov Chains methods and stochastic optimization.

The fact that in the thermodynamic limit the number of degrees of freedom diverges, poses new and challenging mathematical problems when one studies e.g. the ergodic properties or the large deviations in presence of a phase transition, and it requires the development of a new techniques like, for example, infinite dimensional logarithmic Sobolev inequalities for Gibbsian random fields. Moreover a detailed analysis of the dynamics poses new interesting questions in the theory of Gibbsian random fields itself.

This interplay between dynamical and equilibrum problems turned out to be quite fruitful in both directions and, in my opinion, it represents one of the most interesting aspect of the subject.

Although the theory of Gibbsian random fields, *i.e.* equilibrium statistical mechanics, is clearly very important for our purposes, by no means these notes are intended as an introduction to this difficult subject. Rather, I have tried to present (but in general not to derive with the notable exception of a selfcontained and elementary discussion of finite size mixing conditions) as clearly as possible all the results concerning Gibbs measure of lattice spin models needed to study our Markov processes. Then, based on these results, I develop in a selfcontained way the analysis of the ergodic properties of the Glauber dynamics, its behaviour in the presence of a phase transition or when the interactions in the original model are random. Most of the results presented here are not new but some of them are quite recent and many of their proofs have been revisited and, hopefully, considerably simplified.

Unfortunately, due to space–time problems, I have not been able to cover all the interesting aspects of Markovian dynamics in models of classical statistical mechanics. I have in fact left out important and very lively topics like the microscopic description of metastability and the difficult subject of conservative Kawasaki–type dynamics. The choice of the topics presented here is largely based on my own research in the last few years.

I would like to finish this brief introduction with a short overview of the notes. In section 2 I recall some basic facts of lattice Gibbsian random fields and I introduce two important notion of mixing for them. In particular I provide a new, elementary proof that exponential decay of covariences in a finite large enough region implies the same property on any regular (e.g. cubic) region of the lattice.

In section 3 I introduce the Glauber dynamics, its graphical representation and two relevant analytic quantities for the study of its ergodic behaviour, namely the spectral gap and the logarithmic Sobolev constant. For the necessary background in the theory of finite Markov chains, I definitely suggest the excellent notes by Saloff–Coste [SC].

In section 4 I discuss and provide a new proof for the main results concerning the exponential ergodicity in the absence of a phase transition.

In section 5 I discuss without proofs the question of a boundary phase transition for the solid–on–solid approximation of the three dimensional Ising model and its consequences on the rate of exponential convergence to equilibrium for the associated Glauber dynamics.

In section 6 I discuss the behaviour of Glauber dynamics for the two dimensional Ising model in the presence of a phase transition. In particular I provide a new and simpler proof of a sharp lower bound on the spectral gap with plus of free boundary conditions.

In section 7 I apply all the results of section 4 and 6 to analyze the behaviour of the dynamics for a particular random ferromagnet, namely the dilute Ising model.

In section 8 I conclude the notes with a list of open problems.

Acknowledgments

I wish to warmly thank my hosts, the summer school of probability theory and Prof. Bernard in particular, for their very kind hospitality in Saint Flour and the whole staff of the school for providing an excellent environment for the meeting. During the preparation of these notes and also during their presentation in Saint Flour I benefit from the comments and criticism of several colleagues. To all of them go my hearthly thanks.

2. Gibbs Measures of Lattice Spin Models

2.1 Notation

The lattice. We consider the d dimensional lattice $\mathbf{Z}^d$ with *sites* $x = \{x_1, \ldots, x_d\}$ and norm

$$|x| = \max_{i \in \{1, \ldots, d\}} |x_i|$$

The associated distance function is denoted by $d(\cdot, \cdot)$. Sometimes we will need another distance, in the sequel denoted by $d_2(\cdot, \cdot)$, defined by

$$d_2(x, y) = \left(\sum_{i=1}^{d} |x_i - y_i|^2 \right)^{1/2}$$

By Q_L we denote the cube of all $x = (x_1, \ldots, x_d) \in \mathbf{Z}^d$ such that $x_i \in \{0, \ldots, L-1\}$. If $x \in \mathbf{Z}^d$, $Q_L(x)$ stands for $Q_L + x$. We also let B_L be the ball of radius L centered at the origin, *i.e.* $B_L = Q_{2L+1}((-L, \ldots, -L))$.

A finite subset Λ of $\mathbf{Z}^d$ is said to be a *multiple* of Q_L if there exists $y \nu Q_l$ such that Λ is the union of a finite number of cubes $Q_L(x_i + y)$ where $x_i \in L\mathbf{Z}^d$.

If Λ is a finite subset of $\mathbf{Z}^d$ we write $\Lambda \subset\subset \mathbf{Z}^d$. The cardinality of Λ is denoted by $|\Lambda|$. $\mathbf{F}$ is the set of all nonempty finite subsets of $\mathbf{Z}^d$. We define the exterior n-*boundary* as $\partial_n^+ \Lambda = \{x \in \Lambda^c : d(x, \Lambda) \leq n\}$. Given $r \in \mathbf{Z}_+$ we say that a subset V of $\mathbf{Z}^d$ is r-*connected* if, for all $y, z \in V$ there exist $\{x_1, \ldots, x_n\} \subset V$ such that $x_1 = y$, $x_n = z$ and $|x_{i+1} - x_i| \leq r$ for $i = 2, \ldots, n$.

The configuration space. Our *configuration space* is $\Omega = S^{\mathbf{Z}^d}$, where $S = \{-1, 1\}$, or $\Omega_V = S^V$ for some $V \subset \mathbf{Z}^d$. The single spin space S is endowed with the discrete topology and Ω with the corresponding product topology. Given $\sigma \in \Omega$ and $\Lambda \subset \mathbf{Z}^d$ we denote by σ_Λ the natural projection over Ω_Λ. If U, V are disjoint, $\sigma_U \eta_V$ is the configuration on $U \cup V$ which is equal to σ on U and η on V.

If f is a function on Ω, Λ_f denotes the smallest subset of $\mathbf{Z}^d$ such that $f(\sigma)$ depends only on σ_{Λ_f}. f is called *local* if Λ_f is finite. $\mathcal{F}_\Lambda$ stands for the σ-algebra generated by the set of projections $\{\pi_x\}$, $x \in \Lambda$, from Ω to $\{-1, 1\}$, where $\pi_x : \sigma \mapsto \sigma(x)$. When $\Lambda = \mathbf{Z}^d$ we set $\mathcal{F} \equiv \mathcal{F}_{\mathbf{Z}^d}$ and $\mathcal{F}$ coincides with the Borel σ-algebra on Ω with respect to the topology introduced above. By $\|f\|_\infty$ we mean the supremum norm of f. The *gradient* of a function f is defined as

$$(\nabla_x f)(\sigma) = f(\sigma^x) - f(\sigma)$$

where $\sigma^x \in \Omega$ is the configuration obtained from σ, by flipping the spin at the site x. If $\Lambda \in \mathbf{F}$ we let

$$|\nabla_\Lambda f|^2 = \sum_{x \in \Lambda} (\nabla_x f)^2$$

We also define

$$\|\|f\|\| = \sum_{x \in \mathbf{Z}^d} \|\nabla_x f\|_\infty$$

2.2 Gibbs Measures

Definition 2.1. *A finite range, translation–invariant potential* $\{J_\Lambda\}_{\Lambda \in \mathbb{F}}$ *is a real function on the set of all non empty finite subsets of* $\mathbb{Z}^d$ *with the following properties*
(1) $J_A = J_{A+x}$ *for all* $A \in \mathbb{F}$ *and all* $x \in \mathbb{Z}^d$
(2) *There exists* $r > 0$ *such that* $J_A = 0$ *if* $\operatorname{diam} A > r$. r *is called the* range *of the interaction.*
(3) $\|J\| \equiv \sum_{A \ni 0} |J_A| < \infty$

Given a *potential* or *interaction* J with the above two properties and $V \in \mathbb{F}$, we define the Hamiltonian $H_V^J : \Omega \mapsto \mathbb{R}$ by

$$H_V^J(\sigma) = - \sum_{A:\, A \cap V \neq \emptyset} J_A \prod_{x \in A} \sigma(x)$$

For $\sigma, \tau \in \Omega$ we also let $H_V^{J,\tau}(\sigma) = H_V^J(\sigma_V \tau_{V^c})$ and τ is called the *boundary condition*. For each $V \in \mathbb{F}$, $\tau \in \Omega$ the (finite volume) conditional Gibbs measure on $(\Omega, \mathcal{F})$, are given by

$$\mu_V^{J,\tau}(\sigma) = \begin{cases} (Z_V^{J,\tau})^{-1} \exp[-H_V^{J,\tau}(\sigma)] & \text{if } \sigma(x) = \tau(x) \text{ for all } x \in V^c \\ 0 & \text{otherwise.} \end{cases} \tag{2.1}$$

where $Z_V^{J,\tau}$ is the proper normalization factor called partition function. Notice that in (2.1) we have adsorbed in the interaction J the usual inverse temperature factor β in front of the Hamiltonian. In most notation we will drop the superscript J if that does not generate confusion.

Given a measurable bounded function f on Ω, $\mu_V(f)$ denotes the function $\sigma \mapsto \mu_V^\sigma(f)$ where $\mu_V^\sigma(f)$ is just the average of f w.r.t. μ_V^σ. Analogously, if $X \in \mathcal{F}$, $\mu_V^\tau(X) \equiv \mu_V^\tau(\mathbf{1}_X)$, where $\mathbf{1}_X$ is the characteristic function on X. $\mu_V^\tau(f,g)$ stands for the covariance or *truncated correlation* (with respect to μ_V^τ) of f and g. The set of measures (2.1) satisfies the DLR compatibility conditions

$$\mu_\Lambda^\tau(\mu_V(X)) = \mu_\Lambda^\tau(X) \qquad \forall X \in \mathcal{F} \qquad \forall V \subset \Lambda \subset\subset \mathbb{Z}^d \tag{2.2}$$

Definition 2.2. *A probability measure* μ *on* $(\Omega, \mathcal{F})$ *is called a Gibbs measure for* J *if*

$$\mu(\mu_V(X)) = \mu(X) \qquad \forall X \in \mathcal{F} \qquad \forall V \in \mathbb{F} \tag{2.3}$$

Remark. In the above definition we could have replaced the σ–algebra $\mathcal{F}$ with $\mathcal{F}_V$ (see section 2.3.2 in [EFS]).

The set of all Gibbs measures relative to a fixed given potential $\{J\}$ will be denoted by $\mathcal{G}$. It can be proved that $\mathcal{G}$ is a nonempty, convex compact set. We will say that the discrete spin system described by the potential J exhibits a phase transition if $\mathcal{G}$ contains more than one element. The reader is refered to [Geo] and [EFS] for a much more advanced discussion of Gibbs measures.

The most famous example of lattice spin system is certainly the Ising model. For this model the potential J takes the following values

$$J_A = \beta \begin{cases} 1 & \text{if } A = \{x, y\} \text{ with } d_2(x, y) \leq 1 \\ h & \text{if } A = \{x\} \\ 0 & \text{otherwise} \end{cases} \tag{2.4}$$

where β represents the inverse temperature and h the external magnetic field.

For the Ising model in dimension greater than one it is well known (see e.g [Pf]) that there exists a finite value β_c, called the critical point, such that there is a unique Gibbs measure for any $h \neq 0$ or any $\beta < \beta_c$. If instead $h = 0$ and $\beta > \beta_c$ there is a phase transition. In particular the so called spontaneous magnetization is non zero. Namely there exist two Gibbs measures denoted by $\mu_\pm^\beta$, that can be obtained as thermodynamic limit as $L \to \infty$ of the finite volume Gibbs measures $\mu_{B_L}^{\beta,\pm}$ respectively and such that $m^*(\beta) \equiv \mu_+^\beta(\sigma(0)) = -\mu_-^\beta(\sigma(0)) > 0$. Here the supercript $\pm$ in $\mu_{B_L}^{\beta,\pm}$ denotes the special boundary conditions for which all the boundary spins are all equal to either plus or minus one. In this notes we will refer to $\mu_\pm^\beta$ as the *plus and minus phase* respectively.

2.3 Weak and Strong Mixing Conditions

As a next step we define two similar, but at the same time deeply different, notion of *weak dependence of the boundary conditions* for finite volume Gibbs measures (see [MO]). These notion will be denoted in the sequel *weak* and *strong mixing* (not to be confused with the classical notion of strong-mixing for random fields) respectively. They both imply that there exists a unique infinite volume Gibbs state with exponentially decaying truncated correlations functions. Actually the validity of our strong mixing condition on e.g. all squares implies much more, namely analyticity properties of the Gibbs measure, the existence of a convergent cluster expansion (see [O] and [OP]) and good behaviour under the renormalization-group transformation known as the "decimation transformation" (see [MO3] and [MO4]). Moreover, and this is our main motivation, both notion play a key role in the discussion of the exponential ergodicity of a Glauber dynamics for discrete lattice spin systems.

Roughly speaking, the *weak mixing* condition implies that if in a finite volume V we consider the Gibbs state with boundary condition τ, then a local (e.g. in a single site $y \in V^c$) modification of the boundary condition τ has an influence on the corresponding Gibbs measure which decays exponentially fast inside V with the *distance from the boundary* $\partial^+ V$.

Strong mixing condition, instead, implies, in the same setting as above, that the influence of the perturbation decays in V exponentially fast with the distance from the *support of the perturbation* (e.g. the site y).

This distinction is very important since, even if we are in the one phase region with a unique infinite volume Gibbs state with exponentially decaying truncated correlation functions, it may happen that, if we consider the same Gibbs state in a finite volume V, a local perturbation of the boundary condition radically modifies the Gibbs measure close to the boundary while leaving it essentially unchanged in the bulk and this "long range order effect" at the boundary persists even when V becomes arbitrarily large. We will refer to this phenomenon as a "boundary phase transition". It is clear that if a "boundary phase transition" takes place, then our Gibbs measure may satisfy a weak mixing condition but not a strong one.

A "boundary phase transition" is apparently not such an exotic phenomenon since, besides being proved for the so called Czech models [Sh] (in dimension 3 and higher), it is also expected (see section 5) to take place for the three dimensional ferromagnetic Ising model at low temperatures and small enough magnetic field (depending on the temperature). On the contrary, for finite range two dimensional systems and for regular volumes (e.g squares) we do not expect any "boundary phase transition" since the boundary is one–dimensional and, unless the interaction is long range, no phase transition occurs. Thus in two dimensions weak mixing should be equivalent to strong mixing. That is precisely the content of theorem 2.5 below.

We conclude this short introduction with a warning. It may happen, also for very natural model like the Ising model at low temperature and positive external field, that strong mixing holds for "regular" volumes, like all multiples of a given large enough cube, but fails for other sets (see [MO1]). This fact led to a revision of the theory of "completely analytical Gibbsian random fields" (see [DS]) and it plays an important role in the recently much debated issue of pathologies of renormalization group transformations in statistical mechanics (see [EFS]).

Let us now define our two conditions. We first recall that the *variation distance* between two probability measures μ_1, μ_2 on a finite set Y is defined as :

$$\|\mu_1 - \mu_2\| = \frac{1}{2} \sum_{y \in Y} |\mu_1(y) - \mu_2(y)|$$
$$= \sup_{X \subset Y} |\mu_1(X) - \mu_2(X)| \tag{2.5}$$

Given $\Delta \subset V \subset\subset \mathbb{Z}^d$ and a Gibbs measure μ_V^τ on Ω_V, we denote by $\mu_{V,\Delta}^\tau$ the projection of the measure μ_V^τ on Ω_Δ, *i.e.*

$$\mu_{V,\Delta}^\tau(\sigma) = \sum_{\eta; \, \eta_\Delta = \sigma_\Delta} \mu_V^\tau(\eta)$$

We are now in a position to define strong mixing and weak mixing.

Definition 2.3. *We say that the Gibbs measures μ_V satisfy the weak mixing condition in V with constants C and m if for every subset $\Delta \subset V$*

$$\sup_{\tau, \tau'} \|\mu_{V,\Delta}^\tau - \mu_{V,\Delta}^{\tau'}\| \le C \sum_{x \in \Delta, \, y \in \partial_r^+ V} e^{-md(x,y)} \tag{2.6}$$

We denote this condition by $WM(V, C, m)$.

Definition 2.4. *We say that the Gibbs measures μ_V satisfy the strong mixing condition in V with constants C and m if for every subset $\Delta \subset V$ and every site $y \in V^c$*

$$\sup_\tau \|\mu_{V,\Delta}^\tau - \mu_{V,\Delta}^{\tau^y}\| \le Ce^{-md(\Delta,y)} \tag{2.7}$$

We denote this condition by $SM(V, C, m)$.

Remark. It is clear that either one of the above properties becomes interesting when it holds with the *same* constants C and m for an infinite class of finite subsets of $\mathbb{Z}^d$ e.g. all cubes.

It is a relatively easy task to show that strong mixing is more stringent than weak mixing in the sense that, for example, strong mixing for all cubes implies weak mixing for all cubes.

Let us in fact fix a cube Q_L, a subset $\Delta \subset Q_L$ and a pair of boundary conditions τ

and τ'. Let also $\{\tau_i\}_{i=1}^n$ be a sequence of interpolating configurations between τ and τ' such that $n \le \partial_r^+ Q_L$, τ_{i+1} differs from τ_i at exactly one site $y_i \in \partial_r^+ Q_L$, $\tau_1 = \tau$ and τ_n agrees with τ' on $\partial_r^+ Q_L$. Then, using $SM(Q_L, C, m)$, we write

$$\|\mu_{Q_L,\Delta}^\tau - \mu_{Q_L,\Delta}^{\tau'}\| \le \sum_i \|\mu_{Q_L,\Delta}^{\tau_{i+1}} - \mu_{Q_L,\Delta}^{\tau_i}\|$$

$$\le \sum_i Ce^{-md(\Delta, y_i)}$$

$$\le \sum_{x\in\Delta} \sum_{y\in\partial_r^+ Q_L} Ce^{-md(x,y)}$$

i.e. $WM(Q_L, C, m)$.

The converse of the above result, namely weak mixing implies strong mixing, is in general expected to be false in dimension greater than two. In two dimensions we have instead the following (see [MOS])

Theorem 2.5. *In two dimensions, $WM(V, C, m)$ for every $V \subset\subset \mathbf{Z}^d$ implies $SM(Q_L, C', m')$ for every square Q_L, for suitable constants C' and m'.*

Remark. It is very important to notice that it is known, by means of explicit examples, that the above result becomes false if we try to replace in the above theorem *for all squares* with *for all finite subsets of $\mathbf{Z}^2$* (see [MO1]).

We conclude this paragraph by discussing the validity of the above conditions for the Ising model (2.4). In two dimensions it has been proved (see [Hi], [MO1] and particularily [ShSch]) that condition $WM(V, C, m)$ holds true for any set $V \subset\subset \mathbf{Z}^2$ everywhere in the one–phase region *i.e.* whenever the external magnetic field h is different from zero or the inverse temperature β is smaller than the critical value β_c, with constants C and m depending on β and h. Thanks to theorem 2.5 the same holds for $SM(Q_L, C', m')$ for all integers L.

In higher dimensions weak mixing for every finite set V is known to holds for any $\beta < \beta_c$ (see [Hi]) or for large enough β and arbitrary non–zero external field h (see [MO1]). Strong mixing for all cubes has been proved instead for small enough β or large enough βh (see [MO1]). Moreover, as we have already anticipated, there is strong evidence if not a proof that strong mixing for all cubes is false if β is large and h is small enough depending on β (see section 5).

2.4 Mixing properties and bounds on relative densities

In this paragraph we first show that strong mixing is equivalent to the exponential decay of finite–volume covariances, uniformly in the boundary conditions (see condition $SMT(V, C, m)$ below). Then we prove two useful equilibrium results on finite volume Gibbs measures. The first one is what was called in [MO1] "effectiveness" of the strong mixing condition and it implies that in order to verify strong mixing on, e.g. all cubes, it is sufficient to verify it for a finite class of subsets. The second result says that a power–law decay of covariances with a large enough power depending on the dimension implies exponential decay (absence of intermediate rates of decay of covariances). We finally conclude with a simple bound on the relative density between the projection over certain sets of two different Gibbs measures, once one assumes exponential decay of correlations.

Before going on we need the following definition:

Definition 2.6. Given $V \subset\subset \mathbb{Z}^d$, $n, \alpha > 0$, we say that the condition $SMT(V, n, \alpha)$ holds if for all local functions f and g on Ω such that $d(\Lambda_f, \Lambda_g) \geq n$ we have

$$\sup_{\tau \in \Omega} |\mu_V^\tau(f, g)| \leq |\Lambda_f||\Lambda_g| \, \|f\|_\infty \|g\|_\infty \exp[-\alpha \, d(\Lambda_f, \Lambda_g)]$$

With the above notation we have

Theorem 2.7. The following are equivalent
  i) There exist C, m and L_0 such that $SM(\Lambda, C, m)$ holds for all Λ multiples of Q_{L_0}
  ii) There exist l, m and L_0 such that $SMT(\Lambda, l, m,)$ holds for all Λ multiples of Q_{L_0}

Proof. Let us first show that $i) \to ii)$. For this purpose we choose $L \in \mathbb{Z}_+$ in such a way that Q_L is a multiple of Q_{L_0}. Thanks to proposition 2.9 below it is enough to prove condition $SMT(V, L/2, m')$ for some $m' > 0$ independent of L and all sets $\Lambda \subset B_{2L}$ which are multiple of Q_L, provided that L was chosen large enough.

We choose a multiple l_1 of L_0 in such a way that $L^{2/3} \leq l_1 \leq L/8$, we fix a set $\Lambda \subset B_{2L}$ multiple of Q_L and we partition it into disjoint cubes as follows

$$\Lambda = \cup_{i=1}^n B_i, \qquad B_i = Q_{l_1}(x_i) \cap \Lambda; \qquad x_i \in l_1 \mathbb{Z}^d + \bar{x}$$

for a suitable $\bar{x}$.

We also set $\Lambda_I = \cup_{i \in I} B_i \quad \forall I \subset \{1 \ldots n\}$. Given now two arbitrary functions f and g, with $d(\Lambda_f, \Lambda_g) \geq L/2$, let

$$I_f = \{i \in \{1 \ldots n\} : B_i \cap \Lambda_f \neq \emptyset\} \qquad I_f^c = \{1 \ldots n\} \setminus I_f$$

and similarly for g. Notice that, because of our choice of l_1, we have

$$d(\Lambda_{I_f}, \Lambda_{I_g}) \geq L/2 - 2l_1 \geq L/4 \tag{2.8}$$

Now we write, using the DLR equations,

$$\mu_\Lambda^\tau(f, g) = \sum_{(\sigma, \sigma')} \mu_\Lambda^\tau(\sigma) \mu_\Lambda^\tau(\sigma') f(\sigma) [\mu_{\Lambda_{I_f^c}}^\sigma(g) - \mu_{\Lambda_{I_f^c}}^{\sigma'}(g)]$$

$$\leq \|f\|_\infty \|g\|_\infty \sup_{\tau, \tau'} \|\mu_{\Lambda_{I_f^c}, \Lambda_{I_g}}^\tau - \mu_{\Lambda_{I_f^c}, \Lambda_{I_g}}^{\tau'}\| \tag{2.9}$$

so that, using a simple telescopic argument

$$\sup_\tau |\mu_\Lambda^\tau(f, g)| \leq \|f\|_\infty \|g\|_\infty \sum_{y \in \Lambda_{I_f}} \sup_\tau \|\mu_{\Lambda_{I_f^c}, \Lambda_{I_g}}^\tau - \mu_{\Lambda_{I_f^c}, \Lambda_{I_g}}^{\tau^y}\| \tag{2.10}$$

We can now apply property $SM(C, m, \Lambda_{I_j^c})$ to estimate each term in the sum appearing in the r.h.s of (2.10). We get

$$\text{RHS of (2.10)} \leq \|f\|_\infty \|g\|_\infty CL^d e^{-mL/4} \leq \|f\|_\infty \|g\|_\infty |\Lambda_f||\Lambda_g| e^{-md(\Lambda_f, \Lambda_g)/8}$$
(2.11)

namely $SMT(\Lambda, L/2, m/8)$, provided that L is large enough depending on m.

In order to prove the converse, namely $ii) \to i)$, we first need some additional notation.

Given any two measures μ and ν on $(\Omega, \mathcal{F})$, and given $V \in \mathbb{F}$ such that for, each $X \in \mathcal{F}_V$, $\nu(X) = 0$ implies $\mu(X) = 0$, we define the $\mathcal{F}_V$−measurable function

$$\left.\frac{d\mu}{d\nu}\right|_V : \eta \mapsto \frac{\mu\{\sigma \in \Omega : \sigma_V = \eta_V\}}{\nu\{\sigma \in \Omega : \sigma_V = \eta_V\}} \qquad \eta \in \Omega$$
(2.12)

where $0/0$ means 0. We have, of course,

$$\mu(f) = \nu\left(f \left.\frac{d\mu}{d\nu}\right|_V\right) \qquad \forall f \in \mathcal{F}_V$$
(2.13)

With this notation we have the following result

Lemma 2.8. *Let* $\Delta \subset V \subset\subset \mathbb{Z}^d$, *and let* $x \in V^c$ *such that* $d(x, \Delta) > r$. *If* $U \equiv V \backslash \Delta$, *we have*

$$\sup_{\tau \in \Omega} \left\| 1 - \left.\frac{d\mu_V^{\tau^x}}{d\mu_V^\tau}\right|_\Delta \right\|_\infty \leq e^{16\|J\|} \sum_{y \in \Delta} \sup_{\tau \in \Omega} \left| \mu_U^\tau \left(e^{-\nabla_x H_U}, e^{-\nabla_y H_U} \right) \right|$$
(2.14)

Assuming the lemma let us complete the proof of the theorem.
Let V be a multiple of Q_{L_0}, let $\Delta \subset V$ and let $\hat{\Delta}$ be the smallest subset of V containing Δ which is also a multiple of L_0. Thanks to (2.14), property $SMT(V, l, m)$ and the simple bound

$$\left| \mu_U^\tau \left(e^{-\nabla_x H_U}, e^{-\nabla_y H_U} \right) \right| \leq (2r + 1)^{2d} e^{4\|J\|} e^{m(l - d(x,y))}$$

we get

$$\begin{aligned}
\|\mu_{V,\Delta}^\tau - \mu_{V,\Delta}^{\tau^y}\| &\leq \|\mu_{V,\hat{\Delta}}^\tau - \mu_{V,\hat{\Delta}}^{\tau^y}\| \\
&\leq \left\| 1 - \left.\frac{d\mu_V^{\tau^y}}{d\mu_V^\tau}\right|_{\hat{\Delta}} \right\|_\infty \\
&\leq (2r + 1)^{2d} e^{20\|J\|} \sum_{x \in \hat{\Delta}} e^{ml - md(x,y)} \\
&\leq A(2r + 1)^{2d} e^{20\|J\|} \sum_{x \in \Delta} e^{ml - md(x,y)}
\end{aligned}$$
(2.15)

for a suitable constant A depending on L_0 and m. Clearly (2.15) proves $SM(V, C, m)$ with $C = A(2r + 1)^{2d} e^{20\|J\|} e^{ml}$. $\square$

Proof of lemma 2.8 Let $\underline{1} \in \Omega$ be the configuration with all spins equal to $+1$, and let

$$W_{V,\Delta}(\sigma) = \log Z_U^{\sigma} - \log Z_U^{\sigma_{\Delta^c} \underline{1}_{\Delta}}$$

It is easy to show that

$$\left\| 1 - \frac{d\mu_V^{\tau^x}}{d\mu_V^{\tau}}\big|_{\Delta} \right\|_{\infty} \le e^{2\|\nabla_x W_{V,\Delta}\|_{\infty}} \|\nabla_x W_{V,\Delta}\|_{\infty} \qquad \text{for all } \tau \in \Omega$$

which, using the trivial bound $\|\nabla_x W_{V,\Delta}\|_{\infty} \le 4\|J\|$, gives

$$\left\| 1 - \frac{d\mu_V^{\tau^x}}{d\mu_V^{\tau}}\big|_{\Delta} \right\|_{\infty} \le e^{8\|J\|} \|\nabla_x W_{V,\Delta}\|_{\infty} \qquad \text{for all } \tau \in \Omega$$

By proceeding as in Lemma 3.1 of [MO2] one can show that

$$\|\nabla_x W_{V,\Delta}\|_{\infty} \le e^{6\|J\|} \sum_{y \in \Delta} e^{2\|J\|} \sup_{\tau \in \Omega} \left| \mu_U^{\tau}\left(e^{-\nabla_x H_U}, e^{-\nabla_y H_U} \right) \right|$$

which completes the proof of the lemma. □

The next key result (see also [DS], [O], [MO1] and [Yo1]) says that in order to prove exponential decay of covariances for all sets V multiple of a given cube Q_l it is sufficient to verify such property only for a finite class of sets.

Let $\mathcal{F}_l$ be the class of all subsets of $\mathbb{Z}^d$ which are multiple of Q_l and denote by $\mathcal{F}_l^0$ the class of all sets $V \in \mathcal{F}$ such that $V \subset B_{2l}$. Then we have

Proposition 2.9. *Given $\alpha > 0$ there exist positive integers $\bar{l}(d, r, \alpha, \|J\|)$, $\gamma_1(d, r, \|J\|)$, $\hat{m}(d, r, \alpha)$ such that if $l > \bar{l}$ and $SMT(V, l/2, \alpha)$ holds for all $V \in \mathcal{F}_l^0$, then $SMT(V, \gamma_1 l, \hat{m}(d, r, \alpha))$ holds for all $V \in \mathcal{F}_l$.*

Proof. Given a positive integer l define the non–negative, bounded, translation invariant function $f_l(x - y)$ on $\mathbb{Z}^d$ by

$$f_l(x - y) \equiv \sup_{V \in \mathcal{F}_l} \sup_{\tau \in \Omega} |\mu_V^{\tau}(h_x, h_y)| \tag{2.16}$$

where $h_x = e^{-\nabla_x H}$.
The proof is then based on the following key lemma

Lemma 2.10. *Given $\alpha > 0$ there exist $\bar{l}(d, r, \alpha)$ such that, if $l > \bar{l}$ and $SMT(V, l/2, \alpha)$ holds for all $V \in \mathcal{F}_l^0$, then there exists $C(\alpha, d, r, \|J\|)$ such that*

$$f_l(x - y) \le C e^{-\frac{\alpha}{10} d(x,y)} \qquad \forall\, d(x, y) \ge 4l + r$$

Let us assume for a moment the lemma and let us complete the proof of the proposition.
We fix an integer l, a set $V \in \mathcal{F}_l$, a boundary condition τ and two functions f and g with zero $\mu_V^{\tau}(\cdot)$ mean and $d(\Lambda_f, \Lambda_g) \ge 6l + 2r$. Without loss of generality we assume $V = \cup_{i=1}^n Q_l(x_i)$ with $x_i \in l\mathbb{Z}^d$. Let also V_1 (V_2) be the union of those cubes

$\{Q_l(x_i)\}_{i=1}^n$ that intersect the support of f (g). Clearly $d(V_1, V_2) \geq 4l + r$. Then we write

$$|\mu_V^\tau(fg)| = |\mu_V^\tau(f\mu_{V\setminus V_1}(g))|$$
$$\leq \|f\|_\infty \|\mu_{V\setminus V_1}(g)\|_\infty^\tau \tag{2.17}$$

where in $\|\mu_{V\setminus V_1}(g)\|_\infty^\tau$ the supremum over the boundary condition in $\partial_r^+(V \setminus V_1)$ has to be taken only over configurations that agree with τ in $\partial_r^+ V$. In order to bound $\|\mu_{V\setminus V_1}(g)\|_\infty^\tau$ we use the following simple formula valid for any pairs of sets $A \subset B$ and any function f that depends only on the spins in A

$$\sup_\tau \sup_{\sigma; \sigma_{B^c}=\tau_{B^c}} |\mu_A^\sigma(f) - \mu_B^\sigma(f)| = \sup_{\sigma; \sigma_{B^c}=\tau_{B^c}} |\sum_\eta \mu_B^\tau(\eta)[\mu_A^\sigma(f) - \mu_A^\eta(f)]|$$

$$\leq \sup_{\substack{\sigma, \eta; \\ \sigma_{B^c}=\eta_{B^c}}} |\mu_A^\sigma(f) - \mu_A^\eta(f)|$$

$$\leq \sup_\tau \sum_{x \in \partial_r^+ A \cap B} |\mu_A^{\tau^x}(f) - \mu_A^\tau(f)| \tag{2.18}$$

$$= \sup_\tau \sum_{x \in \partial_r^+ A \cap B} \frac{|\mu_A^\tau(h_x, f)|}{\mu_A^\tau(h_x)}$$

$$\leq e^{8\|J\|} \sup_\tau \sum_{x \in \partial_r^+ A \cap B} |\mu_A^\tau(h_x, f)|$$

If we apply (2.18) to (2.17) we get that the r.h.s. of (2.17) is bounded from above by

$$|\mu_V^\tau(fg)| \leq \|f\|_\infty e^{8\|J\|} \sup_\tau \sum_{x \in V_1} |\mu_{V\setminus V_1}^\tau(\bar{h}_x g)| \tag{2.19}$$

where $\bar{h}_x \equiv h_x - \mu_{V\setminus V_1}^\tau(h_x)$. We can at this point repeat the previous steps for each term $\mu_{V\setminus V_1}^\tau(\bar{h}_x g)|$ and write

$$\mu_{V\setminus V_1}^\tau(\bar{h}_x g)| = \mu_{V\setminus V_1}^\tau(g\mu_{V\setminus V_0}(\bar{h}_x))|$$
$$\leq \|g\|_\infty \sup_\tau \sum_{y \in V_2} |\mu_{V\setminus V_0}^\tau(h_x, h_y)| \tag{2.20}$$

where $V_0 = V_1 \cup V_2$.

Notice that, by construction, $V \setminus V_0$ is still a multiple of Q_l. Therefore we can use lemma 2.10 in order to estimate $|\mu_{V\setminus V_0}^\tau(h_x, h_y)|$. We get

$$|\mu_V^\tau(f, g)| \leq \|f\|_\infty \|g\|_\infty e^{16\|J\|} \sum_{\substack{x \in V_1 \\ y \in V_2}} f_l(x - y) \tag{2.21}$$

and property $SMT(V, 5l, \alpha/11)$ follows from lemma 2.10 provided that l is taken large enough. $\square$

Proof of Lemma 2.10. Let

$$F_l(n) \equiv \sup_{x; |x| \geq n} f_l(x)$$

Then the proof follows from the following recursive formula valid for $l > 2r$, any integer $n \geq 4l + r$ and a suitable constant $k(d, r)$:

$$F_l(n) \leq e^{16\|J\|} k(d, r) A(l) l^{d-1} F_l(n - 4l - r) \qquad (2.22)$$

where

$$A(l) = \sup_{V \in \mathcal{F}_l^0} \sup_{x \in Q_l \cap \partial_r^+ Q_l} \sup_{\tau \in \Omega} \sum_{z \in \partial_r^+ B_{2l}} |\mu_V^\tau(h_x, h_z)| \qquad (2.23)$$

Let us in fact assume (2.22). Then it follows immediately that

$$
\begin{aligned}
F_l(n) &\leq \left[k(d, r) e^{16\|J\|} A(l) l^{d-1} \right]^{\lfloor \frac{n}{4l+r} \rfloor} \|F_l\|_\infty \\
&\leq \left[k(d, r) e^{16\|J\|} A(l) l^{d-1} \right]^{\lfloor \frac{n}{4l+r} \rfloor} \|h_0\|_\infty^2
\end{aligned} \qquad (2.24)
$$

where $\lfloor \cdot \rfloor$ denotes the integer part.

Let us now assume $SMT(V, l/2, \alpha)$ for all $V \in \mathcal{F}_l^0$. Then $A(l) \leq e^{-\frac{\alpha}{2} l}$ for all l large enough and therefore it is possible to find $\bar{l}(d, r, \alpha)$ and $C(\alpha, d, r, \|J\|)$ such that for all $l \geq \bar{l}$

$$F_l(n) \leq C e^{-\frac{\alpha}{16} n}$$

Thus the sought exponential decay of $f_l(x)$ follows.

Let us prove (2.22). We fix $n \geq 4l + r$, two sites x, y with $d(x, y) \geq n$, a set $V \in \mathcal{F}_l$ and a boundary configuration τ. Without loss of generality we assume that $V = \cup_{i=1}^n Q_l(x_i)$, $x_i \in l\mathbb{Z}^d$, $x_1 = 0$ and that $x \in Q_l(x_1) \cup \partial_r^+ Q_l(x_1)$. Next we set

$$
\begin{aligned}
A &= V \cap B_{2l} \\
B &= B_{3l} \setminus B_{2l} \\
C &= V \cap B_{3l}^c
\end{aligned}
$$

and we observe that by construction $x \in A$, $y \in C$ and that neither h_x nor h_y depend on the spins in B. Therefore, using the DLR equations together with the Markov property of our Gibbsian field and by denoting with $\bar{h}_x$ the function $h_x - \mu_V^\tau(h_x)$ and similarly for $\bar{h}_y$, we can write

$$
\begin{aligned}
|\mu_V^\tau(h_x, h_y)| &= |\mu_V^\tau(\bar{h}_x \bar{h}_y)| \\
&= |\mu_V^\tau(\mu_C(\bar{h}_y) \mu_A(\bar{h}_x))| \\
&\leq \|\mu_A(\bar{h}_x)\|_\infty^\tau \|\mu_C(\bar{h}_y)\|_\infty^\tau
\end{aligned} \qquad (2.25)
$$

where $\| \cdot \|_\infty^\tau$ means that we take the supremum over all configuration σ that agree with τ outside V. We claim at this point that

$$
\begin{aligned}
\|\mu_A(\bar{h}_x)\|_\infty^\tau &\leq e^{8\|J\|} A(l) \\
\|\mu_C(\bar{h}_y)\|_\infty^\tau &\leq e^{8\|J\|} \sum_{z \in \partial_r^+ C \cap B} f_l(z - y) \\
&\leq e^{8\|J\|} k(d, r) l^{d-1} F_l(n - 4l - r)
\end{aligned} \qquad (2.26)
$$

for a suitable constant $k(d, r)$, so that (2.22) follows from (2.25) and the arbitrariness of x, y, V and τ.

In turn (2.26) follows at once from formula (2.18) and the obvious bound $|B_{3l} \setminus B_{2l}| \leq k(d, r) l^{d-1}$. $\square$

Remark. In the sequel for any given $\alpha > 0$ we will denote by $\widehat{m}(\alpha)$ the constant $\widehat{m}(d, r, \alpha)$ given in the above proposition.

Remark. With some extra work and using once more the Markov property, one can actually replace the basic assumption "$SMT(V, l/2, \alpha)$ holds for all $V \in \mathcal{F}_l^{0}$" of proposition 2.9 with "$SMT(Q_l, l/4, \alpha)$ holds".

Remark. It is clear from the proof of lemma 2.10 that if we would have replaced in the hypothesis of the lemma "for all $V \in \mathcal{F}_l^{0}$" with "for all $V \subset B_{2l}$" we would have derived the exponential decay of the covariances $\mu_V^\tau(h_x, h_y)$, uniformly for all V and τ and not just for all $V \in \mathcal{F}_l$.

It clearly follows from (2.23), (2.24) that in order to get the exponential decay of the function $f_l(x)$ it is enough to assume that $k(d, r) A(l) l^{d-1} < 1$. In particular this is the case if $\sum_{x \in \mathbb{Z}^d} |x|^{(d-1)} f_l(x) < \infty$ since, by the definition of $f_l(x - y)$,

$$A(l) \leq \sup_{x \in Q_l} \sum_{z \in \partial_r^+ B_{2l}} f_l(x - z)$$

Let us formalize all that in the next proposition

Proposition 2.11. *Fix $L_0 < \infty$ and assume that for all sets V multiple of Q_{L_0} and all functions f and g on Ω such that $d(\Lambda_f, \Lambda_g) \geq n$ we have*

$$\sup_{\tau \in \Omega} |\mu_V^\tau(f, g)| \leq |\Lambda_f| |\Lambda_g| \|f\|_\infty \|g\|_\infty \phi(n)$$

with $\lim_{n \to \infty} n^{2(d-1)} \phi(n) = 0$. Then there exists l and m such that $SMT(V, l, m)$ holds for all V multiple of Q_{L_0}.

We conclude this part concerning Gibbsian fields with a last useful result

Proposition 2.12. *For each $m > 0$ there exists $C(d, r, m)$ such that the following holds. Let $A \subset\subset \mathbb{Z}^d$, $A_0 \subset A$ and $B_0 \subset \partial_r^+ A$. Let $\bar{A} = A \cup \partial_r^+ A$ and assume that*

(i) $md_0 \equiv md(A_0, B_0) \geq \max\{C, 100\|J\|, 10(\log|B_0| + 1)\}$
(ii) $SMT(A \setminus A_0, d_0 - 2r, m)$ holds

then for each pair of configurations $\sigma, \tau \in \Omega$ which agree on $\partial_r^+ A \setminus B_0$, we have

$$\left\| 1 - \frac{d\mu_{\bar{A}}^\tau}{d\mu_{\bar{A}}^\sigma} \Big|_{A_0} \right\|_\infty \leq e^{-(m/4)d_0} \tag{2.27}$$

Proof. For each $\eta \in \Omega_{A_0}$, consider the event $F_\eta = \{\sigma \in \Omega : \sigma_{A_0} = \eta\}$. Choose a pair of configurations σ, τ which agree on $\partial_r^+ A \setminus B_0$. Then there exists a sequence of interpolating configurations $\gamma_i \in \Omega$ for $i = 1, \ldots, n$ such that $n \leq |B_0|$, γ_{i+1} differs from γ_i at exactly one site, $\gamma_1 = \sigma$ and γ_n agrees with τ on $\partial_r^+ A$. Thus, for each $\eta \in \Omega_{A_0}$, we can write

$$\left| 1 - \frac{\mu_A^\tau(F_\eta)}{\mu_A^\sigma(F_\eta)} \right| = \left| 1 - \prod_{i=2}^n \frac{\mu_A^{\gamma_i}(F_\eta)}{\mu_A^{\gamma_{i-1}}(F_\eta)} \right| \qquad (2.28)$$

If we define

$$a = \sup_{\zeta \in \Omega, x \in B_0, \eta \in \Omega_{A_0}} \left| 1 - \frac{\mu_A^\zeta(F_\eta)}{\mu_A^{\zeta^x}(F_\eta)} \right|$$

then it is easy to check that, if $a \le \frac{1}{10}$ and $a|B_0| \le 1$ then the RHS of (2.28) cannot exceed $ea|B_0|$, so if we show that, for instance, $a \le e^{-(m/2)d_0}$, the proposition follows.

Let then, for $z \in \mathbb{Z}^d$, $g_z = \exp(-\nabla_z H_{A \setminus A_0})$. By lemma 2.8, and the SMT property given in the hypotheses, we find

$$a \le \sup_{x \in B_0} e^{14\|J\|} \sum_{y \in A_0} e^{2\|J\|} |\Lambda_{g_x}| |\Lambda_{g_y}| \, \|g_x\|_\infty \|g_y\|_\infty \, e^{-md(\Lambda_{g_x}, \Lambda_{g_y})} \le$$

$$\le (2r+1)^{2d} e^{20\|J\|} \sup_{x \in B_0} \sum_{y \in A_0} e^{-m(|x-y|-2r)}$$

In the second inequality we have used the fact that Λ_{g_x} is contained in a ball of center x and radius r, and the fact that $\|g_x\|_\infty \le \exp(2\|J\|)$. Finally, using the hypothesis on d_0, we easily get

$$a \le e^{-(m/2)d_0}$$

$\square$

3. The Glauber Dynamics

In this section we define the Markov process that will be the main object of our investigation. In what follows the interaction J is fixed and, whenever confusion does not arise, it will not appear in our notation.

3.1 The Dynamics in Finite Volume.

The stochastic dynamics we want to study is determined by the Markov generators L_V, $V \subset\subset \mathbf{Z}^d$, defined by

$$(L_V f)(\sigma) = \sum_{x \in V} c_J(x, \sigma)(\nabla_x f)(\sigma) \qquad \sigma \in \Omega \tag{3.1}$$

The nonnegative real quantities $c_J(x, \sigma)$ are the *transition rates* for the process.

The general assumptions on the transition rates are

(H1) *Finite range interactions.* If $\sigma(y) = \sigma'(y)$ for all y such that $d(x, y) \le r$, then $c_J(x, \sigma) = c_J(x, \sigma')$

(H2) *Detailed balance.* For all $\sigma \in \Omega$ and $x \in \mathbf{Z}^d$,

$$\exp\left[-H_{\{x\}}(\sigma)\right]c_J(x, \sigma) = \exp\left[-H_{\{x\}}(\sigma^x)\right]c_J(x, \sigma^x) \tag{3.2}$$

(H3) *Positivity and boundedness.* There exist positive real numbers c_m and c_M such that

$$0 < c_m \le \inf_{x,\sigma} c_J(x, \sigma) \quad \text{and} \quad \sup_{x,\sigma} c_J(x, \sigma) \le c_M \tag{3.3}$$

(H4) *Translation invariance.* If, for some $k \in \mathbf{Z}^d$, $\sigma'(y) = \sigma(y + k)$ for all $y \in \mathbf{Z}^d$ then $c_J(x, \sigma') = c_J(x + k, \sigma)$ for all $x \in \mathbf{Z}^d$

Three cases one may want to keep in mind are

$$c_J(x, \sigma) = \min\left\{e^{-(\nabla_x H_{\{x\}})(\sigma)}, 1\right\} \tag{3.4}$$

$$c_J(x, \sigma) = \mu^\sigma_{\{x\}}(\sigma^x) = \left[1 + e^{(\nabla_x H_{\{x\}})(\sigma)}\right]^{-1} \tag{3.5}$$

$$c_J(x, \sigma) = \frac{1}{2}\left[1 + e^{-(\nabla_x H_{\{x\}})(\sigma)}\right] \tag{3.6}$$

The first two examples correspond to the so called Metropolis and heat–bath dynamics respectively.

We denote by L_V^τ the operator L_V acting on $L^2(\Omega, d\mu_V^\tau)$ (this amounts to choose τ as the boundary condition). Assumptions (1), (2) and (3) guarantee that there exists a unique Markov process whose generator is L_V^τ, and whose semigroup we denote by $\{T_V^\tau(t)\}_{t \ge 0}$. L_V^τ is a bounded selfadjoint operator on $L^2(\Omega, d\mu_V^\tau)$. The process has a unique invariant measure given by μ_V^τ which is moreover *reversible* for the process.

3.2. Infinite Volume Dynamics.

Let μ be a Gibbs measure for the interaction J. Since the transition rates are bounded and of finite range, the infinite volume generator L obtained by choosing $V = \mathbf{Z}^d$ in (3.1) is well defined on the set of functions f such that $\|\|f\|\|$ is finite. We can then take the closure of L in $C(\Omega)$, the metric space of all continuous functions on Ω with the sup–distance, and get a Markov generator (see, for instance Theorem 3.9 in Chapter I in [L]) or take the closure in $L^2(\Omega, d\mu)$ and get a selfadjoint Markov generator in $L^2(\Omega, \mu)$ (see Theorem 4.1 in Chapter IV of [L]) that will be denoted by L. In the latter case, since the generator is self–adjoint on $L^2(\Omega, d\mu)$ the associated Markov process is reversible w.r.t. the Gibbs measure μ. We conclude with a general result relating the set of invariant measures of the infinite volume Glauber dynamics with the set of Gibbs meaures for the given interaction J (see Theorems 2.14, 2.15, 2.16, 5.12 and 5.14 in [L])

Theorem 3.1. *Assume (H1)...(H4). Then*
  a) *If $d = 1$ or $d = 2$ the set of invariant measures for the above Markov process coincides with the set of Gibbs measures $\mathcal{G}$*
  b) *If $d \geq 3$ then*
      i) *any invariant measure which is also translation invariant is a Gibbs measure*
     ii) *the set of Gibbs measures coincides with the set of reversible invariant measures*
    iii) *If the process is attractive (see paragraph 3.4 below) then the process is ergodic if and only if there is no phase transition*

3.3. Graphical Construction.

We describe here a very convenient way, introduced in [Sch], to realize simultaneously on the same probability space all Markov processes whose generator is L_V^τ, as the initial and the boundary conditions vary in Ω. As a byproduct of the construction we will get, in a rather simple way, a key result which shows that "information" propagates through the system at finite speed.

We associate to each site $x \in \mathbb{Z}^d$ two independent Poisson processes, each one with rate c_M, and we assume independence as x varies in $\mathbb{Z}^d$. We denote the successive arrival after time $t = 0$ of the two processes by $\{t_{x,n}^+\}_{n=1,2...}$ and $\{t_{x,n}^-\}_{n=1,2...}$. We say that at time t there has been an *upmark* at x if $t_{x,n}^+ = t$ for some n and similarly for a *downmark*. Notice that with probability one all the arrival times are different. Next we associate to each arrival time $t_{x,n}^*$, where $*$ stands for either $+$ or $-$, a random variable $U_{x,n}^*$ uniformly distributed in the interval $[0,1]$. We assume that these random variables are mutually independent and independent from the Poisson processes. This completes the construction of the probability space. The corresponding probability measure and expectation are denoted by $\mathbb{P}$ and $\mathbb{E}$ respectively.

Given now $V \subset\subset \mathbb{Z}^d$, a boundary condition $\tau \in \Omega$ and an initial condition $\eta \in \Omega_V$, we construct a Markov process $\{\sigma_t^{V,J,\tau,\eta}\}_{t \geq 0}$ on the above probability space according to the following updating rules. Let us suppose that $t = t_{x,n}^*$ for some $x \in V$ and $n \in \mathbb{Z}_+$ and that the configuration immmediately before t was σ. Then:
  i) The spin $\sigma(y)$ with $y \neq x$ does not change
  ii) If $\sigma(x) = -1$ and the mark was a downmark then $\sigma(x)$ does not change. Similarly if $\sigma(x) = +1$ and the mark was an upmark
  iii) If $\sigma(x) = -1$ and the mark was an upmark then we flip $\sigma(x)$ if and only if $c_J^\tau(x, \sigma) \geq U_{x,n}^+ c_M$. Similarly if $\sigma(x) = +1$ and the mark was a downmark.

One can easily check that the above continuous Markov chain on Ω_V has indeed the correct flip rates $c_J^\tau(x, \sigma)$ so that the above construction represents a global coupling among all processes generated by L_V^τ as the boundary condition τ and the initial condition η vary. In order to investigate how the process $\sigma_t^{V,J,\tau,\eta}(x)$ at site x is affected by a faraway change either in the boundary condition τ or in the initial configuration η, let us introduce, for any given integer l, the event $E(x, t, l)$ that there exists a collection of sites $\{x_0, \ldots x_n\}$ and times $\{t_0, \ldots t_n\}$ such that

i) $0 < t_0 < \ldots < t_n \leq t$ and at each time t_i there is a mark at site x_i
ii) $d(x_0, x) \geq l$, $d(x_i, x_{i+1}) \leq r$ and $x_n = x$, where r denotes the range of the local flip rates.

By construction, if the event $E(x, t, l)^c$ occurred, then the spin $\sigma_t^{V,J,\tau,\eta}(x)$ does not depend on the spin $\eta(y)$ if $d(x, y) > l$ and similarly for $\tau(y)$. The probability of the event $E(x, t, l)$ can be estimated from above by

$$\mathbb{P}(E(x, t, l)) \leq \sum_{n \geq [l/r]} (2r + 1)^{dn} \, \mathbb{P}(Z \geq n) \tag{3.7}$$

where Z is a Poisson random variable with mean $2tc_M$. An elementary calculation shows that

$$\mathbb{P}(Z \geq n) \leq e^{-n(\log(n/2c_M t) - 1)} \tag{3.8}$$

As a consequence we get that the r.h.s of (3.7), as a function of l, for any given fixed time t tends to zero, as $l \to \infty$, faster than an exponential. Moreover there exists a constant k_0, depending only on the dimension d, the range r and c_M, such that the r.h.s of (3.7) is smaller than e^{-t} for all $l \geq k_0 t$.
Let us formalize all what we said in a lemma.

Lemma 3.2. *There exists a constant k_0 depending on d, r and c_M, such that for all local function f and all $t \geq 0$*
i) for all pairs $V_1 \subset\subset \mathbb{Z}^d$ and $V_2 \subset\subset \mathbb{Z}^d$, with $d(V_i^c, \Lambda_f) \geq k_0 t$, $i = 1, 2$,

$$\sup_{\tau_1, \tau_2 \in \Omega} \|T_{V_1}^{\tau_1}(t)f - T_{V_2}^{\tau_2}(t)f\|_\infty \leq \||f\|| e^{-t}$$

ii) for all $V \subset\subset \mathbb{Z}^d$ with $d(V^c, \Lambda_f) \geq k_0 t$ and all $\sigma_1, \sigma_2 \in \Omega_V$, with $\sigma_1(x) = \sigma_2(x)$ for all x such that $d(x, \Lambda_f) \leq k_0 t$,

$$\sup_{\tau \in \Omega} |T_V^\tau(t)f(\sigma_1) - T_V^\tau(t)f(\sigma_2)| \leq \||f\|| e^{-t}$$

Proof. We use the global coupling defined above. More precisely we write

$$T_{V_1}^{\tau_1}(t)f(\sigma_1) - T_{V_2}^{\tau_2}(t)f(\sigma_2) = \mathbb{E}\big(f(\sigma_t^{V_1, J, \tau_1, \sigma_1}) - f(\sigma_t^{V_2, J, \tau_2, \sigma_2})\big) \tag{3.9}$$

and notice that, by a simple telescoping argument,

$$\sup_{\sigma, \eta} |f(\sigma) - f(\eta)| \leq \sum_x \|\nabla_x f\|_\infty \chi(\sigma(x) \neq \eta(x))$$

Therefore the absolute value of the r.h.s of (3.9) can be bounded from above by

$$\||f\|| \sup_{x \in \Lambda_f} \mathbb{P}\big(\sigma_t^{V_1, J, \tau_1, \sigma_1}(x) \neq \sigma_t^{V_2, J, \tau_2, \sigma_2}(x)\big) \tag{3.10}$$

Using the hypotheses of the lemma we get that each term in the sum appearing in the r.h.s of (3.10) can be bounded from above by

$$\mathbb{P}\big(E(x, t, l)\big)$$

with $l \geq k_0 t$. Thus the lemma follows from (3.7) and (3.8). $\quad\square$

We conclude by observing that, in our particular framework, one could construct directly the infinite volume dynamics using the global coupling constructed above together with lemma 3.2, without appealing to the much more general results of [L].

3.4 Attractive Dynamics for Ferromagnetic Interactions.

Let us introduce a partial order on the configuration space Ω by saying that $\sigma \leq \eta$ iff $\sigma(x) \leq \eta(x) \; \forall x \in \mathbb{Z}^d$. A function $f : \Omega_V \mapsto \mathbb{R}$ is called *monotone increasing (decreasing)* if $\sigma \leq \sigma'$ implies $f(\sigma) \leq f(\sigma')$ $(f(\sigma) \geq f(\sigma'))$. An event is called *positive (negative)* if its characteristic function is increasing (decreasing). Given two probability measures μ, μ' on Ω_V we write $\mu \leq \mu'$ if $\mu(f) \leq \mu'(f)$ for all increasing functions f (with $\mu(f)$ we denote the expectation with respect μ).

Then the dynamics defined by the transition rates $c_J(x, \sigma)$ is said to be *attractive* if $\sigma \leq \eta$ and $\sigma(x) = \eta(x)$ imply that

$$\eta(x) c_J(x, \eta) \leq \sigma(x) c_J(x, \sigma)$$

An example of attractive dynamics are the Metropolis and Heat Bath dynamics for a (generalized) Ising model, in which the interaction $J = \{J_A\}_{A \in F}$ is a one and two-body ferromagnetic interaction of the form

$$J_A = \begin{cases} J \geq 0 & \text{if } A = \{x, y\} \text{ with } d_2(x, y) \leq r \\ h & \text{if } A = \{x\} \\ 0 & \text{otherwise} \end{cases}$$

An very useful consequence of attractivity is the following *monotonicity* property in the boundary condition and in the initial configuration of the Markov process $\sigma_t^{V, J, \tau, \eta}$ given by the graphical construction

$$\mathbb{P}\big(\sigma_t^{V, J, \tau_1, \eta_1} \leq \sigma_t^{V, J, \tau_2, \eta_2}\big) = 1 \quad \forall \, \tau_1 \leq \tau_2 \text{ and } \forall \, \eta_1 \leq \eta_2 \tag{3.11}$$

Such property follows immediately from the updating rules which govern the evolution of the Markov process $\sigma_t^{V, J, \tau, \eta}$. In turn (3.11) implies

(1) If f is an increasing function on Ω_V then $T_V^\tau(t)f$ is also increasing for all $t \geq 0$
(2) If ρ_1, ρ_2 are two probability measures on Ω_V such that $\rho_1 \leq \rho_2$ then
$\rho_1 T_V^\tau(t) \leq \rho_2 T_V^\tau(t)$ for all $t \geq 0$

Moreover (3.11) allows us to define a *standard* coupling between two finite volume Gibbs measures which preserve the order of the b.c. Take in fact two boundary conditions $\tau_1 \leq \tau_2$ and let $\nu_V^{\tau_1,\tau_2}$ be the unique invariant measure of the coupled process $(\sigma_t^{V,J,\tau_1}, \sigma_t^{V,J,\tau_2})$ on the set $\{(\sigma,\eta): \sigma \leq \eta\}$. Then we have

(1) $\nu_V^{\tau_1,\tau_2}\{(\sigma_1,\sigma_2): \sigma_1 = \sigma\} = \mu_V^{\tau_1}(\sigma)$ for all $\sigma \in \Omega_V$

(2) $\nu_V^{\tau_1,\tau_2}\{(\sigma_1,\sigma_2): \sigma_2 = \sigma\} = \mu_V^{\tau_2}(\sigma)$ for all $\sigma \in \Omega_V$

(3) $\nu_V^{\tau_1,\tau_2}\{(\sigma_1,\sigma_2): \sigma_1 \leq \sigma_2\} = 1$ so that $\mu_V^{\tau_1} \leq \mu_V^{\tau_2}$.

In particular, from (3) it follows that any finite volume Gibbs measure μ_V^{τ} is an FKG–measure i.e. satisfies (see chapter II.2 of [L])

$$\mu_V^{\tau}(fg) \geq \mu_V^{\tau}(f)\mu_V^{\tau}(g)$$

for any increasing functions f and g.

3.5 Spectral Gap and Logarithmic Sobolev Constant

A fundamental quantity associated with the dynamics of a reversible system is the gap of the generator, *i.e.*

$$\text{gap}(L_V^{\tau}) = \inf \text{spec}\left(-L_V^{\tau} \restriction \mathbf{1}^{\perp}\right)$$

where $\mathbf{1}^{\perp}$ is the subspace of $L^2(\Omega, d\mu_V^{\tau})$ orthogonal to the constant functions. The gap can also be characterized as

$$\text{gap}(L_V^{\tau}) = \inf_{f \in L^2(\Omega, d\mu_V^{\tau})} \frac{\mathcal{E}_V^{\tau}(f,f)}{\text{Var}_V^{\tau}(f)} \tag{3.12}$$

where $\mathcal{E}$ is the Dirichlet form associated with the generator L,

$$\mathcal{E}_V^{\tau}(f,f) = \frac{1}{2} \sum_{\sigma \in \Omega_V} \sum_{x \in V} \mu_V^{\tau}(\sigma)\, c(x,\sigma)\, [(\nabla_x f)(\sigma)]^2 \tag{3.13}$$

and Var_V^{τ} is the variance relative to the probability measure μ_V^{τ}. When the transition rates are chosen as in (3.6), it is easy to verify that the Dirichlet form takes a particularly simple form

$$\mathcal{E}_V^{\tau}(f,f) = \frac{1}{2}\mu_V^{\tau}(|\nabla_V f|^2) \tag{3.14}$$

By simple spectral theory, for any $f \in L^2(\Omega, d\mu_V^{\tau})$, one gets the following bound on the Markov semigroup $T_V^{\tau}(t)$ in terms of the spectral gap

$$\|T_V^{\tau}(t)f - \mu_V^{\tau}(f)\|_{2,\mu_V^{\tau}}^2 \leq e^{-2\,\text{gap}(L_V^{\tau})t}\|f\|_{2,\mu_V^{\tau}}^2$$

where $\|f\|_{2,\mu_V^{\tau}}$ denotes the $L^2(\Omega, d\mu_V^{\tau})$ norm.

Quite bad is the analogous bound in the $\|\cdot\|_{\infty}$ norm. In finite volume and for general f, the best one can get using only the spectral gap is

$$\begin{aligned}
\|T_V^{\tau}(t)f - \mu_V^{\tau}(f)\|_{\infty} &\leq \left[\inf_{\sigma}\mu_V^{\tau}(\sigma)\right]^{-1/2}\|T_V^{\tau}(t)f - \mu_V^{\tau}(f)\|_{2,\mu_V^{\tau}} \\
&\leq \left[\inf_{\sigma}\mu_V^{\tau}(\sigma)\right]^{-1/2}e^{-\text{gap}(L_V^{\tau})t}\|f\|_{2,\mu_V^{\tau}}
\end{aligned} \tag{3.15}$$

Notice that the factor $\left[\inf_\sigma \mu_V^\tau(\sigma)\right]^{-1/2}$ is usually exponentially large in the size of V and therefore the above bound becomes meaningful only for extremely large times. For local functions the above estimate can be slightly improved. Let in fact f be a local function. Then, using lemma 3.2, one can safely replace the original set V by a "ball" of radius $k_0 t$ around the support of f, Λ_f (see (3.20)...(3.21) below for details). However the prefactor multiplying the negative exponential is still very large, of the order of $\exp(Ct^d)$ so that the resulting bound is practically useless.

Much more powerful are the estimates obtained through "hypercontractive bounds" on the Markov semigroup. For this purpose, we define the *logarithmic Sobolev constant* $c_s(L_V^\tau)$ associated with the generator L_V^τ as the infimum over all c such that, for all positive functions f

$$\mu_V^\tau(f^2 \log f) \leq c\mathcal{E}_V^\tau(f, f) + \mu_V^\tau(f^2) \log \sqrt{\mu_V^\tau(f^2)} \qquad (3.16)$$

More generally, given a probability measure ν on Ω, we denote by $c_s(\nu)$ the smallest constant c such that, for all positive local functions f

$$\nu(f^2 \log f) \leq \frac{c}{2} \nu(|\nabla_V f|^2) + \nu(f^2) \log \sqrt{\nu(f^2)} \qquad (3.17)$$

Notice that if we take $\nu = \mu_V^\tau$ then $c_s(\mu_V^\tau)$ coincides with the logarithmic Sobolev constant associated with the generator L_V^τ corresponding to the rates (3.6). Even if the transition rates are different from those given in (3.6) we can estimate $c_s(L_V^\tau)$ in terms of $c_s(\mu_V^\tau)$ as follows:

$$\frac{c_m}{c_M} c_s(L_V^\tau) \leq c_s(\mu_V^\tau) \leq \frac{c_M}{c_m} c_s(L_V^\tau) \qquad (3.18)$$

Finally we set

$$\begin{aligned} c_s(L_V) &= \sup_\tau c_s(L_V^\tau) \\ c_s(\mu_V) &= \sup_\tau c_s(\mu_V^\tau) \end{aligned} \qquad (3.19)$$

The role of the logarithmic Sobolev constant in the analysis of the exponential ergodicity of the dynamics is clarified by the following theorem

Theorem 3.3. *Assume*

$$\sup_L \sup_\tau c_s(L_{B_L}^\tau) < c_s^\infty < +\infty$$

Then:
  a) *there exists a positive constant m and for any local function f there exists a constant C_f such that*

$$\sup_L \sup_\tau \|T_{B_L}^\tau(t)f - \mu_{B_L}^\tau(f)\|_\infty \leq C_f e^{-mt}$$

  b) *There exist positive α and l_0 such that $SMT(B_L, l_0, \alpha)$ holds for all L*

c) *The infinite volume dynamics is exponentially ergodic in the $\|\cdot\|_\infty$−norm and there exists a unique Gibbs measure μ with covariances decaying exponentially fast.*

Proof.

a) Let us fix a cube B_L, a boundary condition $\tau \in \Omega$ and a local function f such that $\mu^\tau_{B_L}(f) = 0$. Without loss of generality we assume that $0 \in \Lambda_f$. For any $t \geq 0$ let also $l_t = \min\{l : d(\Lambda_f, B_l^c) \geq k_0 t\}$ where k_0 is the constant appearing in lemma 3.2 and let $\Lambda_t \equiv B_{l_t \wedge L}$. Then we write

$$\|T^\tau_{B_L}(t)f\|_\infty \leq \|T^\tau_{B_L}(t)f - T^\tau_{\Lambda_t}(f)\|_\infty + \|T^\tau_{\Lambda_t}(t)f\|_\infty \tag{3.20}$$

The first term in the r.h.s of (3.20) is smaller than $\||f\|| e^{-t}$ because of lemma 3.2. In order to estimate from above the second term we use the following trivial bound valid for any $q \geq 1$, any $V \subset\subset \mathbf{Z}^d$, any positive probability measure ν on Ω_V and any function f

$$\|f\|_\infty \leq \left[\inf_\sigma \nu(\sigma)\right]^{-\frac{1}{q}} \|f\|_{q,\nu} \tag{3.21}$$

where $\|f\|_{q,\nu}$ denotes the $L^q(\Omega_V, d\nu)$ norm.

Given $\varepsilon \in (0,1)$, we take ν as the marginal on Ω_{Λ_t} of the Gibbs measure $\mu^\tau_{B_L}$ and apply (3.21) to $T^\tau_{\Lambda_t}(t)f$. Using once more lemma 3.2 and the trivial bound $\inf_\sigma \mu^\tau_V(\sigma) \geq e^{-2\|J\|_\infty|V|}$, we get

$$\begin{aligned}
\|T^\tau_{\Lambda_t}(t)f\|_\infty &\leq A(t,q)\|T^\tau_{\Lambda_t}(t)f\|_{q,\mu^\tau_{B_L}} \\
&\leq A(t,q)\|T^\tau_{B_L}(t)f\|_{q,\mu^\tau_{B_L}} + A(t,q)\||f\|| e^{-t} \\
&= A(t,q)\|T^\tau_{B_L}((1-\varepsilon)t)f_t\|_{q,\mu^\tau_{B_L}} + A(t,q)\||f\|| e^{-t}
\end{aligned} \tag{3.22}$$

where

$$A(t,q) = e^{2\|J\|_\infty|\Lambda_t|/q} \qquad \text{and} \qquad f_t = T^\tau_{B_L}(\varepsilon t)f$$

Notice that so far we never used the key assumption of the theorem, namely boundedness of the logarithmic Sobolev constant. It is at this stage that such hypothesis becomes crucial. Thanks in fact to Gross's integration lemma (see e.g [DeSt]) we have that the Markov semigroup $T^\tau_V(t)$ maps $L^2(\Omega, d\mu^\tau_V)$ contractively into $L^q(\Omega, d\mu^\tau_V)$ for any $2 \leq q \leq 1 + e^{t/c_s(L^\tau_V)}$. Thus

$$\begin{aligned}
A(t,q)\|T^\tau_{B_L}((1-\varepsilon)t)f_t\|_{q,\mu^\tau_{B_L}} &\leq A(t,q)\|f_t\|_{2,\mu^\tau_{B_L}} \\
&\leq A(t,q)\|f\|_{2,\mu^\tau_{B_L}} e^{-\operatorname{gap}(L^\tau_{B_L})\varepsilon t}
\end{aligned} \tag{3.23}$$

for all $2 \leq q \leq 1 + e^{(1-\varepsilon)t/c_s(L^\tau_{B_L})}$. Since $\sup_\tau \sup_L c_s(L^\tau_{B_L}) < c_s^\infty < +\infty$ we can apply the above inequality with $q = q(t) = 1 + e^{(1-\varepsilon)t/c_s^\infty}$ and get that $\lim_{t\to\infty} A(t, q(t)) = 1$. In order to conclude the proof of the first part of the theorem it is enough to recall a well known result (see e.g [SC]) that says that

$$\operatorname{gap}(L^\tau_V) \geq c_s(L^\tau_V)^{-1}$$

b) To prove exponential decay of covariances we proceed as in Lemma 3.1 of [SZ]. Let fix a box B_L and a boundary condition τ and let f and g be two local functions with zero $\mu_{B_L}^\tau$–mean and supports Λ_f, Λ_g contained in B_L. Let $l = \lfloor d(\Lambda_f, \Lambda_g)/2 \rfloor$ and let $t = \frac{l}{k_0}$, k_0 being the constant appearing in lemma 3.2. Let also $\Lambda_1 = \{x \in B_L : d(x, \Lambda_f) \le l\}$ and take $\Lambda_2 = B_L \setminus \Lambda_1$. Then, using reversibility, we write

$$
\begin{aligned}
|\mu_{B_L}^\tau(fg)| &= |\mu_{B_L}^\tau\big(T_{B_L}^\tau(t)(fg)\big)| \\
&\le |\mu_{B_L}^\tau\big(g\, T_{B_L}^\tau(2t)f\big)| + \|T_{B_L}^\tau(t)(fg) - T_{B_L}^\tau(t)f\, T_{B_L}^\tau(t)g\|_\infty
\end{aligned}
\tag{3.24}
$$

The first term in the r.h.s of (3.24) is bounded from above by

$$
\|g\|_{2,\mu_{B_L}^\tau} \|f\|_{2,\mu_{B_L}^\tau} e^{-t \operatorname{gap}(L_{B_L}^\tau)}
\tag{3.25}
$$

Since $\operatorname{gap}(L_{B_L}^\tau) \ge 1/c_s^\infty$ and $\|g\|_{2,\mu_{B_L}^\tau} \le \|g\|_\infty$ we get, remembering the definition of t, that (3.25) is bounded from above by

$$
\|g\|_\infty \|f\|_\infty e^{-\frac{l}{k_0 c_s^\infty}}
\tag{3.26}
$$

The second term in the r.h.s of (3.24), thanks to lemma approx and our choice of time t is bounded from above by

$$
\begin{aligned}
\|T_{B_L}^\tau(t)fg &- T_{\Lambda_1}^\tau(t)f\, T_{\Lambda_2}^\tau(t)g\|_\infty + \|T_{B_L}^\tau(t)f - T_{\Lambda_1}^\tau(t)f\|_\infty \|g\|_\infty \\
&+ \|T_{B_L}^\tau(t)g - T_{\Lambda_2}^\tau(t)g\|_\infty \|f\|_\infty \\
&\le \big(\|\!|fg|\!\| + \|g\|_\infty \|\!|f|\!\| + \|f\|_\infty \|\!|g|\!\|\big)e^{-t} \\
&\le 4\|g\|_\infty \|f\|_\infty |\Lambda_f||\Lambda_g| e^{-\frac{l}{k_0}}
\end{aligned}
\tag{3.27}
$$

Take now l_0 such that

$$
e^{-\frac{l_0}{2k_0 c_s^\infty}} + 4e^{-\frac{l_0}{2k_0}} \le 1
$$

and set $\alpha = \min[(2k_0 c_s^\infty)^{-1}, (2k_0)^{-1}]$. With this choice and using (3.26), (3.27) the r.h.s of (3.24) is bounded from above by

$$
\|g\|_\infty \|f\|_\infty |\Lambda_f||\Lambda_g| e^{-\alpha l}
$$

if $l \ge l_0$. Thus $SMT(B_L, l_0, \alpha)$ follows.

c) To prove exponential ergodicity of the infinite volume dynamics we fix a local function f with $0 \in \Lambda_f$ and set, for any time t, $\Lambda_t = B_{l_t}$ where l_t was defined at the beginning of the proof. Then, thanks once more to lemma 3.2, for any $\tau \in \Omega$ we have

$$
\begin{aligned}
\sup_{\sigma,\eta} |T(t)f(\sigma) - T(t)f(\eta)| &\le \sup_{\sigma,\eta} |T_{\Lambda_t}^\tau(t)f(\sigma) - T_{\Lambda_t}^\tau(t)f(\eta)| + 2\|\!|f|\!\| e^{-t} \\
&\le 2C_f e^{-mt} + 2\|\!|f|\!\| e^{-t}
\end{aligned}
$$

where in the last step we used the bound (3.23). Thus the infinite volume dynamics is exponentially ergodic and therefore (see theorem 3.1) there in no phase transition. $\square$

Remark. Notice that, in the proof of the exponential decay of covariances, we could have assumed the (apparently) weaker condition

$$\inf_L \inf_\tau \mathrm{gap}(L^\tau_{B_L}) > 0$$

It is remarkable that such a weaker hypotheses implies the stronger one used in the theorem above (see [MO1], [MO2] and [SZ]). This issue will be illustrated in the next section (see theorem 4.6).

Remark. It is worthwhile to observe that exponential ergodicity in the $\|\cdot\|_\infty$−norm for all local functions f, *i.e.*

$$\|T(t)f - \mu(f)\|_\infty \le C_f e^{-mt}$$

for some finite constant C_f and some $m > 0$ independent of f, implies that the spectral gap of the infinite volume generator L on $L^2(\Omega, d\mu)$ is larger or equal to m. Here μ denotes the unique invariant measure of the Glauber dynamics and therefore the unique Gibbs state for the interaction J.

Let us in fact assume exponential ergodicity in the $\|\cdot\|_\infty$−norm with rate $m > 0$ for all local functions and let $\delta \in (0, 1)$. Let also $P_{[0,\delta m]}$ be the spectral projection of the interval $[0, \delta m]$ associated to the selfadjoint generator L. Then necessarily $P_{[0,\delta m]}f = 0$ for all local functions f with zero μ−mean. If not we would get the contradiction

$$\|P_{[0,\delta m]}f\|_{2,\mu} e^{-\delta mt} \le \|T(t)f\|_{2,\mu} \le C_f e^{-mt} \qquad \forall t \ge 0$$

for any local function f with zero μ−mean. To finish the argument it is sufficient to observe that, since the continuous functions are dense in $L^2(\Omega, d\mu)$, for any $f \in L^2(\Omega, d\mu)$ there exists a sequence $f^{(n)}$ of local functions such that $\limsup_{n\to\infty} \|f - f^{(n)}\|_{2,\mu} = 0$.

3.6 From Single Spin Dynamics to Block Dynamics.

We will also consider a more general version of the finite volume dynamics discussed so far in which more than one spin can flip at once. Let $\mathcal{D} = \{V_1, \ldots, V_n\}$ be an arbitrary collection of finite sets $V_i \in \mathbb{F}$ and let $V = \cup_i V_i$. Then we will denote by *block dynamics* with blocks $\{V_1, \ldots, V_n\}$ the continuos time Markov chain in which each block waits an exponential time of mean one and then the configuration inside the block is replaced by a new configuration distributed according to the Gibbs measure of the block given the previous configuration outside the block. More precisely, the generator of the Markov process corresponding to $\mathcal{D}$ is defined as

$$L^\tau_{\mathcal{D}} f = \sum_{i=1}^n (\mu^\tau_{V_i}(f) - f) \tag{3.28}$$

From the DLR condition (2.2) it follows that $L^\tau_{\mathcal{D}}$ is self-adjoint on $L^2(\Omega, d\mu^\tau_V)$, *i.e.* the block dynamics is reversible w.r.t. the Gibbs measure μ^τ_V, so that μ^τ_V represents its unique (in finite volume) invariant measure.

3.7 General Results on the Spectral Gap

In this section we collect some technical results to be used in the next sections. Although most of the results presented here are rather simple and some of them can actually be found in the literature, we thought however useful, also for future purposes, to put them together in a sort of primitive tool–box for the subject.

We begin by proving a simple, but important result relating the spectral gap for the single–site Glauber dynamics to the spectral gap of a block–dynamics defined on the same set V. This simple, but in our approach very important result can be roughly understood as follows. The relaxation time ($\equiv$ inverse of the spectral gap) of a single–site Glauber dynamics for a given finite volume Gibbs measure is not larger than the largest among the relaxation times of the same dynamics restricted to each of the blocks of some block–dynamics for the same Gibbs measure times the relaxation time of the block–dynamics itself. More precisely

Proposition 3.4. *Let* $\mathcal{D} = \{V_1, \ldots, V_n\}$ *be an arbitrary collection of finite sets* $V_i \in \mathbb{F}$ *and let* $V = \cup_i V_i$. *For any given boundary condition* $\tau \in \Omega$, *let* L_V^τ *be given by (3.1) with transition rates satisfying* $(H1), (H2), (H3)$ *and let* $L_{\mathcal{D}}^\tau$ *be given by (3.28). Then for any* $f \in L^2(\Omega, d\mu_V^\tau)$ *we have*

$$\mathrm{Var}_V^\tau(f) \leq$$

$$\left(\mathrm{gap}(L_{\mathcal{D}}^\tau)^{-1} \inf_i \inf_{\varphi \in \Omega} \mathrm{gap}(L_{V_i}^\varphi)\right)^{-1} \frac{1}{2} \sum_{\sigma \in \Omega} \mu_V^\tau(\sigma) \sum_{x \in V} N_x c_J(x, \sigma) \left[f(\sigma^x) - f(\sigma)\right]^2$$

(3.29)

where $N_x = \#\{i : V_i \ni x\}$. *In particular*

$$\mathrm{gap}(L_V^\tau) \geq \mathrm{gap}(L_{\mathcal{D}}^\tau) \inf_i \inf_{\varphi \in \Omega} \mathrm{gap}(L_{V_i}^\varphi) \left(\sup_{x \in V} \#\{i : V_i \ni x\}\right)^{-1}$$

(3.30)

Proof. Let

$$g = \inf_i \inf_{\varphi \in \Omega} \mathrm{gap}(L_{V_i}^\varphi)$$

Thanks to (3.12), (3.30) follows immediately from (3.29). In turn (3.29) is proven if we can show that

$$\mathcal{E}_{\mathcal{D}}^\tau(f, f) \leq g^{-1} \frac{1}{2} \sum_{\sigma \in \Omega} \mu_V^\tau(\sigma) \sum_{x \in V} \#\{i : V_i \ni x\} c_J(x, \sigma) \left[f(\sigma^x) - f(\sigma)\right]^2$$

(3.31)

for all $f \in L^2(\Omega, d\mu_V^\tau)$. But, using (3.28), we find

$$\mathcal{E}_{\mathcal{D}}^\tau(f, f) = \sum_{\sigma' \in \Omega} \mu_V^\tau(\sigma') \sum_i \mathrm{Var}_{V_i}^{\sigma'}(f)$$

$$\leq g^{-1} \sum_{\sigma' \in \Omega} \mu_V^\tau(\sigma') \sum_i \mathcal{E}_{V_i}^{\sigma'}(f, f)$$

(3.32)

On the other side, again because of (2.2),

$$\sum_{\sigma' \in \Omega} \mu_V^\tau(\sigma') \sum_i \mathcal{E}_{V_i}^{\sigma'}(f, f) =$$

$$= \frac{1}{2} \sum_{\sigma' \in \Omega} \mu_V^\tau(\sigma') \sum_i \sum_{\sigma \in \Omega} \mu_{V_i}^{\sigma'}(\sigma) \sum_{x \in V_i} c_J(x, \sigma) \left[f(\sigma^x) - f(\sigma) \right]^2$$

$$\leq \frac{1}{2} \sum_{\sigma \in \Omega} \mu_V^\tau(\sigma) \sum_{x \in V} \#\{ i : V_i \ni x \} c_J(x, \sigma) \left[f(\sigma^x) - f(\sigma) \right]^2$$

(3.31) and the proposition follows. $\square$

Next we provide three lower bounds on the spectral gap of the block dynamics with just two blocks.

Proposition 3.5. *Let* $V \subset\subset \mathbf{Z}^d$, *and let* A, B *be two (possibly intersecting) subsets of* V *such that* $V = A \cup B$. *Let* $\mathcal{D} = \{A, B\}$. *Assume that*

$$\sup_{\tau \in \Omega} \left\| 1 - \frac{d\mu_A^\tau}{d\mu_V^\tau} \Big|_{\partial_r^+ B} \right\|_\infty \leq \varepsilon < 1 \tag{3.33}$$

Then the gap for the block dynamics on $\mathcal{D}$ *satisfies*

$$\inf_{\tau \in \Omega} \mathrm{gap}(L_\mathcal{D}^\tau) \geq 1 - \sqrt{\varepsilon}$$

Proof. The action of the semigroup $T_\mathcal{D}(t)$ associated to the block dynamics is given by

$$T_\mathcal{D}(t) f = \sum_{n=0}^\infty \frac{t^n}{n!} (L_\mathcal{D})^n f$$

Using the explicit expression for $L_\mathcal{D}$ and some elementary combinatorics, it is not difficult to show that

$$T_\mathcal{D}(t) f = \sum_{n=0}^\infty \frac{(2t)^n}{n!} e^{-2t} \frac{1}{2^n} \sum_{X \in \{A, B\}^n} \mu_{X_1} \cdots \mu_{X_n}(f) \tag{3.34}$$

Since $(\mu_A)^2 = \mu_A$ (and similarly for B) the last summation (over X) in (3.34) can be written as

$$\sum_{k=0}^{n-1} \binom{n-1}{k} (\widehat{A}_{k+1} + \widehat{B}_{k+1}) f \tag{3.35}$$

where

$$\widehat{A}_k = (\mu_A \circ \mu_B)^{\lfloor k/2 \rfloor} \circ \mu_A^{k - 2\lfloor k/2 \rfloor} \qquad \widehat{B}_k = (\mu_B \circ \mu_A)^{\lfloor k/2 \rfloor} \circ \mu_B^{k - 2\lfloor k/2 \rfloor}$$

If now g is an arbitrary bounded measurable function on Ω, such that $\mu_V(g) = 0$, we get

$$\| \mu_A \mu_B \mu_A g \|_\infty \leq \| \mu_V \mu_B \mu_A g \|_\infty + \| \mu_V \mu_B \mu_A g - \mu_A \mu_B \mu_A g \|_\infty \tag{3.36}$$

By the DLR property (2.2) the first term on the RHS of (3.36) is equal to $\mu_V(g) = 0$. Furthermore, since the interaction has range r, the function $h \equiv \mu_B \mu_A g$ is $\mathcal{F}_{V^c \cup \partial_r^+ B}$ measurable. This fact together with hypothesis (3.33) and the trivial observation that μ_A and μ_V agree on $\mathcal{F}_{V^c}$ implies

$$\|\mu_V \mu_B \mu_A g - \mu_A \mu_B \mu_A g\|_\infty \leq \varepsilon \|\mu_B \mu_A g\|_\infty \leq \varepsilon \|\mu_A g\|_\infty \tag{3.37}$$

Iterating this inequality we get, for each bounded measurable f with $\mu_V(f) = 0$,

$$\|\widehat{A}_k f\|_\infty \leq (\sqrt{\varepsilon})^{k-3} \|f\|_\infty \qquad \|\widehat{B}_k f\|_\infty \leq (\sqrt{\varepsilon})^{k-3} \|f\|_\infty \tag{3.38}$$

Thus, we get that the sup norm of (3.35) is not greater than

$$\|f\|_\infty \frac{2}{\varepsilon^{3/2}} (1 + \sqrt{\varepsilon})^{n-1}$$

which, inserted back into (3.34) yields

$$\|T_{\mathcal{D}}(t)f\|_\infty \leq \|f\|_\infty 4\varepsilon^{-3/2} e^{-2t} \sum_{n=0}^{\infty} \frac{t^n}{n!} (1 + \sqrt{\varepsilon})^n = \|f\|_\infty 4\varepsilon^{-3/2} e^{-(1-\sqrt{\varepsilon})t} \qquad \square$$

Corollary 3.6. *Let V, A and B be as in Proposition 3.5. Let also $A_0 = A \cap \partial_s^+ B$, with $s \geq r$, $B_0 = B \cap \partial_r^+ A$ and $\bar{A} = A \cup \partial_r^+ A$. For each $m > 0$ there exists $C(d, r, m)$ such that if*

(i) $md_0 \equiv md(A_0, B_0) \geq \max\{ C, 100\|J\|, 10(\log|B_0| + 1) \}$
(ii) $SMT(A\backslash A_0, d_0 - 2r, m)$ *holds*

then

$$\inf_{\tau \in \Omega} \text{gap}(L_{\{A,B\}}^\tau) \geq \frac{1}{2}$$

Proof. Thanks to Proposition 3.5 it is sufficient to show that

$$\sup_{\tau \in \Omega} \left\|1 - \frac{d\mu_A^\tau}{d\mu_V^\tau}\Big|_{A_0}\right\|_\infty \leq \frac{1}{4} \tag{3.39}$$

By the DLR property (2.2) we have

$$\text{LHS of (3.39)} \leq \sup_{\tau, \sigma \in \Omega \,:\, \tau_{V^c} = \sigma_{V^c}} \left\|1 - \frac{d\mu_A^\tau}{d\mu_A^\sigma}\Big|_{A_0}\right\|_\infty \tag{3.40}$$

At this point we can use Proposition 2.12 and obtain the result. $\square$

Remark. It is important to observe that, while it is very likely that a block dynamics with few large blocks will have a spectral gap larger than the spectral gap of a single spin flip dynamics, precisely the opposite may happen for the logarithmic Sobolev constant $c_s(L_{\mathcal{D}}^\tau)$. To see this fix a finite set V and consider the covering $\mathcal{D}$ with just one block eqaul to V itself. Then clearly $\text{gap}(L_{\mathcal{D}}^\tau) = 1$ since $\mathcal{E}_V^\tau(f, f) = \text{Var}_V^\tau(f)$ but $c_s(L_{\mathcal{D}}^\tau) \geq \sup_\eta -\frac{1}{2} \log(\mu_V^\tau(\eta_V))(1 - \mu_V^\tau(\eta_V))$. To prove it let us consider, for any $\eta \in \Omega$, the test function

$$f_\varepsilon(\sigma) = \begin{cases} (\mu_V^\tau(\eta_V))^{-1/2} & \text{if } \sigma_V = \eta_V \\ \varepsilon & \text{otherwise} \end{cases}$$

where $\varepsilon > 0$ is a positive number that eventually will be sent to zero. Then an explicit computation shows that

$$\lim_{\varepsilon \to 0} \mathrm{Var}_V^\tau(f_\varepsilon) = 1 - \mu_V^\tau(\eta_V)$$

$$\lim_{\varepsilon \to 0} \mu_V^\tau\big(f_\varepsilon^2 \log(f_\varepsilon)\big) = -\frac{1}{2} \log\big(\mu_V^\tau(\eta_V)\big)$$

$$\lim_{\varepsilon \to 0} \mu_V^\tau(f_\varepsilon^2) = 1$$

so that, necessarily, $c_s(L_D^\tau) \geq -\frac{1}{2} \log\big(\mu_V^\tau(\eta_V)\big)\big(1 - \mu_V^\tau(\eta_V)\big)$ for all $\eta \in \Omega$. We conclude by observing that, for non pathological models, the above lower bound is of the order of $\log(|V|)$, in agreement with the general uppr bound proved in proposition 3.9.

Proposition 3.7. *Let V, A, B be as in Proposition 3.5. Let*

$$N = |\partial_r^+ A \cap B| \wedge |\partial_r^+ B \cap A|$$

Then there exists $k = k(d, r)$ such that

$$\inf_{\tau \in \Omega} \mathrm{gap}(L_{\{A,B\}}^\tau) \geq \exp[-k\|J\|_V N]$$

Proof. We can assume $N = |\partial_r^+ A \cap B|$. Consider a new interaction J^0 such that A and $V \backslash A$ are decoupled, *i.e.*

$$J_X^0 = \begin{cases} J_X & \text{if } X \cap \partial_r^+ A \cap B = \emptyset \\ 0 & \text{otherwise} \end{cases}$$

We have clearly $\sum_{X \ni x} \|J_X - J_X^0\| = 0$ unless x is in a neighborhood of radius r of $\partial_r^+ A \cap B$, hence

$$\sum_{x \in V} \sum_{X \ni x} \|J_X - J_X^0\| \leq k_1 N \|J\|$$

for some k_1 which depends on d and r. This implies that for all functions f on Ω,

$$\mathcal{E}_{\{A,B\}}^{J,\tau}(f,f) \geq \exp(-4k_1\|J\|N)\, \mathcal{E}_{\{A,B\}}^{J^0,\tau}(f,f) \tag{3.41}$$

$$\mathrm{Var}_V^{J,\tau}(f) \leq \exp(4k_1\|J\|N)\, \mathrm{Var}_V^{J^0,\tau}(f) \tag{3.42}$$

From (3.41), (3.42) and the variational characterization of the gap (3.12), it follows that

$$\mathrm{gap}(L_{\{A,B\}}^{J,\tau}) \geq \exp[-8k_1\|J\|N]\, \mathrm{gap}(L_{\{A,B\}}^{J^0,\tau}) \tag{3.43}$$

In order to estimate the gap for the block–dynamics with couplings J^0, we just notice that the hypotheses of Proposition 3.5 are satisfied with $\varepsilon = 0$, and thus $\mathrm{gap}(L_{\{A,B\}}^{J^0,\tau}) \geq 1$. $\square$

The second important result is a very general lower bound on the spectral gap of Glauber dynamics in an *arbitrary* set $V \subset\subset \mathbf{Z}^d$. It says that the spectral gap is always larger than a negative exponential of $|V|^{\frac{d-1}{d}}$. Notice that if V is cube then $|V|^{\frac{d-1}{d}}$ is simply its surface. In this case the bound is certainly optimal, at least in our general setting, since it is known that for several models of lattice *discrete spins* in the phase coexistence region, the activation energy between different stable phases is proportional to the *surface* of the region in consideration (see [M] and [CGMS] for more precise statements for the Ising model). Apparently the situation for *continuous* spins systems can be very different. For Heisenberg models, in fact, it is believed on the basis of spin–wave theory (see [B1], [B2]) that, at least for cubic regions, the spectral gap does not go to zero faster than the inverse of the volume. It is a challenging problem to actually prove it!

Theorem 3.8. *Let $d \geq 2$. There exist $k(d, r, \kappa_1)$, such that, for each $\Lambda \subset\subset \mathbb{Z}^d$ and for each $\tau \in \Omega$, we have (c_m was defined in (3.3))*

$$\mathrm{gap}(L_\Lambda^\tau) \geq c_m \exp\left[-k \|J\| |\Lambda|^{\frac{d-1}{d}} \right] \tag{3.44}$$

Proof. For each non–negative integer n, let

$(K_n) = $ the inequality (3.44) holds for all $\Lambda \in \mathbb{F}$ such that $|\Lambda| \leq (3/2)^n$

We want to show, that (K_n) holds for all $n \in \mathbb{Z}_+$, by proving that there exists $n_0(d, r) \in \mathbb{Z}_+$ such that (K_{n_0}) holds, and such that, for all $n \geq n_0$, (K_n) implies (K_{n+1}).

Assume then that K_{n-1} holds, and take any Λ such that $(3/2)^{n-1} < |\Lambda| \leq (3/2)^n$. Let $v = |\Lambda|$. Using a geometric lemma proved in [CMM1] (see Proposition A1.1 there), it is possible to write Λ as the disjoint union of two subsets X and Y, such that

(a) $v/2 - k_1 v^{\frac{d-1}{d}} \leq |X| \leq v/2$
(b) $|\delta_r(X, Y)| \leq k_1 v^{\frac{d-1}{d}}$

where

$$\delta_r(A, B) = (\partial_r^+ A \cap B) \cup (\partial_r^+ B \cap A)$$

and k_1 depends only on d and r. There exists then $n_0(d, r)$ such that if $n > n_0$ (and thus $v > (3/2)^{n_0-1}$), then $|Y| \leq (2/3)|\Lambda|$. So we can apply the inductive hypothesis to both X and Y. Moreover proposition 3.4 shows that

$$\inf_{\tau \subset \Omega} \mathrm{gap}(L_\Lambda^\tau) \geq \inf_{\tau \in \Omega} \inf_{W \in \{X,Y\}} \mathrm{gap}(L_W^\tau) \inf_{\tau \in \Omega} \mathrm{gap}(L_{\{X,Y\}}^\tau) \tag{3.45}$$

where, as usual, the last term refers to the block dynamics. By Proposition 3.7 we know that

$$\inf_{\tau \in \Omega} \mathrm{gap}(L_{\{X,Y\}}^\tau) \geq e^{-k_2 \|J\| v^{\frac{d-1}{d}}}$$

for some $k_2 = k_2(d, r)$. Together with the inductive hypothesis on X and Y, this gives

$$\inf_{\tau \in \Omega} \mathrm{gap}(L_\Lambda^\tau) \geq c_m \exp\left[-k\|J\| |Y|^{\frac{d-1}{d}} - k_2 \|J\| v^{\frac{d-1}{d}} \right] \tag{3.46}$$

Since $|Y| \leq (2/3)v$, we have

$$\inf_{\tau \in \Omega} \mathrm{gap}(L_\Lambda^\tau) \geq c_m \exp\left[-k\|J\| v^{\frac{d-1}{d}} \right] \qquad \text{if } k \geq \frac{k_2}{1 - (2/3)^{\frac{d-1}{d}}}$$

In this way we have shown that (K_n) implies (K_{n+1}) for all $n \geq n_0(d,r)$. All is left is to prove (K_{n_0}). For this purpose we observe that

$$e^{-2\|J-J'\|_\Lambda|\Lambda|} \leq \frac{\mu_\Lambda^{J,\tau}(\sigma)}{\mu_\Lambda^{J',\tau}(\sigma)} \leq e^{2\|J-J'\|_\Lambda|\Lambda|} \qquad \text{for all } \tau, \sigma, \Lambda, J, J' \qquad (3.47)$$

Choose now any Λ with volume not exceeding $(3/2)^{n_0}$ and let $\widetilde{L}_\Lambda$ be the generator of the heat-bath dynamics with $J = 0$, i.e.

$$\widetilde{L}_\Lambda = \sum_{x \in \Lambda} \widetilde{L}_{\{x\}} = \sum_{x \in \Lambda} (\mu_{\{x\}}^{J=0} - \mathbf{1})$$

Since all $\widetilde{L}_{\{x\}}$ commute, it follows that $\text{gap}(\widetilde{L}_\Lambda) = \text{gap}(\widetilde{L}_{\{x\}}) = 1$ (the last equality can be checked via an explicit calculation). From (3.3), (3.12) and (3.47), it now follows that

$$\text{gap}(L_\Lambda^{J,\tau}) \geq \text{gap}(\widetilde{L}_\Lambda) \, e^{-6\|J\|\|\Lambda\|} \, c_m e^{-\kappa_1\|J\|} \geq c_m \exp\left[-(6 + \kappa_1)\|J\|(3/2)^{n_0}\right]$$

which implies (K_{n_0}) (and then (3.44)), if we take $k \geq (6 + \kappa_1)(3/2)^{n_0}$. $\square$

3.8 General Results on the Logarithmic Sobolev Constant

Here we recall two well known results concerning the logarithmic Sobolev constant.

Proposition 3.9. *For each $\Lambda \subset\subset \mathbf{Z}^d$ we have*

$$c_s(L_\Lambda^\tau) \leq \left[4 + (4\|J\| + 2\log 2)|\Lambda|\right](\text{gap}(L_\Lambda^\tau))^{-1}$$

Proof. The proposition follows from (3.12), Proposition 3.10 below, and from a trivial estimate on $\inf_\sigma \mu_\Lambda^\tau(\sigma)$.
$\square$

Proposition 3.10. *Let Ω be a finite set, let μ be a probability measure on $(\Omega, 2^\Omega)$ and assume*

$$\mu_0 \equiv \inf_{x \in \Omega} \mu(x) > 0$$

Then, for each positive function f on Ω, we have

$$\mu(f^2 \log f) \leq (4 + 2\log \mu_0^{-1}) \text{Var}(f) + \mu(f^2) \log \sqrt{\mu(f^2)}$$

Proof. We can assume $\mu(f^2) = 1$. If we let $f = \mu(f)(1 + g)$, we find $\mu(g) = 0$ and $\mu(g^2) = \text{Var}(f)/\mu(f)^2$. Let A be the set of all $x \in \Omega$ such that $|g(x)| < 1$. We can then write

$$\mu(f^2 \log f) = \mu(f^2 \log f \, \mathbf{I}_A) + \mu(f^2 \log f \, \mathbf{I}_{A^c}) \qquad (3.48)$$

Let's denote by X_1 respectively X_2 the first and the second term in the RHS of (3.48). Using the inequalities $\log(1 + g) \leq g$ and $\log \mu(f) \leq \log \mu(f^2) \leq 0$, we get

$$\begin{aligned} X_1 &\leq \mu(f)^2 \mu[(g + 2g^2 + g^3)\mathbf{I}_A] \\ &\leq \mu(f)^2\left[3\mu(g^2) + \mu(g\mathbf{I}_A)\right] = 3\text{Var}(f) + \mu(f)^2\mu(g\mathbf{I}_A) \end{aligned} \qquad (3.49)$$

To take care of the last term we remember that $\mu(g) = 0$, so $\mu(g\mathbb{1}_A) = -\mu(g\mathbb{1}_{A^c})$ which implies

$$|\mu(g\mathbb{1}_A)| \leq \mu(|g|\mathbb{1}_{A^c}) \leq (\mu(g^2)\mu(\mathbb{1}_{A^c}))^{1/2} \leq \mu(g^2)$$

Thus we get $X_1 \leq 4\operatorname{Var} f$. As for X_2, we write

$$X_2 \leq (\sup_{x\in\Omega} \log f(x))\,\mu(f^2\mathbb{1}_{A^c}) \leq \log\|f\|_\infty \mu(f^2\mathbb{1}_{A^c}) \tag{3.50}$$

Finally we observe that $\|f\|_\infty$ is bounded by $(\mu(f^2)/\mu_0)^{1/2} = \mu_0^{-(1/2)}$ while

$$\mu(f^2\mathbb{1}_{A^c}) = \mu(f)^2\mu((1 + 2g + g^2)\mathbb{1}_{A^c}) \leq 4\mu(f)^2\mu(g^2) = 4\operatorname{Var}(f)$$

This concludes the proof. $\square$

Proposition 3.11. *Given a finite set $V \subset\subset \mathbb{Z}^d$ and two probability measures on Ω_V, ν_1 and ν_2, assume $\|\frac{d\nu_1}{d\nu_2}\|_\infty \vee \|\frac{d\nu_2}{d\nu_1}\|_\infty \leq \varepsilon$. Then*

$$c_s(\nu_1) \leq \varepsilon^2 c_s(\nu_2) \tag{3.51}$$

Proof. Inequality (3.51) follows immediately from the bound

$$\nu_2(|\nabla_V f|^2) \leq \varepsilon \nu_1(|\nabla_V f|^2)$$

and the identity

$$\nu_1(f^2 \log f^2) - \nu_1(f^2) \log(\nu_1(f^2)) = \min_{a>0} \nu_1\big(f^2 \log f^2 - \log(a)f^2 - f^2 + a\big)$$

together with the observation that the expression $f^2 \log f^2 - \log(a)f^2 - f^2 + a$ is non–negative for all positive f and a (see [HS]). $\square$

3.9 Possible Rates of Convergence to Equilibrium

We conclude this general section on the dynamics with a discussion of the possible rates of convergence to equilibrium when the infinite volume process is ergodic. The flavour of the results discussed below is that, for local functions f, either the convergence of $T(t)f$ to $\mu(f)$ is very slow (e.g a small inverse power of the time) or it is exponentially fast.

Let us first consider the attractive case. Define

$$\rho(t) = \mathbb{P}\big(\sigma_t^+(0) \neq \sigma_t^-(0)\big) \tag{3.52}$$

where $\mathbb{P}$ is the global coupling constructed in section 3.3, σ_t^+ denotes the dynamics in infinite volume starting from the configuration identically equal to $+1$ and similarily for σ_t^-. It is easy to see that if $\rho(t)$ tends to zero as $t \to \infty$ then the infinite volume dynamics is ergodic. In fact, for any local function f, we can write

$$\sup_{\sigma,\eta} |T_t(f)(\sigma) - T_t(f)(\eta)| = \sup_{\sigma,\eta} |\mathbb{E}(f(\sigma_t^\sigma) - f(\sigma_t^\eta))|$$

$$\leq \||f\|| \sup_{x\in\Lambda_f} \mathbb{P}\big(\sigma_t^+(x) \neq \sigma_t^-(x)\big)$$

$$= \||f\|| \rho(t)$$

where we used (3.10) and (3.11) together with translation invariance.

The main result on the absence of intermediate speed of relaxation to equilibrium in the attractive case can then be formulated as follows (see [Ho1])

Theorem 3.12. *There exists two constants ε and t_0 depending only on the flip rates and on the range of the interaction r such that, $t^d \rho(t) \leq \varepsilon$ for some $t \geq t_0$ implies that $\rho(t)$ decays to zero exponentially fast.*

Proof. The proof is based on the following key inequality valid for all $t \geq 0$:

$$\rho(2t) \leq Ct^d \rho(t)^2 + e^{-t} \tag{3.53}$$

for a suitable constant C.

Let us in fact assume (3.53) and define

$$u(t) \equiv 2^d (Ct^d + 1)\big(\rho(t) \vee e^{-t/2}\big)$$

Then, by explicit calculation, $u(2t) \leq u(t)^2$ and thus, if there exists a time t_0 such that $u(t_0) < 1$, then $u(t)$ tends to zero exponentially fast. The existence of a time t_0 with the above property is precisely the assumption appearing in the theorem.

To prove (3.53) let $\nu_t^{+,-}$ be the joint distribution of σ_t^+ and σ_t^- given by the graphical construction. Because of (3.11) $\nu_t^{+,-}$ is above the diagonal i.e. $\nu_t^{+,-}((\xi, \eta); \ \xi \geq \eta) = 1$. Let now χ_t be the characteristic function of the event that $\sigma_t^+(x) = \sigma_t^-(x)$ for all $|x| \leq k_0 t$, where k_0 appears in lemma 3.2. Then, using the Markov property, we can write $\rho(2t)$ as:

$$
\begin{aligned}
\rho(2t) &= \mathbb{P}\big(\sigma_{2t}^+(0) \neq \sigma_{2t}^-(0)\big) \\
&= \int d\nu_t^{+,-}(\xi, \xi') \, \chi_t \, \mathbb{P}\big(\sigma_t^\xi(0) \neq \sigma_t^{\xi'}(0)\big) \\
&\quad + \int d\nu_t^{+,-}(\xi, \xi')(1 - \chi_t)\, \mathbb{P}\big(\sigma_t^\xi(0) \neq \sigma_t^{\xi'}(0)\big)
\end{aligned}
\tag{3.54}
$$

The first term in the r.h.s of (3.54) is bounded from above by e^{-t} because of lemma 3.2. To bound the second term we observe that, because of monotonicity in the starting configuration,

$$\mathbb{P}\big(\sigma_t^\xi(0) \neq \sigma_t^{\xi'}(0)\big) \leq \rho(t)$$

for any $\xi \geq \xi'$. Moreover

$$\int d\nu_t^{+,-}(\xi, \xi')(1 - \chi_t) \leq \frac{1}{2}(2k_0 t + 1)^d \rho(t)$$

because of translation invariance and the definition of $\rho(t)$. Thus (3.53) follows and the theorem is proved. □

Let us now examine the more difficult, non-attractive case. The analogous of theorem 3.12 reads as follows (see [Yo1])

Theorem 3.13. *Assume that for some L_0, some positive non–increasing function $\phi(t)$ satisfying $\sum_n n^{2(d-1)}\phi(n) < \infty$, any set V multiple of Q_{L_0} and any local function f*

$$\sup_\tau \mu_V^\tau\left(|T_V^\tau(t)f - \mu_V^\tau(f)|\right) \leq \|f\|_\infty |\Lambda_f|\,\phi(t) \tag{3.55}$$

Then there exists L_0', a positive constant m and for any local function f there exists a constant $C'(f) = C'(|\Lambda_f|, \|f\|_\infty)$ such that for any set V multiple of $Q_{L_0'}$

$$\sup_\tau \|T_V^\tau(t)f - \mu_V^\tau(f)\|_\infty \leq C'(f)\,e^{-mt}$$

Proof. The idea of the proof is to show that the assumption of the theorem is sufficient to guarantee that there exists L_0' such that property $SMT(V, l, m)$ for some l and m and all sets V which are multiple of $Q_{L_0'}$ holds. Once the exponential decay of covariances is available then we get, thanks to theorem 4.6 below, that there exits a finite constant c_s^∞ such that

$$\sup_\tau c_s(L_V^\tau) < \infty$$

for all sets V multiple of $Q_{L_0'}$. Thus the thesis follows from theorem 3.3.

In order to prove strong mixing we first establish some form of decay for covariances of the Gibbs measure μ_V^τ. For this purpose let f and g be two local functions with supports contained in the set V multiple of Q_{L_0} and let $l = d(\Lambda_f, \Lambda_g)$. Formula (3.24) written for μ_V^τ, together with (3.55), gives the following bound

$$|\mu_V^\tau(f; g)| \leq \|g\|_\infty \|f\|_\infty |\Lambda_f|\phi(2k_0^{-1}l) + \||f\|| \,\||g\|| \,|\Lambda_f| \,|\Lambda_g| e^{-k_0^{-1}l}$$

i.e. the decay of covariances for large values of l is governed by the function $\phi(2l) \wedge e^{-l}$. To conclude the proof we appeal to proposition 2.11. $\square$

The above result indicates that a power law decay to equilibrium in the L^1 sense is enough to get exponential ergodicity. One possibility to establish such a weaker decay is through generalized Nash inequalities (see e.g [SC] and [BZ]).

Let $c_N(L_0, \alpha)$, $\alpha > 0$ and $L_0 < \infty$, be the smallest constant c such that the following inequality holds for all sets V multiple of Q_{L_0} and all local function f:

$$\mathrm{Var}_V^\tau(f)^{1+\alpha} \leq c\,\mathcal{E}_V^\tau(f, f)\,\||f\||^{2\alpha} \qquad \forall\,\tau \in \Omega \tag{3.56}$$

With the above notation we have the following result

Theorem 3.14. *Assume that $c_N(L_0, \alpha) < \infty$ for some finite L_0 and some $\alpha < (6d - 2)^{-1}$. Then the conclusions of theorem 3.13 hold.*

Proof. Let us fix a set V multiple of Q_{L_0}, a boundary condition $\tau \in \Omega$ and a local function f with zero mean w.r.t. the Gibbs measure μ_V^τ.

Thanks to theorem 3.13 it is sufficient to bound the decay rate to equilibrium of $T_V^\tau(t)f$ in the $L^1(d\mu_V^\tau)$–sense, uniformly in V. For this purpose define

$$f_t = T_V^\tau(t)f$$
$$u_t = \mathrm{Var}_V^\tau(f_t)$$

An explicit computation gives

$$\dot{u}_t = -2\mathcal{E}_V^\tau(f_t, f_t) \leq -\frac{2}{c_N(\alpha)\||f_t\||^{2\alpha}} u_t^{1+\alpha} \tag{3.57}$$

for some $\alpha < (6d - 2)^{-1}$, because of (3.56) together with the assumption of the theorem.

Unfortunately, contrary to the L^p–norms, the seminorm $\||f_t\||$ is not obviously bounded from above by $\||f\||$. However, thanks to lemma 3.2, it is quite simple to check that

$$\||f_t\|| \leq At^d\||f\||$$

for a suitable constant A. We have in fact that

$$\||f_t\|| = \sum_y \sup_\eta |\,\mathbf{E}(f(\sigma_t^{V,J,\tau,\eta^y}) - f(\sigma_t^{V,J,\tau,\eta}))|$$

$$\leq \sum_y \sum_x \|\nabla_x f\|_\infty \sup_\eta \mathbf{P}(\sigma_t^{V,J,\tau,\eta^y}(x) \neq \sigma_t^{V,J,\tau,\eta}(x))$$

$$\leq (kt)^d\||f\|| + \sum_x \|\nabla_x f\|_\infty \sum_{y:\, d(x,y)\geq kt} \sup_\eta \mathbf{P}(\sigma_t^{V,J,\tau,\eta^y}(x) \neq \sigma_t^{V,J,\tau,\eta}(x))$$

$$\leq 2(kt)^d\||f\||$$

if the constant k is chosen large enough. In the last step we used the bound (3.8) in order to estimate $\mathbf{P}(\sigma_t^{V,J,\tau,\eta^y}(x) \neq \sigma_t^{V,J,\tau,\eta}(x))$ when $d(x,y) \geq kt$.

Thus (3.57) reads

$$\dot{u}_t \leq -2\big(c_N(\alpha)A\||f\||^{2\alpha}\big)^{-1} t^{-2d\alpha} u_t^{1+\alpha}$$

Since $2d\alpha < 1$, this differential inequality implies

$$u_t \leq C_f t^{-(1-2d\alpha)/\alpha} \tag{3.58}$$

for a suitable constant C_f depending only on $\|f\|_\infty$ and $|\Lambda_f|$.

To conclude the proof it is enough to write

$$\mu_V^\tau\big(|T_V^\tau(t)f|\big) \leq u_t^{1/2} \leq \big[C_f t^{-(1-2d\alpha)/\alpha}\big]^{1/2} \tag{3.59}$$

and notice that the r.h.s of (3.59) satisfies the hypothesis of theorem 3.13. $\square$

We conclude with an example of a spin system whose infinite volume dynamics is ergodic but with an algebraic decay to equilibrium (see [Ho2]).

Let us consider the two dimensional Ising model (see (2.4)) at the critical temperature $i.e.$ we take $d = 2$ and a two-body ferromagnetic interaction of the form

$$J_A = \begin{cases} \beta_c & \text{if } A = \{x, y\} \text{ with } d_2(x,y) \leq 1 \\ 0 & \text{otherwise} \end{cases}$$

where $\beta_c = \frac{1}{2}\log(1 + \sqrt{2})$.

It is well known (see e.g [MCoy]) that the above system has a unique infinite volume Gibbs measure μ with polynomially decaying correlations functions. More precisely

$$\lim_{x_1 \to \infty} \frac{\log\big[\mu\big(\sigma((0,0))\sigma((x_1,0))\big)\big]}{\log(x_1)} = -\frac{1}{4}$$

where $\sigma((x_1, x_2))$ denotes the spin at the site $x = (x_1, x_2)$. Given now a time t, let $x = (\lfloor k_0 t \rfloor, 0)$. Then, using reversibility together with lemma 3.2, we can write

$$
\begin{aligned}
\mu\big(\sigma(0)\sigma(x)\big) &= \mu\big(T(t/2)\sigma(0)\sigma(x)\big) \\
&\leq \mu\big((T(t/2)\sigma(0))\,(T(t/2)\sigma(x))\big) + 2e^{-t} \\
&\leq \|T(t/2)\sigma(0)\|_{2,\mu}\,\|T(t/2)\sigma(x)\|_{2,\mu} + 2e^{-t} \\
&= \|T(t/2)\sigma(0)\|_{2,\mu}^2 + 2e^{-t}
\end{aligned}
$$

i.e.

$$\lim_{t \to \infty} \frac{\log(\|T(t/2)\sigma(0)\|_{2,\mu})}{\log(t)} \geq -\frac{1}{4}$$

Thus the relaxation to equilibrium is not exponential.

4. One Phase Region

In this section we discuss the main results on the exponential ergodicity in the absence of a phase transition. In this case equilibrium is reached by an homogeneous process: far apart regions equilibrate in finite time without exchanging almost any information, very much like an infinite collection of non–interacting continuous time ergodic Markov chains.

We first discuss the attractive case and we prove exponential ergodicity of the infinite volume dynamics under the weak mixing condition (see definition 2.3). Then we turn to the general case and prove exponential ergodicity uniformly in all cubes under the strong mixing condition (see definition 2.4). We refer the reader to [MO1], [SZ], [LY], [La], [MO2], [Yo1] and references therein to have a complete overview of related results under similar assumptions.

4.1 The Attractive Case

In this first paragraph we assume that the dynamics is attractive. In this case it is not difficult to check that condition $WM(V, C, m_0)$ for all $V \subset\subset \mathbb{Z}^d$ is equivalent to the following

There exist positive constants C and m such that, for any integer L

$$\mu^+_{B_L}(\sigma(0)) - \mu^-_{B_L}(\sigma(0)) \leq Ce^{-mL} \tag{4.1}$$

Indeed, by the very definition of variation distance, $WM(V, C, m)$ for all $V \subset\subset \mathbb{Z}^d$ trivially implies (4.1). To see the converse, let $\nu^{+,-}_V$ be the joint representation above the diagonal of the two Gibbs measures μ^+_V and μ^-_V given by attractivity (see section 3.4) and set, for any $x \in V$, $l_x = d(x, \partial^+_r V)$. Then, thanks to attractivity and finite volume ergodicity, we can write

$$
\begin{aligned}
\sup_{\tau_1, \tau_2} \|\mu^{\tau_1}_{V,\Delta} - \mu^{\tau_2}_{V,\Delta}\| &= \sup_{\tau_1,\tau_2} \sup_{X \in \mathcal{F}_\Delta} |\mu^{\tau_1}_{V,\Delta}(X) - \mu^{\tau_2}_{V,\Delta}(X)| \\
&\leq \sup_{\tau_1,\tau_2} \lim_{t\to\infty} \mathbb{P}(\sigma^{V,\tau_1,+}_t(x) \neq \sigma^{V,\tau_2,+}_t(x) \text{ for some } x \in \Delta) \\
&\leq \lim_{t\to\infty} \sum_{x \in \Delta} \mathbb{P}(\sigma^{V,+,+}_t(x) \neq \sigma^{V,-,+}_t(x)) \\
&= \sum_{x \in \Delta} \nu^{+,-}_V(\sigma(x) \neq \eta(x)) \\
&\leq \sum_{x \in \Delta} \nu^{+,-}_{B_{l_x}+x}(\sigma(x) \neq \eta(x)) \\
&\leq \sum_{x \in \Delta} Ce^{-md(x,\partial^+_r V)} \\
&\leq \sum_{x \in \Delta, y \in \partial^+_r V} Ce^{-md(x,y)}
\end{aligned}
$$

Thus $WM(V, C, m)$ holds.

Theorem 4.1. *In the attractive case the following are equivalent:*
i) $WM(V, C, m)$ *for all* $V \subset\subset \mathbb{Z}^d$
ii) *There exists a positive constant m and for any local function f there exists a constant C_f such that:*

$$\|T_t(f) - \mu(f)\|_\infty \leq C_f e^{-mt}$$

Proof. Let us first show that i) → ii). For this purpose and as in section 3.9 we define

$$\rho(t) = \mathbb{P}\big(\sigma_t^+(0) \neq \sigma_t^-(0)\big)$$

As in theorem 3.12, if $\rho(t)$ decays exponentially fast to zero then the theorem follows. Moreover, thanks to theorem 3.12, the exponential decay of $\rho(t)$ follows once one is able to show that $\rho(t)$ goes to zero faster than t^{-d}. In order to prove such a weaker decay of $\rho(t)$ let us first show, as a preliminary step, that under our assumption $\rho(t) \to 0$ as $t \to \infty$.

Lemma 4.2. *Under the same assumption of theorem 4.1 $\rho(t) \to 0$ as $t \to \infty$.*

Proof. Using the attractivity of the dynamics, for any $V \subset\subset \mathbb{Z}^d$ such that $0 \in V$ and any $t \geq 0$

$$\rho(t) \leq \mathbb{P}(\sigma_t^{V,+,+}(0) \neq \sigma_t^{V,-,-}(0)) \tag{4.2}$$

where $\sigma_t^{V,+,+}$ denotes the dynamics in V with plus boundary condition and initial condition identically equal to $+1$ and similarily for $\sigma_t^{V,-,-}$. If we pass to the limit $t \to \infty$ in (4.2) and use finite volume ergodicity, we get

$$\limsup_{t \to \infty} \rho(t) \leq 2\big[\mu_V^+(\sigma(0)) - \mu_V^-(\sigma(0))\big] \tag{4.3}$$

Finally, by taking V as a large cube centered at the origin and using weak mixing, we get that the r.h.s of (4.3) can be made arbitrarily small. $\square$

As a second step, using weak mixing, we will establish a powerful recursive inequality satisfied by $\rho(t)$ which, combined with lemma 4.2, will show that $t^d \rho(t) \to 0$ as $t \to \infty$.

Lemma 4.3. *Under the same assumption of theorem 4.1, for all $t \geq 0$ and all $L \in \mathbb{Z}_+$*

$$\rho(2t) \leq 2(2L+1)^d \rho(t)^2 + 2C e^{-mL}$$

Remark. Although our inequality is apparently very similar to the one appearing in the proof of theorem 3.12 on the absence of intermediate speed of convergence to equilibrium, the differences between the two are actually quite deep and mainly consists in the freedom of the parameter L that in the proof of theorem 3.12 was fixed and of the order of the time t.

Proof. We write $\rho(2t)$ as:

$$\rho(2t) \; = \; \int d\mu(\eta) \left[\mathbb{P}(\sigma_{2t}^+(0) - \sigma_{2t}^\eta(0))\right] \; + \; \int d\mu(\eta) \left[\mathbb{P}(\sigma_{2t}^\eta(0)) - \sigma_{2t}^-(0))\right] \qquad (4.4)$$

and we show that each one of the two integrals is bounded by a half of the r.h.s. of the recursive inequality.

Let $\nu_t^{+,\eta}$ be the joint distribution of σ_t^+ and σ_t^η given by the graphical construction. Because of (3.11) $\nu_t^{+,\eta}$ is above the diagonal i.e. $\nu_t^{+,\eta}((\xi,\eta); \; \xi \geq \eta) \; = \; 1$. Then, using the Markov property, we can write:

$$\int d\mu(\eta) \left[\mathbb{P}(\sigma_{2t}^+(0) \neq \sigma_{2t}^\eta(0))\right] = \int d\mu(\eta) \int d\nu_t^{+,\eta}(\xi,\xi') \, \chi_L \, \mathbb{P}(\sigma_t^\xi(0) \neq \sigma_t^{\xi'}(0))$$

$$+ \int d\mu(\eta) \int d\nu_t^{+,\eta}(\xi,\xi')(1 - \chi_L)\mathbb{P}(\sigma_t^\xi(0) \neq \sigma_t^{\xi'}(0))$$

$$(4.5)$$

where χ_L is the characteristic function of the event that $\xi(x) = \xi'(x)$ for all $|x| \leq L$. By attractivity and traslation invariance, the second term in the r.h.s. of (4.5) can be bounded from above by:

$$(2L + 1)^d \rho(t) \int d\mu(\eta)\nu_t^{+,\eta}(\xi(0) \neq \xi'(0)) \; \leq \; (2L+1)^d \rho(t)^2 \qquad (4.6)$$

Let now, for any given $\tau \in \Omega_{B_L}$, $\hat{\chi}_{L,\tau}$ be the characteristic function of the event :

$$\xi(x) = \xi'(x) = \tau(x) \qquad \forall x \in Q_L$$

Thus

$$\chi_L \; = \; \sum_{\tau \in \Omega_{B_L}} \hat{\chi}_{L,\tau}$$

and therefore the first term in the r.h.s. of (4.6) can be written as:

$$\int d\mu(\eta) \sum_{\tau \in \Omega_{B_L}} \int d\nu_t^{+,\eta}(\xi,\xi') \, \hat{\chi}_{L,\tau} \, \mathbb{P}(\sigma_t^\xi(0) \neq \sigma_t^{\xi'}(0)) \qquad (4.7)$$

Attractivity allows us to bound the quantity $\mathbb{E}(\sigma_t^\xi(0) \neq \sigma_t^{\xi'}(0))$ by imposing extra plus and minus boundary conditions outside the cube B_L. More precisely :

$$\mathbb{P}(\sigma_t^\xi(0) \neq \sigma_t^{\xi'}(0)) \leq \mathbb{P}(\sigma_t^{B_L,+,\xi}(0) \neq \sigma_t^{B_L,-,\xi'}(0)) \qquad (4.8)$$

Thus (4.7) is bounded from above by :

$$\int d\mu(\eta) \sum_{\tau \in \Omega_{B_L}} \int d\nu_t^{+,\eta}(\xi,\xi')\hat{\chi}_{L,\tau}\, \mathbb{P}(\sigma_t^{B_L,+,\tau}(0) \neq \sigma_t^{B_L,-,\tau}(0)) \leq$$

$$\int d\mu(\eta) \sum_{\tau \in \Omega_{B_L}} \nu_t^{+,\eta}(\xi'(j) = \tau(j) \, \forall j \in Q_L)\left[\mathbb{P}(\sigma_t^{B_L,+,\tau}(0) \neq \sigma_t^{B_L,-,\tau}(0))\right] \qquad (4.9)$$

We use at this point attractivity to write

$$\mathbb{P}(\sigma_t^{B_L,+,\tau}(0) \neq \sigma_t^{B_L,-,\tau}(0)) = 2\,\mathbb{E}(\sigma_t^{B_L,+,\tau}(0) - \sigma_t^{B_L,-,\tau}(0))$$

and we notice that, since $f(\tau) \equiv \mathbb{E}(\sigma_t^{B_L,+,\tau}(0))$ is an increasing function of τ and $\mu \leq \mu_{B_L}^+$, then

$$\int d\mu(\eta) \int d\nu_t^{+,\eta}(\xi,\xi')f(\xi') = \int d\mu(\eta)f(\eta)$$
$$\leq \int d\mu_{B_L}^+(\eta)\,\mathbb{E}(f(\sigma_t^{B_L,+,\eta}))$$
$$= \int d\mu_{B_L}^+(\eta)\,\mathbb{E}(\sigma_{2t}^{B_L,+,\eta}(0))$$
$$= \mu_{B_L}^+(\sigma(0))$$

where we used the Markov property and the reversibility of $\sigma_t^{B_L,+,\eta}$ w.r.t $\mu_{B_L}^+$. A similar reasoning shows that

$$\int d\mu(\eta) \int d\nu_t^{+,\eta}(\xi,\xi')f(\xi') \geq \mu_{B_L}^-(\sigma(0))$$

Thus the r.h.s of (4.9) is bounded from above by

$$\mu_{B_L}^+(\sigma(0)) - \mu_{B_L}^-(\sigma(0)) \leq Ce^{-mL} \tag{4.10}$$

because of the weak mixing assumption. $\square$

As a final step we combine together lemma 4.2 and lemma 4.3.

Lemma 4.4. *If, for some positive m and C, $\rho(2t) \leq 2(2L+1)^d\rho(t)^2 + 2Ce^{-mL}$ for all $t \geq 0$ and all $L \in \mathbb{Z}_+$, and if $\lim_{t\to\infty} \rho(t) = 0$, then*

$$\lim_{t\to\infty} t^d\rho(t) = 0$$

Proof. Assume that t is so large that

$$2(-\frac{4}{m}\log(\rho(t)) + 1)^d + Ce^m)\sqrt{\rho(t)} \leq 1 \tag{4.11}$$

and set $L(t) = \left[-\frac{2}{m}\log(\rho(t)) \right]$. Then

$$\rho(2t) \leq 2\left(-\frac{4}{m}\log(\rho(t)) + 1\right)^d\rho(t)^2 + 2Ce^{2\log(\rho(t))+m} \leq \rho(t)^{\frac{3}{2}}$$

Let now t_0 be so large that $\rho(t_0) < 1$ and (4.11) holds for all $t \geq t_0$. One gets at once that for all $n \in \mathbb{Z}_+$

$$\rho(2^n t_0) \leq \rho(t_0)^{(\frac{3}{2})^n}$$

from which the lemma follows. $\square$

Once lemma 4.4 is available then we can apply theorem 3.12 and conclude that $\rho(t)$ must decay exponentially fast.

ii) → i).

Clearly *ii*) together with attractivity imply that the infinite volume dynamics is ergodic with a unique invariant measure μ (see Corollary 2.4 in chapter III of [L]). Thus we write:

$$\mu^+_{B_L}(\sigma(0)) - \mu^-_{B_L}(\sigma(0)) = [\mu^+_{B_L}(\sigma(0)) - \mu(\sigma(0))] + [\mu(\sigma(0)) - \mu^-_{B_L}(\sigma(0))] \tag{4.12}$$

Let us estimate $\mu^+_{B_L}(\sigma(0)) - \mu(\sigma(0))$. By adding and subtracting the term $\mathbb{E}(\sigma^+_t(0))$ and using the exponential convergence to equilibrium together with attractivity, we get

$$0 \leq \mu^+_{B_L}(\sigma(0)) - \mu(\sigma(0)) \leq Ce^{-mt} + \mu^+_{B_L}(\sigma(0)) - \mathbb{E}(\sigma^+_t(0)) \tag{4.13}$$

We now choose the time t as $t = k_0^{-1}L$ where the constant k_0 appears in lemma 3.2 and add and subtract the term $\mathbb{E}(\sigma_t^{B_L,+,+}(0))$. Using lemma 3.2 the r.h.s. of (4.13) can be bounded from above by :

$$Ce^{-k_0^{-1}mL} + 2e^{-k_0^{-1}L} + \mu^+_{B_L}(\sigma(0)) - \mathbb{E}(\sigma_t^{B_L,+,+}(0)) \leq Ce^{-k_0^{-1}mL} + 2e^{-k_0^{-1}L}$$

since, by attractivity, $\mu^+_{B_L}(\sigma(0)) \leq \mathbb{E}(\sigma_t^{B_L,+,+}(0))$.

In conclusion we have shown that $\mu^+_{B_L}(\sigma(0)) - \mu(\sigma(0))$ is smaller than $Ce^{-k_0^{-1}mL} + 2e^{-k_0^{-1}L}$. The same argument applies also to the other term in the r.h.s. of (4.12) $\mu(\sigma(0)) - \mu^-_{B_L}(\sigma(0))$. $\square$

4.2 The General Case: Recursive Analysis

Here we prove exponential ergodicity under a strong mixing assumption on the family of all parallelepipeds with ratio between the smallest and the largest side greater than a given fixed constant. For simplicity we carry out our analysis in two dimensions but the extension to higher dimension is straightforward.

Let $R(l_1, l_2)$ denote the rectangle

$$R(l_1, l_2) = \{(x_1, x_2) \in \mathbb{Z}^2; \ 0 \leq x_1 \leq l_1 - 1, \ 0 \leq x_2 \leq l_2 - 1\} ; \qquad l_1, l_2 \in \mathbb{Z}_+$$

and let $\mathcal{R}_L$ be the family of "fat" rectangles with "size" smaller than L, namely those rectangles $R(l_1, l_2, x) \equiv R(l_1, l_2) + x$, $x \in \mathbb{Z}^2$, with $l_1 \wedge l_2 \geq 0.1(l_1 \vee l_2)$ and $l_1 \vee l_2 \leq L$.
Let also

$$g(L) = \min_{R \in \mathcal{R}_L} \min_{\tau} \text{gap}(L_R^\tau)$$

$$c_s(L) = \sup_{R \in \mathcal{R}_L} c_s(\mu_R^J)$$

where $\text{gap}(L_R^\tau)$ and $c_s(\mu_R^J)$ have been defined in (3.12) and (3.19) respectively. With the above notation we will prove the following key recursive bounds.

Theorem 4.5. *Assume* $SMT(R, l, m)$ *for all* $R \in \bigcup_{L \geq l} \mathcal{R}_L$. *Then there exists a positive constant* $k = k(d, r, \|J\|)$ *such that*

$$g(2L) \geq (1 - \frac{k}{\sqrt{L}})g(L)$$

for all L *large enough. In particular* $\inf_L g(L) > 0$.

Theorem 4.6. *Assume* $SMT(R, l, m)$ *for all* $R \in \bigcup_{L \geq l} \mathcal{R}_L$. *Then there exists a positive constant* $k = k(d, r, \|J\|)$ *such that*

$$c_s(2L) \leq (1 + \frac{k}{\sqrt{L}})c_s(L)$$

for all L *large enough. In particular* $\sup_L c_s(L) < \infty$.

Proof of theorem 4.5. The second part of the theorem is a straightforward consequence of the recursive bound. Thus we concentrate on what happens to the gap when one doubles the length scale. In order to have a clear understanding of what will follow let us consider first the trivial case of zero interaction, $J_A = 0 \, \forall \, A \subset\subset \mathbf{Z}^d$, in e.g two dimensions. Let us take a cube Q_{2L} and divide it into two equal vertical non–overlapping rectangles $R_i(L, 2L)$. Then formula (3.30) naturally allows us to compute the ratio between the spectral gap in Q_{2L} and the spectral gap in $R_i(L, 2L)$, $i = 1, 2$, in terms of the spectral gap of the block dynamics in Q_{2L} with blocks $R_i(L, 2L)$, $i = 1, 2$. Since the interaction is zero this last quantity is exactly equal to one either by exact computation or because of proposition 3.5, and we get that spectral gap in Q_{2L} is equal to that in either one of the two rectangles. We can at this point repeat the whole procedure for each one of the two rectangles and prove finally that the gap does not change when we double the scale. When the interaction is present and strong mixing holds one could be tempted to proceed in exactly the same way. However in the interacting case a lower bound on the spectral gap of the block dynamics with just two blocks is no longer that simple unless the two blocks overlap a little bit and one can appeal to proposition 3.5. On the other hand if the two rectangles overlap then an error term due precisely to the overlap appears in the formula comparing the single site dynamics with the block dynamics (see (3.29) and (3.30)). In particular, without any particular effort, one would get immediately a bound like

$$g(2L) \geq \frac{1}{16}g(L)$$

which already implies that $g(L) \geq L^{-\alpha}$ for a suitable $\alpha > 0$ and any L large enough. The way out to show that the factor $\frac{1}{16}$ can be actually replace by $1 - \frac{k}{\sqrt{L}}$ is to "spread out" over a non–zero fraction of the starting volume Q_{2L} the error term in (3.29). This appears to be a new idea and in our opinion leads to a rather simple proof of both theorems 4.5, 4.6. Let us now explain the details.

If $g(L) = g(2L)$ there is nothing to prove so we suppose that $g(2L) < g(L)$. Let us consider a rectangle $R \equiv R(l_1, l_2) \in \mathcal{R}_{2L}$ and assume, without loss of generality, that the longest side is l_2. Since we are assuming that $g(2L) < g(L)$ necessarily $L < l_2 \leq 2L$.

We set $d = [\sqrt{L}]$ and, for any integer $1 \le n \le \frac{L}{10d}$, we cover R with the following two rectangles

$$R_n^{top} = \{x \in R; \; l_2/2 + (n-1)d < x_2 \le l_2 - 1\}$$
$$R_n^{bot} = \{x \in R; \; 0 \le x_2 \le l_2/2 + nd\}$$

Notice that, since $l_1 \ge 0.1 l_2$ and $l_2 > L$, we can use our assumption and apply proposition 2.12 to get that

$$\sup_{\tau \in \Omega} \left\| 1 - \frac{d\mu_{R_n^{top}}^{\tau}}{d\mu_R^{\tau}} \Big|_{\partial_r^+ R_n^{bot} \cap R_n^{top}} \right\|_{\infty} \le \sup_{\tau, \eta \in \Omega: \, \tau_{R^c} = \eta_{R^c}} \left\| 1 - \frac{d\mu_{R_n^{top}}^{\tau}}{d\mu_{R_n^{top}}^{\eta}} \Big|_{\partial_r^+ R_n^{bot} \cap R_n^{top}} \right\|_{\infty}$$
$$\le e^{-2C\sqrt{L}}$$

for any L large enough and a suitable positive constant C independent of L. Thus, thanks to (3.12), (3.29), (3.33) and proposition 3.5, for any $f \in L^2(\Omega, d\mu_V^{\tau})$ we get

$$\mathrm{Var}_R^{\tau}(f) \le \lambda(\tau, L, R_n^{top}, R_n^{bot}) \Big\{ \mathcal{E}_R^{\tau}(f, f) + \frac{1}{2} \mu_R^{\tau} \Big(\sum_{x \in R_n^{top} \cap R_n^{bot}} c_J(x, \sigma) |\nabla_x f|^2 \Big) \Big\} \quad (4.14)$$

where

$$\lambda(\tau, L, R_n^{top}, R_n^{bot}) = (1 - e^{-C\sqrt{L}})^{-1} \big(\mathrm{gap}(L_{R^{top}}^{\tau}) \wedge \mathrm{gap}(L_{R_n^{bot}}^{\tau}) \big)^{-1}$$

Notice that, as n varies, the strips $R_n^{top} \cap R_n^{bot}$ are disjoint. Therefore, if we average (4.14) over $n \in [1, \frac{L}{10d}]$ and use the trivial bound

$$\sum_{n=1}^{[L/10d]} \frac{1}{2} \mu_R^{\tau} \Big(\sum_{x \in R_n^{top} \cap R_n^{bot}} c_J(x, \sigma) |\nabla_x f|^2 \Big) \le \mathcal{E}_R^{\tau}(f, f)$$

we get

$$\mathrm{Var}_R^{\tau}(f) \le \big(1 + [\tfrac{L}{10d}]^{-1}\big) \sup_{1 \le n \le \frac{L}{10d}} \lambda(\tau, L, R_n^{top}, R_n^{bot}) \mathcal{E}_R^{\tau}(f, f) \quad (4.15)$$

i.e

$$\mathrm{gap}(L_R^{\tau}) \ge \big(1 - \frac{C_1}{\sqrt{L}}\big) \min_{1 \le n \le [L/10d]} \big(\mathrm{gap}(L_{R^{top}}^{\tau}) \wedge \mathrm{gap}(L_{R_n^{bot}}^{\tau}) \big) \quad (4.16)$$

for a suitable positive constant C_1.

Let us now fix $n \in [1, \frac{L}{10d}]$ and examine the spectral gap of the bottom rectangle R_n^{bot}, the reasoning being similar for the top one. There are two cases to analyze:

a) $l_1 \leq \frac{3}{2}L$. In this case one easily verifies that $R_n^{bot} \in \mathcal{R}_{\frac{3}{2}L}$ and thus, by definition,

$$\text{gap}(L_{R_n^{bot}}^\tau) \geq g(\tfrac{3}{2}L)$$

b) $l_1 > \frac{3}{2}L$. In this case $R_n^{bot} \in \mathcal{R}_{2L}$ but now the *longest* side is l_1 and the *shortest* one is smaller than $\frac{l_2}{2} + nd + 1$ which in turn is smaller than $1.2L$ since $l_2 \leq 2L$ and

$$nd \leq \tfrac{L}{10}$$

In conclusion we obtain that the r.h.s of (4.16) is larger than

$$[1 - \frac{C_1}{\sqrt{L}}]\left(g(\tfrac{3}{2}L) \wedge \min_{\substack{R(l_1,l_2) \in \mathcal{R}_{2L} \\ l_1 \leq 1.2L, l_2 \geq 3/2L}} \text{gap}(L_{R(l_1,l_2)}^\tau)\right) \tag{4.17}$$

In (4.17), without loss of generality, we kept the longest side always along the second coordinate direction.

Let us finally apply the lower bound (4.16) to an arbitrary element $R(l_1, l_2) \in \mathcal{R}_{2L}$ such that $l_1 \leq 1.2L$ and $l_2 \geq 3/2L$ and let us repeat the previous two–cases analysis. Since $l_1 \leq 1.2L$, case b) above is no longer possible and thus we get immediately from (4.16) that

$$\min_{\substack{R(l_1,l_2) \in \mathcal{R}_{2L} \\ l_1 \leq 1.2L, l_2 \geq 3/2L}} \text{gap}(L_{R(l_1,l_2)}^\tau)) \geq (1 - \frac{C_1}{\sqrt{L}})g(\tfrac{3}{2}L) \tag{4.18}$$

By combining (4.17) with (4.18) we finally get

$$g(2L) = \min_{R \in \mathcal{R}_{2L}} \text{gap}(L_R^\tau) \geq (1 - \frac{C_1}{\sqrt{L}})^2 g(\tfrac{3}{2}L) \tag{4.19}$$

In order to conclude the proof of the theorem it is enough to iterate two more times (4.19). $\square$

Proof of theorem 4.6. We proceed exactly as in the proof of theorem 4.5 but instead of using block–dynamics with two blocks in order to relate the logarithmic Sobolev constant on scale $2L$ to that on scale L we use conditional expectation. The reason for this is that, contrary to what happens for the spectral gap, the Dirichlet form of a block dynamics with few large blocks may have a logarithmic Sobolev constant much larger than the logarithmic Sobolev constant of a corresponding single spin flip dynamics (see the remark at the end of the proof of corollary 3.6).

If $c_s(2L) = c_s(L)$ there is nothing to prove so we suppose that $c_s(2L) > c_s(L)$. Let us consider a rectangle $R \equiv R(l_1, l_2) \in \mathcal{R}_{2L}$ and assume, without loss of generality, that the longest side is l_2. Since we are assuming that $c_s(2L) > c_s(L)$ necessarily $L < l_2 \leq 2L$. Let also d, R_n^{top}, R_n^{bot} be as in the proof of proposition 4.5. Then, for any non–negative $f \in L^2(\Omega, \mu_R^\tau)$ such that $\mu_R^\tau(f^2) = 1$, we write

$$\mu_R^\tau(f^2 \log f) = \mu_R^\tau\left(\mu_{R_n^{bot}}^\tau(f^2 \log f)\right)$$

$$\leq \left[\sup_{n \in [1, L/10d]} c_s(\mu_{R_n^{bot}}^\tau)\right]\frac{1}{2}\mu_R^\tau\left(|\nabla_{R_n^{bot}} f|^2\right) + \mu_R^\tau(g_n^2 \log g_n) \tag{4.20}$$

where $g_n \equiv \left(\mu^\tau_{R^{bot}_n}(f^2)\right)^{\frac{1}{2}}$.

Notice that $\Lambda_{g_n} \subset R^{top}_n \setminus R^{bot}_n$ and that, as in the proof of theorem 4.5, our assumption implies

$$\sup_{\tau \in \Omega} \left\| 1 - \frac{d\mu^\tau_{R^{top}_n}}{d\mu^\tau_R} \right|_{R^{top}_n \setminus R^{bot}_n} \right\|_\infty \leq e^{-2C\sqrt{L}}$$

for any L large enough and a suitable positive constant C independent of L.

Thus we can use proposition 3.11 to estimate from above the second term in the r.h.s of (4.20) by

$$\mu^\tau_R(g^2_n \log g_n) \leq (1 + e^{-C\sqrt{L}})c_s(\mu^\tau_{R^{top}_n})\frac{1}{2}\mu^\tau_R\left(|\nabla_{R^{top}_n} g_n|^2\right) + \mu^\tau_R(g^2_n) \log\left(\sqrt{\mu^\tau_R(g^2_n)}\right) \tag{4.21}$$

Notice that the last term in the r.h.s of (4.21) is zero because of the identity $\mu^\tau_R(g^2_n) = \mu^\tau_R(f^2) = 1$.

In conclusion the r.h.s. of (4.20) is bounded from above by

$$\left[\sup_{n \in [1, L/10d]} c_s(\mu^\tau_{R^{bot}_n})\right]\frac{1}{2}\mu^\tau_R\left(|\nabla_{R^{bot}_n} f|^2\right) + (1 + e^{-C\sqrt{L}})\left[\sup_{n \in [1, L/10d]} c_s(\mu^\tau_{R^{top}_n})\right]\frac{1}{2}\mu^\tau_R\left(|\nabla_{R^{top}_n} g_n|^2\right) \tag{4.22}$$

In order to conclude the proof we need an estimate of the term $|\nabla_{R^{top}_n} g_n|^2$ in term of quantities like $|\nabla_{R^{top}_n} f|^2$. This estimate is provided by the following key lemma.

Lemma 4.7. *In the same hypotheses of theorem 4.6 there exist two positive constants k_1 and k_2 such that*

$$|\nabla_{R^{top}_n} g_n|^2 \leq \mu^\tau_{R^{bot}_n}(|\nabla_{R^{top}_n} f|^2) + k_1\mu^\tau_{R^{bot}_n}(|\nabla_{R^{bot}_n \cap R^{top}_n} f|^2) + e^{-k_2\sqrt{L}}\mu^\tau_{R^{bot}_n}(|\nabla_{R^{bot}_n} f|^2)$$

for all L large enough.

Let us postpone the proof of lemma 4.7 and let us conclude the proof of proposition 4.6. If we insert the bound of lemma 4.7 into (4.22) we get our final estimate

$$\mu^\tau_R(f^2 \log f) \leq$$
$$(1 + e^{-k_3\sqrt{L}}) \sup_{n \in [1, L/10d]} \left[c_s(\mu^\tau_{R^{bot}_n}) \vee c_s(\mu^\tau_{R^{top}_n})\right]\frac{1}{2}\mu^\tau_R(|\nabla_R f|^2) + k_4\left[\sup_{n \in [1, L/10d]} c_s(\mu^\tau_{R^{top}_n})\right]\frac{1}{2}\mu^\tau_R\left(|\nabla_{R^{bot}_n \cap R^{top}_n} f|^2\right) \tag{4.23}$$

for suitable constants k_3, k_4 and all L large enough independent of f.

By averaging inequality (4.23) over $n \in [1, L/10d]$ as in the proof of theorem 4.5, we get our final bound

$$\mu^\tau_R(f^2 \log f) \leq (1 + \frac{k_5}{\sqrt{L}}) \sup_{n \in [1, L/10d]} \left[c_s(\mu^\tau_{R^{bot}_n}) \vee c_s(\mu^\tau_{R^{top}_n})\right]\frac{1}{2}\mu^\tau_R(|\nabla_R f|^2) \tag{4.24}$$

for a suitable constant k_5, any L large enough and any non–negative $f \in L^2(\Omega, \mu_R^\tau)$ with $\mu_R^\tau(f^2) = 1$. By definition, such inequality implies that

$$c_s(2L) \leq (1 + \frac{k_5}{\sqrt{L}}) \sup_{n \in [1, L/10d]} [c_s(\mu_{R_n^{bot}}^\tau) \vee c_s(\mu_{R_n^{top}}^\tau)] \tag{4.25}$$

for any L large enough.

At this point we continue exactly as in the proof of theorem 4.5 after (4.16) and conclude that

$$c_s(2L) \leq (1 + \frac{k_7}{\sqrt{L}})c_s(\frac{3}{2}L) \tag{4.26}$$

for any L large enough and a suitable constant k_7. Two more iterations of (4.26) suffice to conclude the proof of the theorem. $\square$

In order to conclude our analysis of the logarithmic Sobolev constant we have to prove lemma 4.7. This seemingly technical point is however a key part of the whole approach and its proof requires a new recursive argument that we present in the following lemma.

Lemma 4.8. *Given $R \in \mathcal{R}_L$ and $m > 0$, define $k_m(R)$ as the smallest constant k such that*

$$|\nabla_x \sqrt{\mu_R^\tau(f^2)}|^2 \leq k(\mu_R^\tau(|\nabla_x f|^2) + \sum_{y \in R} e^{-m|x-y|} \mu_R^\tau(|\nabla_y f|^2)) \tag{4.27}$$

for all $\tau \in \Omega$, $x \in \partial_r^+ R$ and $f \in L^2(\Omega, d\mu_R^\tau)$, r being the range of the interaction. Let also $k_m(L) \equiv \sup_{R \in \mathcal{R}_L} k_m(R)$.
Then, in the same hypothesis of theorem 4.6 there exist two positive constants m_0 and a such that

$$k_m(2L) \leq (1 + e^{-aL})k_m(L)$$

for any $m \leq m_0$ and any L large enough.

Corollary 4.9. *In the same hypothesis of theorem 4.6 there exist two positive constants k and m such that, for any rectangle $R(l_1, l_2)$ with $l_1 \wedge l_2 \geq 0.1 \, l_1 \vee l_2$, for all $\tau \in \Omega$, $x \in \partial_r^+ R$ and $f \in L^2(\Omega, d\mu_R^\tau)$*

$$|\nabla_x \sqrt{\mu_R^\tau(f^2)}|^2 \leq k(\mu_R^\tau(|\nabla_x f|^2) + \sum_{y \in R} e^{-m|x-y|} \mu_R^\tau(|\nabla_y f|^2)) \tag{4.28}$$

Proof of the Corollary. It follows immediately from lemma 4.8 that the $\sup_L k_m(L) < \infty$ if m is small enough. Thus the statement for "fat" rectangles with large enough size follows from lemma 4.8. The statement for "fat" rectangles with small size is obviously true if we take the constant k large enough. $\square$

Proof of Lemma 4.7. Take $x \in R_n^{top}$. Then, if $x \notin \partial_r^+ R_n^{bot}$, we compute directly $|\nabla_x g_n|^2$ to found that

$$|\nabla_x g_n|^2 = \mu_{R_n^{bot}}^\tau(|\nabla_x f|^2)$$

If instead $x \in \partial_r^+ R_n^{bot}$ we apply corollary 4.9 and get

$$|\nabla_x g_n|^2 \le k\big(\mu_{R_n^{bot}}^\tau(|\nabla_x f|^2) + \sum_{y \in R_n^{bot}} e^{-m|x-y|}\mu_{R_n^{bot}}^\tau(|\nabla_y f|^2)\big)$$

for a suitable constant k and m. By summing over $x \in R_n^{top}$ we get the sought bound. $\square$

Proof of Lemma 4.8. Given $m > 0$ and L large enough, we suppose that $k_m(2L) > k_m(L)$. Let us consider a rectangle $R \equiv R(l_1, l_2) \in \mathcal{R}_{2L}$ and assume, without loss of generality, that the longest side is l_2. Since we are assuming that $k_m(2L) > k_m(L)$ necessarily $L < l_2 \le L$. Let us also fix $x \in \partial_r^+ R$ and $f \in L^2(\Omega, d\mu_R^\tau)$. A simple computation (see e.g [Z] or [MO2]) shows that

$$|\nabla_x f^2| = 2|\langle f \rangle_x| \, |\nabla_x f| \tag{4.29}$$

where $\langle g \rangle_x = \frac{1}{2}\big(g(\sigma^x) + g(\sigma)\big)$.

Thus, if

$$|\nabla_x \mu_R^\tau(f^2)| \le 2|\langle \sqrt{\mu_R^\tau(f^2)} \rangle_x| \, A$$

then

$$|\nabla_x \sqrt{\mu_R^\tau(f^2)}| \le A \tag{4.30}$$

Let us compute $|\nabla_x \mu_R^\tau(f^2)|$. For this purpose let R_i, $i = 1, 2, 3$ be the rectangles in R defined by

$$R_1 = \big\{(x_1, x_2) \in \mathbf{Z}^2; \ 0 \le x_1 \le l_1 - 1, \ 0 \le x_2 \le \frac{l_2}{2}\big\}$$

$$R_2 = \big\{(x_1, x_2) \in \mathbf{Z}^2; \ 0 \le x_1 \le l_1 - 1, \ \frac{l_2}{4} \le x_2 \le \frac{3l_2}{4}\big\}$$

$$R_3 = \big\{(x_1, x_2) \in \mathbf{Z}^2; \ 0 \le x_1 \le l_1 - 1, \ \frac{l_2}{2} \le x_2 \le l_2\big\}$$

Then, depending whether the second coordinate x_2 of the site x lies in the interval $I_1 = (-\infty, \frac{3}{8}l_2]$, $I_2 = (\frac{3}{8}l_2, \frac{5}{8}l_2]$ or $I_3 = (\frac{5}{8}l_2, \infty)$, we write, using (2.2),

$$|\nabla_x \mu_R^\tau(f^2)| = |\nabla_x \mu_R^\tau(\mu_{R_i}^\tau(f^2))| \quad \text{if} \quad x_2 \in I_i \quad i = 1, 2, 3 \tag{4.31}$$

Let us consider for shortness only the first case, $x_2 \in I_1$, the other ones being similar. A simple computation shows that

$$|\nabla_x \mu_R^\tau(\mu_{R_1}^\tau(f^2))| \le |\mu_R^\tau((h_x - 1)\mu_{R_1}^{\tau^x}(f^2))| + |\mu_R^\tau(\nabla_x \mu_{R_1}^\tau(f^2))| \tag{4.32}$$

where $h_x \equiv \frac{d\mu_R^{\tau^x}}{d\mu_R^\tau}\Big|_{R \setminus R_1}$. Notice that, since $d(x, R \setminus R_1) \ge 0.1 l_2 \ge 0.1 L$, for any L large enough our assumption together with proposition 2.12 implies

$$\|h_x - 1\|_\infty \le e^{-CL}$$

for a suitable constant C independent of τ, L and x. Thus the first term in the r.h.s of (4.32) can be bounded from above by

$$\sqrt{2}e^{-CL}\sqrt{\mu_R^\tau(\mu_{R_1}^{\tau^x}(f^2))}\sqrt{\operatorname{Var}_R^\tau(\sqrt{\mu_{R_1}^{\tau^x}(f^2)})}$$

$$\leq ke^{-CL}\sqrt{\mu_R^\tau(f^2)}\sqrt{\operatorname{Var}_R^{\tau^x}(\sqrt{\mu_{R_1}^{\tau^x}(f^2)})}$$

$$\leq ke^{-CL}\langle\sqrt{\mu_R^\tau(f^2)}\rangle_x\{\operatorname{Var}_R^\tau(f)\}^{\frac{1}{2}} \tag{4.33}$$

$$\leq \bar{k}e^{-CL}\langle\sqrt{\mu_R^\tau(f^2)}\rangle_x\{\mu_R^\tau(|\nabla_R f|^2)\}^{\frac{1}{2}}$$

for a suitable constant k. In the above steps we used the simple bounds valid for any non–negative g

$$|\mu_R^\tau(h;g)| \leq \|h\|_\infty\sqrt{2\mu_R^\tau(g)}\sqrt{\operatorname{Var}_R^\tau(\sqrt{g})}$$

$$\|\frac{d\mu_R^{\tau^x}}{d\mu_R^\tau}\|_\infty \leq e^{2\|J\|_\infty}$$

$$\operatorname{Var}_R^{\tau^x}(\sqrt{\mu_{R_1}^{\tau^x}(f^2)}) \leq \operatorname{Var}_R^{\tau^x}(f)$$

$$\operatorname{Var}_R^{\tau^x}(g) \leq \|\frac{d\mu_R^{\tau^x}}{d\mu_R^\tau}\|_\infty^2\operatorname{Var}_R^\tau(g)$$

(3.12) and the fact that $\sup_L\sup_{R\in\mathcal{R}_L}\operatorname{gap}(L_R^\tau)^{-1} < \infty$.

Let us bound from above the second term in the r.h.s of (4.32). We observe that each R_i, $i = 1, 2, 3$, either belongs to $\mathcal{R}_{\frac{3}{2}L}$ or it belongs to $\mathcal{R}_{2L}$ but with the shortest side smaller than L, depending whether l_1 is smaller or larger than $\frac{3}{2}L$. Thus we can use the definition of $k_m(L)$, $k_m(R)$ together with (4.29) to write

$$|\nabla_x\mu_{R_1}^\tau(f^2)|$$
$$\leq 2\langle\sqrt{\mu_{R_1}^\tau(f^2)}\rangle_x\sqrt{k_m'(L)}\{\mu_{R_1}^\tau(|\nabla_x f|^2) + \sum_{y\in R_1}e^{-m|x-y|}\mu_{R_1}^\tau(|\nabla_y f|^2)\}^{\frac{1}{2}} \tag{4.34}$$

with

$$k_m'(L) = \begin{cases} \sup_{\substack{R(l_1,l_2)\in\mathcal{R}_{2L} \\ l_1\leq L}} k_m(R) & \text{if } l_1 > \frac{3}{2}L \\ k_m(\frac{3}{2}L) & \text{otherwise} \end{cases}$$

Above, as in the proof of theorem 4.5, without loss of generality we kept the longest side always in the second direction. A final Schwartz inequality together with (2.2) gives

$$\mu_R^\tau(|\nabla_x\mu_{R_1}^\tau(f^2)|)$$
$$\leq 2\langle\sqrt{\mu_R^\tau(f^2)}\rangle_x\sqrt{k_m'(L)}\{\mu_R^\tau(|\nabla_x f|^2) + \sum_{y\in R_1}e^{-m|x-y|}\mu_R^\tau(|\nabla_y f|^2)\}^{\frac{1}{2}} \tag{4.35}$$

In conclusion, if we combine together (4.30), (4.33) and (4.34) we get

$$|\nabla_x \sqrt{\mu_R^\tau(f^2)}|^2 \leq$$

$$\left\{ \bar{k}e^{-CL}\{\mu_R^\tau(|\nabla_R f|^2)\}^{\frac{1}{2}} + \sqrt{k_m'(L)}\{\mu_R^\tau(|\nabla_x f|^2) + \sum_{y \in R} e^{-m|x-y|}\mu_R^\tau(|\nabla_y f|^2)\}^{\frac{1}{2}} \right\}^2$$

$$\leq 3\bar{k}e^{-CL}\mu_R^\tau(|\nabla_R f|^2) + (1 + \bar{k}e^{-CL})k_m'(L)\{\mu_R^\tau(|\nabla_x f|^2)$$

$$+ \sum_{y \in R} e^{-m|x-y|}\mu_R^\tau(|\nabla_y f|^2)\}$$

$$(4.36)$$

where, without any harm, we have extended the summation over $y \in R_1$ to the whole rectangle R.

Observe that

$$3\bar{k}e^{-CL}\mu_R^\tau(|\nabla_R f|^2) \leq e^{-(C/2)L} \sum_{y \in R} e^{-m|x-y|}\mu_R^\tau(|\nabla_y f|^2)$$

if m is small enough compared to the constant C and CL is large enough. Thus the r.h.s of (4.36) is smaller than

$$k_m'(L)(1 + e^{-(C/3)L})\{\mu_R^\tau(|\nabla_x f|^2) + \sum_{y \in R} e^{-m|x-y|}\mu_R^\tau(|\nabla_y f|^2)\}$$

i.e.

$$k_m(R) \leq k_m'(L)(1 + e^{-(C/3)L}) \tag{4.37}$$

If we now reapply the whole procedure to an arbitrary rectangle in $\mathcal{R}_{2L}$ but with the shortest side smaller than L, we get from (4.37) that

$$k_m'(L) \leq (1 + e^{-(C/3)L})\, k_m(\frac{3}{2}L)$$

In conclusion

$$k_m(R) \leq (1 + e^{-(C/3)L})^2 k_m(\frac{3}{2}L) \qquad \forall R \in \mathcal{R}_{2L} \tag{4.38}$$

provided that L is large enough and m small enough.
Two iterations of (4.38) suffices to conclude the proof. $\square$

We conclude by stating one final result, whose proof can be given along the same lines of the previous ones, that will turn out to be quite useful when discussing systems with random interactions.

Proposition 4.10. *Fix L_0 and let Λ be a multiple of Q_{L_0}. Assume that there exists $m > 0$ such that $SMT(V, L_0/2, m)$ holds for all subsets V of Λ that are multiples of Q_{L_0}. Then there exists two positive constants δ and $\varepsilon = \varepsilon(m)$ such that, if $L_0^d e^{-m\sqrt{L_0}} \leq \varepsilon$, then*

$$\inf_\tau \text{gap}(L_\Lambda^\tau) \geq \frac{1}{2}\inf_\tau \text{gap}(L_{Q_{L_0}}^\tau)$$

Moreover, if $\mathrm{gap}(L^\tau_{Q_{L_0}}) \geq e^{-m\delta L_0}$, then

$$\sup_\tau c_s(\mu^\tau_\Lambda) \leq 2\sup_\tau c_s(\mu^\tau_{Q_{L_0}})$$

5. Boundary Phase Transitions

In this section we analyze more closely the problem of the exponential ergodicity in the uniform norm of the dynamics for three dimensional systems in the one phase region.

Already in [MO1] (see also [MOS]) it was discussed the possibility that, even if the interaction is such that there is no phase transition and weak mixing holds, in a finite large enough cube Q_L one could have some sort of "boundary phase transition" which could slow down dramatically the approach to equilibrium.

We will provide here a substantial evidence that this phenomenon happens in the three dimensional Ising model at low temperature and nonzero magnetic field h (see (2.4)) for which weak mixing holds. More precisely if β is large enough we will argue that it is possible to tune the magnetic field as a function of β in such a way that the gap in the spectrum of the generator of the dynamics in a finite cube with suitably chosen boundary conditions becomes *exponentially small* in the side of the cube. Notice that, for the same value of the thermodynamic parameters weak mixing holds *i.e.* theorem 4.1 applies and the gap of any attractive *infinite volume* Glauber dynamics is bounded away from zero. Actually the same arguments that proved theorem 4.1 could be used to show that the thermodynamic limit of the gap in finite cubes with + boundary conditions is bounded away from zero [SchYo] (see also section 5.3 below).

In order to clarify the discussion, let us consider the three dimensional Ising model in a cube Q_L of side L, at low temperature and small, positive external field h. As boundary conditions we take $-$ on the bottom face V of the cube and free (*i.e.* absent) on the other faces. Since the magnetic field is positive and the temperature is very low, the typical configurations of the systems will be mostly plus spins (plus phase) away from the bottom face. Thus there will be, with large probability, a unique Peierls contour (*i.e.* a connected union of dual plaquettes separating nearest neighbor spins of opposite sign) Γ separating the plus bulk phase from the minus boundary conditions on the bottom face, and it is quite clear that the statistical properties of such a contour will play an important role in the mixing properties of the Gibbs measure of the system and on the rate of approach to equilibrium of an associated Glauber dynamics. Unfortunately a detailed analysis of the probability distribution of the contour Γ and its dynamical propertie requires quite deep ideas and technology like the Pirogov-Sinai theory (see e.g. [Zha]). Such analysis becomes simpler in of uses the so called solid-on-solid (SOS) approximation for Γ.

5.1 The Solid-on-Solid Approximation

We approximate the contour Γ with a two dimensional surface $\varphi = \{\varphi(x)\}_{x \in V}$, where the random variable $\varphi(x) \in \mathbb{N}$ represents the height of the surface at x above the bottom face of the cube, and we assume that the probability distribution of the surface with ψ boundary condition outside V is that of the solid–on–solid model (SOS)

$$\mu_V^{h,\psi}(\varphi) = \frac{1}{Z_V^{h,\psi}} \exp\left[-H_V^{h,\psi}\right]$$

where

$$H_V^{h,\psi}(\varphi) = \frac{J}{2} \sum_{\substack{x,y \in V;\\ |x-y|=1}} |\varphi(x) - \varphi(y)| + h \sum_{x \in V} \varphi(x) + J \sum_{\substack{x \in V\ y \in V^c\\ |x-y|=1}} |\varphi(x) - \psi(y)| \quad (5.1)$$

The SOS approximation corresponds to taking the original Gibbs measure and conditioning on the event that for any $x = (x_1, x_2, -1)$ in V there exists a non–negative integer $\varphi(x)$ such that

$$\sigma(x_1, x_2, n) = \begin{cases} -1 & \text{if } n \le \varphi(x) \\ +1 & \text{if } n > \varphi(x) \end{cases}$$

Although strictly speaking the above event has a very small probability w.r.t the original Ising Gibbs measure, the SOS model is considered a reasonable approximation at very low temperatures (large values of J).

A kinetic version of the model is readily obtained by considering a Glauber dynamics for it, namely a single spin Markov process on the configuration space $\Omega = \mathbf{N}^V$, reversible with respect to $\mu_V^\psi(\varphi)$ and such that each move consists in replacing at some site x, $\varphi(x)$ with $\varphi(x) \pm 1$ (but the new φ has to be non–negative) with rates that satisfy H1, H3, H4 and are reversible w.r.t the Gibbs measure $\mu_V^{h,\psi}$.

Two cases one may want to keep in mind are (χ is the characteristic function, $s = \pm$ and $\varphi^{x,s} = \varphi(x) + s$)

$$c(x, \varphi, s) = \min\{e^{-J\Delta_{x,s}H_V^{h,\psi}(\varphi)}, 1\} \chi\{\varphi^{x,s} \in \Omega\}$$

and

$$c(x, \varphi, s) = \left[1 + \exp(J\Delta_{x,s}H_V^{h,\psi}(\varphi))\right]^{-1} \chi\{\varphi^{x,s} \in \Omega\}$$

where

$$\Delta_{x,s}H_V^{h,\psi}(\varphi) = H_{\{x\}}^\varphi(\varphi(x) + s) - H_{\{x\}}^\varphi(\varphi(x)).$$

As before we will denote by $L_V^{h,\psi}$ the corresponding generator in the region V with boundary conditions ψ. It is worthwhile to mention that, with above choice of the jump rates, the dynamics becomes attractive.

For the SOS model described by (5.1), and variants of it, it has long been realized (see e.g. [BEF], [FP1], [FP2] and more recently [LM], [BDZ], [DM]) that the relevant statistical properties of the surface φ are determined by the competition between the attraction to the wall due to the external field h and the "entropic repulsion" from the wall. Although the techniques to rigorously study these competing effects are quite involved, nevertheless the basic ideas can be understood in simple terms.

Imagine that the relevant configurations of the SOS Gibbs measure are "flat" rigid surfaces at certain height, with isolated small fluctuations that we assume to be only up or down parallelepipeds of unit base and certain length. Let us denote by $\Phi(k)$ the set of such surfaces which sit at height $k \in \mathbf{N}$. Clearly, if $\varphi \in \Phi(k)$, the downward fluctuations cannot be longer than k itself because of the presence of the wall under the surface. Thus, if $k < k'$, the cardinality of $\Phi(k')$ is larger than that of $\Phi(k)$. Having this in mind one can then try to compute the value of k such that $\mu_V^{h,\psi}(\Phi(k))$ takes its maximum, or, what is the same, compute the ratio

$$\lambda(k) \equiv \frac{\mu_V^{h,\emptyset}(\Phi(k))}{\mu_V^{h,\emptyset}(\Phi(k-1))}$$

where $\emptyset$ means free (i.e. absent) boundary conditions. One easily gets from (5.1) that $\lambda(k)$ is approximately (neglecting the boundary conditions) given by

$$\lambda(k) \approx \exp(-hL^2 + L^2 e^{-4Jk})$$

where the term $-hL^2$ takes into account the energy one has to pay in order to lift up by one unit a rigid flat interface, while the term $L^2 e^{-4Jk}$ represents the entropy gain obtained by increasing by one the height. Notice that the factor e^{-4Jk} is the energy one has to pay in order to create a downward fluctuation of length k, namely a fluctuation which can be present in a surface $\varphi \in \Phi(k)$ and not in a surface $\varphi \in \Phi(k-1)$.

The above computation suggests that, when L is very large and the magnetic field h is decreased from value $e^{-4J(k-1)}$ to the new value e^{-4Jk}, the average height of the surface should jump from $k-1$ to k. Thus there should exist a critical value $h_{k-1}^*(J)$ between $e^{-4J(k-1)}$ and e^{-4Jk} such that, for $h = h_{k-1}^*(J)$, the surface is undecided whether to stay at height $k-1$ or k. It is such a situation that we call "layering phase transition".

In [DM] the phenomenon of "layering phase transition" was proved for the first time. Their result was then extended and improved in [CM1]. Let us recall the main result of [CM1], since it plays an essential role in our analysis of the non equilibrium case.

Theorem 5.1. *There exists J_0 such that for all $J \geq J_0$ there are positive numbers $\{h_k^*(J)\}_{k=1}^{k_{max}}$, with $k_{max} = \lfloor e^{\frac{J}{20000}} \rfloor$, such that the following holds for $k = 1, \ldots, k_{max}$*
  i) $\frac{1}{4}e^{-4Jk} \leq h_k^*(J) \leq 4e^{-4Jk}$

ii) if $h_k^*(J) < h < h_{k-1}^*(J)$ (define $h_0^*(J) = +\infty$) then
   a) there exists a unique Gibbs measure
   b) there exist $m(J,h) > 0$, $C(J,h) > 0$ such that for any $L \geq \lfloor 8/h + 1 \rfloor$

$$\sup_{\psi,\psi' \in \Omega} |\mu_{Q_L}^{h,\psi}(\varphi(x_0)) - \mu_{Q_L}^{h,\psi'}(\varphi(x_0))| \leq C(J,h)e^{-m(J,h)L}$$

where $\mu_{Q_L}^{h,\psi}(\varphi(x_0))$ denotes the expected value of the height of the surface at the center x_0 of the square Q_L with boundary conditions ψ.
iii) if $h = h_k^*(J)$ there are at least two distinct extreme Gibbs measures.

Remark.
  a) Notice that, although the spins $\varphi(x)$ are unbounded, ii.b) above is nothing else but weak mixing in the attractive case (see section 3.1).
  b) In [LMaz] the restriction $k \leq k_{max}$ was removed and the proof of ii.b) was considerably simplified.

A byproduct of the results of [CM1] is that if $h_k^*(J) < h < h_{k-1}^*(J)$ then the typical configurations of the Gibbs measure $\mu_{Q_L}^{h,\psi}$ are, in a sense that can be made precise, "flat" surfaces at height k. On the other hand, at $h = h_k^*(J)$ and with free boundary condition, in [CM1] it was proved that the ± 1 random variables

$$\sigma(x) = \text{sign}(\varphi(x) - k - \frac{1}{2})$$

behave roughly as a *two dimensional* Ising model at low temperature with *zero* external field and free boundary conditions, i.e a system in the phase coexistence region.

The associated Glauber dynamics in a square Q_L of side L, with boundary conditions ψ and magnetic field h, was then studied in [CM2]. The main result of [CM2] is

Theorem 5.2. *In the same setting as in Theorem [CM1] we have, for all $k = 1, \ldots, k_{max}$,*

i) if $h_k^*(J) < h < h_{k-1}^*(J)$ then there exist $L_0(J, h)$, $\kappa(J, h) > 0$ such that

$$\inf_{L \geq L_0} \inf_{\psi \in \Omega} \text{gap}(L_{Q_L}^{h, \psi}) \geq \kappa(J, h)$$

ii) if $h = h_k^*(J)$, then there exist positive constants $C_1(J, h)$, $C_2(J, h)$ such that for all $L > 10/h$

$$C_1(J, h) \, e^{-100JkL} \leq \text{gap}(L_{Q_L}^{h, \emptyset}) \leq C_2(J, h) \, e^{-\frac{1}{40}JL}$$

where $\emptyset$ means free boundary conditions.

Proof. The proof of i) goes as follows. Since the SOS system is two dimensional one can adapt the arguments of [MOS] to transform the weak mixing condition ii.b) of theorem 5.1 into the strong mixing condition of section 2.3 and then use the recursive analysis of the proof of theorem 4.5 to get i).

To prove part ii), namely to find an upper bound on $\text{gap}(L_{Q_L}^{h_k^*(J), \emptyset})$, one uses the "look for the bottleneck" approach, *i.e.* one takes advantage of the variational characterization for the gap (3.12) and choose an appropriate test function f which illustrates how the system, in order to relax to equilibrium, has to make an excursion to a very unlikely region of the phase space.

Given $\varphi \in \Omega$ we set

$$\sigma(x) = \text{sign}(\varphi(x) - k - 1/2)$$

and, for $U \subset\subset \mathbb{Z}^2$

$$M_U(\sigma) = M_U(\sigma(\varphi)) = \sum_{x \in U} \sigma(x)$$
$$m_U(\sigma) = m_U(\sigma(\varphi)) = |U|^{-1} M_U(\sigma).$$

In analogy with the solution of the similar problem for the two dimensional Ising model in the phase coexistence region, we take as a test function

$$f(\varphi) = \chi\{M_{Q_L}(\varphi) > 0\} - \chi\{M_{Q_L}(\varphi) < 0\}.$$

Using (3.3), we find

$$\mathcal{E}_{Q_L}^{h_k^*,\theta}(f,f) \leq \sup_{\varphi \in \Omega} \sum_{s=\pm} c(x,\varphi,s)\, \mu_{Q_L}^{h_k^*,\theta}\{|M_{Q_L}(\varphi)| \leq 2\}$$
$$\leq 2c_M(J,h_k^*)\, \mu_{Q_L}^{h_k^*,\theta}\{|M_{Q_L}(\varphi)| \leq 2\}, \tag{5.2}$$

while

$$\mathrm{Var}_{Q_L}^{h_k^*,\theta}(f) = \mu_{Q_L}^{h_k^*,\theta}\{M_{Q_L}(\varphi) \neq 0\} - \left(\mu_{Q_L}^{h_k^*,\theta}\{M_{Q_L}(\varphi) > 0\} - \mu_{Q_L}^{h_k^*,\theta}\{M_{Q_L}(\varphi) < 0\}\right)^2 \tag{5.3}$$

The sougth upper bound now follows from the following two estimates valid for any L large enough. These bounds should be compared with the analogous ones for the two dimensional Ising model with free boundary conditions in the phase coexistence region (see (6.24) and (6.30)).

$$\mu_{Q_L}^{h_k^*,\theta}\{|M_{Q_L}(\varphi)| \leq 2\} \leq e^{-\frac{1}{12}JL}.$$

and

$$\mu_{Q_L}^{h_k^*,\theta}\{\varphi \in \Omega_{Q_L} : M_{Q_L}(\sigma(\varphi)) < 0\} \geq e^{-\frac{1}{20}JL}$$
$$\mu_{\Lambda}^{h_k^*(\beta),\theta}\{\varphi \in \Omega_{Q_L} : M_{Q_L}(\sigma(\varphi)) > 0\} \geq e^{-\frac{1}{20}JL}$$

5.2 Back to the Ising Model

Going back to the 3D Ising model discussed above, we can conclude that there is a very good evidence that for three dimensional systems satisfying the weak mixing condition it is possible to have at the same time

$$\inf_{\Lambda,\tau} \mathrm{gap}(L_{\Lambda}^{\tau}) = 0 \quad \text{and} \quad \mathrm{gap}(L_{\mathbb{Z}^d}^{J}) > 0$$

due to the occurrence of some sort of "boundary phase transition". Theorem 4.5 or theorem 4.6 imply that in this situation the system cannot satisfy strong mixing.

It is interesting to better understand how the expected value of some local observable over the dynamics in a finite cube Q_L at time t is affected by the phenomenon of boundary phase transition. As before we consider this problem for the three dimensional Ising model (2.4) in Q_L with the boundary conditions described at the beginning of this paragraph, $\beta \gg 1$ and $h \equiv h_k^*(\beta)$. Here, in analogy with the SOS approximation, the critical value $h_k^*(\beta)$ should be such that the height of the Peierls contour Γ above the bottom face V is undecided between level k and $k+1$. Without loss of generality we will consider the average magnetization in the center x_0 of the cube.

Our first claim is that up to time $T = k_0^{-1}L$, k_0 being the constant appearing in lemma 3.2

$$\|T_{Q_L}^{\tau}(t)\,\sigma(x_0) - \mu_{Q_L}^{\tau}(\sigma(x_0))\|_{\infty} \leq Ce^{-t} \tag{5.4}$$

with C independent of L.
To prove (5.4) we write

$$\|T_{Q_L}^\tau(t)\sigma(x_0) - \mu_{Q_L}^\tau(\sigma(x_0))\|_\infty \leq \quad \|T_{Q_L}^\tau(t)\sigma(x_0) - T^J(t)\sigma(x_0)\|_\infty$$
$$+ \|T^J(t)\sigma(x_0) - \mu^J(\sigma(x_0))\|_\infty \qquad (5.5)$$
$$+ |\mu^J(\sigma(x_0)) - \mu_{Q_L}^\tau(\sigma(x_0))|$$

where μ^J is, as before, the unique infinite volume Gibbs state. Because of lemma 3.2

$$\|T_{Q_L}^\tau(t)\sigma(x_0) - T^J(t)\sigma(x_0)\|_\infty \leq e^{-t} \qquad \forall\, t \leq k_0^{-1}L$$

Using weak mixing together with theorem 3.3, the second term in the r.h.s of (5.5) is bounded from above by $C_1 e^{-m_1 t}$ for all $t \geq 0$ and suitable constants C_1 and m_1. Finally, thanks again to weak mixing, the third term in the r.h.s of (5.5) is bounded from above by $C_2 e^{-m_2 L} \leq C_2 e^{-m_2 k_0 t}$ and (5.4) follows.

For times t larger than $k_0^{-1}L$ but smaller than $e^{\epsilon L}$, where ϵ is a small positive constant, we conjecture that

$$T_{Q_L}^\tau(t)\sigma(0)(+) - T_{Q_L}^\tau(t)\sigma(0)(-) \geq e^{-m'L} \qquad (5.6)$$

for a suitable $m' > 0$ independent of L, where $+$ and $-$ denote the two configurations identically equal to $+1$ and -1 respectively. In order to justify (5.6), we observe that the analysis of the SOS model (see [CM2]) for $h = h_k^*(J)$ shows that, with very high probability (larger than $1 - e^{-cL}$, $c > 0$), any interface Γ starting below (above) height k does not jump to height $k+1$ (k) before a time which is exponentially large in L. Therefore, the difference in the l.h.s of (5.6) should be bounded from below, using monotonicity, by the difference of magnetizations at $x = x_0$ computed with the Gibbs states in the parallelepiped $V \times [k, L-1]$ with plus and minus b.c. on the bottom face and free b.c. on the remaining faces. Such a difference can be shown to be exactly of the order of the r.h.s of (5.6).

5.3 Recent Progresses

We conclude by discussing some recent progresses on the relaxation properties of the d-dimensional Ising model at low temperature and very small positive magnetic field h. Although our discussion indicates that one may have

$$\inf_\tau \mathrm{gap}(L_{Q_L}^\tau) \approx e^{-CL}$$

for the three dimensional Ising model at low enough temperature and suitably chosen small magnetic field h, it is also clear that for many natural boundary conditions, e.g plus b.c. of free b.c., such a patology should not occur, since no large contour is present in the system. The next theorem (see [Yo2] and also [SchYo]) goes exactly in this direction.

Theorem 5.3. *Consider the standard d-dimensional Ising model given by (2.4). There exists $\beta_0(d)$ such that for any $\beta \geq \beta_0$ and any $h_0 > 0$ there exists a $c > 0$ such that if $h > 0$ and*

$$\min_{x \in Q_L} \left\{ h + \sum_{\substack{v \notin Q_L \\ d_2(x,v)=1}} \tau(y) \right\} \geq h_0$$

then $\mathrm{gap}(L^\tau_{Q_L}) \geq c$. *Moreover, if the Glauber dynamics is attractive, there exist two finite positive constants C_1, C_2 independent of L such that*

$$\|T^\tau_{Q_L}(t)f - \mu^\tau_{Q_L}(f)\|_\infty \leq C_1 \|\|f\|\| e^{-C_2 t}$$

The reader may at this point be quite confused about what is known and what it remains to prove in the absence of a phase transition even for the simplest model like the Ising model. We thus summarize the results for the latter in dimension greater than one in the one phase region, *i.e.* when either the magnetic field h is different from zero or the inverse temperature β is strictly smaller than the critical value β_c.

i) The simplest case is two dimensions. Here weak mixing holds and therefore, thanks to theorem 4.1, we have infinite volume exponential ergodicity in the $\| \cdot \|_\infty$ norm. Moreover, thanks to theorems 2.5 and 2.7, the assumptions of theorem 4.6 hold and therefore, thanks to theorem 3.3, exponential ergodicity in any finite cube holds with constants independent of the size of the cube. One can say that in two dimensions the problem of the exponential ergodicity in the one phase region is fully solved.

ii) In higher dimension one still knows that weak mixing, and thus exponential ergodicity in infinite volume, holds for any $\beta < \beta_c$ and any h (see [Hi]) or for any large enough β (much larger than the critical value β_c) and any $h \neq 0$ (see [MO1]). However there is no result concerning weak mixing in the region of non-zero h and β just above β_c. Analogously strong mixing and thus the strong form of exponential ergodicity given by theorems 4.6 and 3.3 for cubes or "fat" parallelepipeds, has been proved only deep inside the one phase region *i.e.* at high temperature ($\beta << \beta_c$) or for $h\beta$ large enough. Moreover, contrary to the two dimensional case, we conjecture (but not prove) that strong mixing fails for β large enough and suitably small h. For these special values of the thermodinamic parameters, the relaxation time in a large cube should drastically depend on the boundary conditions.

6. Phase Coexistence

So far we have discussed the relaxation to equilibrium of the Glauber dynamics in the one phase region. A natural question arises as to what happens when the thermodynamic parameters (e.g inverse temperature and the external magnetic field in the Ising model) are such that we do have a phase transition in the thermodynamic limit.

To be more concrete, let us consider the usual Ising model (2.4) in d-dimensions $d \geq 2$ without external field and let us suppose that the inverse temperature β is larger than the critical value β_c. Then, thanks to theorem 3.1, any associated infinite volume Glauber dynamics is not ergodic and it is rather natural to ask how this absence of ergodicity is reflected if we look at the dynamics in a finite, but large cube Q_L of side L, where ergodicity is never broken.

A first partial answer was provided in [Th] few years ago for very low temperatures. In [Th] it was proved that, if the boundary conditions are free, *i.e.* absent, then the relaxation time to equilibrium, that in a first approximation can be taken equal to the inverse of the gap in the spectrum of the generator $L_{V_L}^{J,\tau}$ of the dynamics, diverges, as $L \to \infty$, at least as an exponential of the *surface* L^{d-1}.

The reason for such a result is the presence of a rather tight "bottleneck" in the phase space. When in fact the boundary conditions are either free or periodic, the energy landscape determined by the function $H_{Q_L}^{J,\tau}(\sigma)$ has only two absolute minima, corresponding to the two configurations identically equal to either $+1$ or to -1. Thus the dynamics started e.g. from all minuses, in order to relax to equilibrium, has to reach the neighboorood of the opposite minimum by necessarily crossing the set of configurations of zero magnetization. Since the Gibbs measure gives to the latter a very small weight, of the order of a negative exponential of the *surface* of Q_L (see e.g. [P]), a kind of bottleneck is present and the result follows by rather simple arguments (see section 6.5 below).

The same reasoning also suggests that if the double well structure of the Gibbs measure is completely removed by the boundary conditions, e.g. by fixing equal to $+1$ all spins outside Q_L, or if we measure the relaxation to equilibrium of a function f which is *even* w.r.t a global spin flip $\sigma \to -\sigma$, then the relaxation time should be much shorter than in the previous case since there are no bottlenecks to cross. Actually in this case the interesting but unproven conjecture, is that, at least in two dimensions with plus boundary conditions, the relaxation time will diverge, as $L \to \infty$, like L^2. The proof of the above conjecture would have some very nice consequences on the analysis of the speed of relaxation to equilibrium for the infinite volume dynamics started in one of the pure phases of the system, e.g the plus phase, for which it has been predicted (see [FH]) a stretched exponential decay of the form $e^{-\sqrt{t}}$ in $d = 2$ and a pure exponential law in $d = 3$.

In this section, following [M], [CGMS] and [MM], we consider the above and other related questions for the two dimensional Ising model at inverse temperature β above the critical one β_c and without external field. We will first prove a *lower* bound on the gap in the spectrum of the generator $L_{Q_L}^{J,+}$ of the Glauber dynamics with $+$ boundary conditions of the form:

$$\text{gap}(L_{Q_L}^{J,+}) \geq e^{-\delta L} \tag{6.1}$$

where $\delta = \delta(L)$ is such that $\lim_{L\to\infty} \delta(L) = 0$. Such a bound, although quite far from the conjectured L^2 law, is in any case much larger than the *upper bound* obtained in [Th] with free boundary conditions. We will also show how to use (6.1) to derive an upper bound on the time autocorrelation functions of the infinite volume dynamics started in the plus phase.

Then we will compute exactly the asymptotics of the gap with open boundary conditions and show that

$$\lim_{L\to\infty} -\frac{1}{\beta L} \log(\text{gap}(L_{Q_L}^{J,\emptyset})) = \tau_\beta \qquad (6.2)$$

where τ_β denotes the surface tension in the direction of e.g. the horizontal axis (see e.g [Pf] for a precise definition) and $\emptyset$ denotes free (*i.e.* absent) boundary conditions.

In this case, the picture of the relaxation behaviour to the Gibbs equilibrium measure that comes out of our analysis (see also [MM] for an important improvement), is the following: the system first relaxes rather rapidly to one of the two phases and then it creates, via a large fluctuation, a thin layer of the opposite phase along one of the sides of Λ. Such a process requires already a time of the order of $e^{\beta\tau_\beta L}$. After that, the opposite phase invades the whole system by moving, in a much shorter time scale the interface to the side opposite to the initial one and equilibrium is finally reached. The time required for this final process can be computed to be of the order of L^3 at least in the SOS approximation described in section 5 (see [Po]).

Once this picture is established it is not difficult to show that, under a suitable stretching of the time by a factor $a(L) \approx e^{\beta\tau_\beta(0)L}$, the magnetization in the square Q_L behaves in time as a continuous Markov chain with state space $\{-m^*(\beta), +m^*(\beta)\}$ and unitary jump rates, where $m^*(\beta)$ is the spontaneous magnetization (see Theorem 6.1 of [M]).

The key ingredients of our analysis (sse [M] and [CGMS]) are
  (i) a geometric bound on the gap in the spectrum of the generator of the dynamics in a rectangle
 (ii) some detailed equilibrium estimates related to the large fluctuations of an horizontal interface of length L
(iii) a precise estimate of the equilibrium probability of having anomalous magnetization
   $m_\Lambda(\sigma) \in (-m^*(\beta), m^*(\beta))$.

The first one (i) is borrowed from a clever technique to bound from below the gap of symmetric Markov chains on complicated graphs introduced in [JS1], [JS2] in the framework of hard computational problems. It is important for us that its validity is independent of β. The second, at least for the Ising model, is nowdays available for any temperature below the critical one (see [CGMS]) thanks to the powerful methods of duality (see [Pf]). The third, after the extension of [Io1] and [Io2] of the basic results of [DKS], is also available for any $\beta > \beta_c$ (see [CGMS]).

6.1. Some Preliminary Key Equilibrium Results

Given the two dimensional lattice $\mathbb{Z}^2$ let $\mathbb{Z}_*^2 = \mathbb{Z}^2 + (1/2, 1/2)$ denotes its dual. For $x, y \in \mathbb{R}^2$ $[x,y]$ is the *closed segment* with x, y as its endpoints. The *edges* of $\mathbb{Z}^2$ ($\mathbb{Z}_*^2$) are those $e = [x,y]$ with x, y nearest neighbors in $\mathbb{Z}^2$ ($\mathbb{Z}_*^2$). Given an edge e of $\mathbb{Z}^2$,

e^* is the unique edge in $\mathbf{Z}_*^2$ that intersects e. We denote by $\mathcal{E}_\Lambda$ the set of all edges such that both endpoints are in Λ and by $\bar{\mathcal{E}}_\Lambda$ the set of all edges with at least one endpoint in Λ.

Given $\Lambda \subset \mathbf{Z}^2$ we define Λ^* as the set of all $x \in \mathbf{Z}_*^2$ such that $d(x, \Lambda) = 1$. The set of the dual edges is defined as

$$\mathcal{E}_\Lambda^* = \{ e^* : e \in \bar{\mathcal{E}}_\Lambda \}$$

Notice that, in general, $\mathcal{E}_\Lambda^* \subset \mathcal{E}_{\Lambda^*}$ (the equality holds, for instance, in the case of rectangles). We also define the *boundary* $\delta\Lambda = \{ e^* : e \in \bar{\mathcal{E}}_\Lambda \backslash \mathcal{E}_\Lambda \}$

We will often consider our model on a $(2L+1) \times (2M+1)$ rectangle

$$R_{L,M} = \{ (x_1, x_2) \in \mathbf{Z}^2 : -L \leq x_1 \leq L, \; -M \leq x_2 \leq M \}$$

and thus $R_{L,L}$ coincides with B_L.

Given $V \subset\subset \mathbf{Z}^2$ and some *boundary condition* (b.c.) $\tau \in \Omega$, we consider the generalized Ising hamiltonian

$$H_V^{J,\tau}(\sigma) = -\frac{1}{2} \sum_{\substack{x,y \\ [x,y] \in \mathcal{E}_V}} J(x,y) \big(\sigma(x)\sigma(y) \big) - \sum_{\substack{x,y \\ [x,y]^* \in \delta V}} J(x,y) \big(\sigma(x)\tau(y) \big) \qquad (6.3)$$

The coupling J have been introduced for technical reasons, but our main result is for $J = \beta$, $\beta > \beta_c$ (see section 2). We always assume $0 \leq J(x,y)$ for all x, y.

For further convenience we define the hamiltonian with *free boundary conditions*

$$H_V^{J,\emptyset}(\sigma) = -\frac{1}{2} \sum_{\substack{x,y \\ [x,y] \in \mathcal{E}_V}} J(x,y)\, \sigma(x)\sigma(y) \qquad (6.4)$$

When $J(x,y) = \beta$ for all x, y, we substitute the superscript J with β.

When V is a rectangle $V = R_{L,M}$ and $\eta \in \Omega$, we define the $[k]_\eta$ boundary condition by (let $x = (x_1, x_2)$)

$$[k]_\eta(x) = \begin{cases} \eta(x) & \text{if } x_2 \geq M - k + 1 \\ +1 & \text{if } x_2 \leq M - k \end{cases} \qquad (6.5)$$

so, in particular, $[0]_\eta$ b.c. means η on the top side of the rectangle and $+1$ on the remaining three sides. A rectangle V has a δ–boundary δV consisting of a top, bottom, left and right side, that we denote respectively with $\delta_t V$, $\delta_b V$, $\delta_l V$ and $\delta_r V$. So, for instance

$$\delta_t R_{L,M} = \{ e = [x,y]^* : [x,y] = [(j,M),(j,M+1)] \; j = -L, \ldots, L \}$$

In the following we will often choose $J = \beta$ everywhere with the exception of one or more sides of a certain rectangle where we take $J = \varepsilon\beta$, $\varepsilon \langle 1$. So, we introduce the notation

$$J_\varepsilon^\square(V; x, y) = \beta \begin{cases} \varepsilon & \text{if } [x, y]^* \in \delta\Lambda \\ 1 & \text{otherwise} \end{cases}$$

$$J_\varepsilon^\sqcup(V; x, y) = \beta \begin{cases} \varepsilon & \text{if } [x, y]^* \in \delta\Lambda \backslash \delta_t\Lambda \\ 1 & \text{otherwise} \end{cases}$$

$$J_\varepsilon^\parallel(V; x, y) = \beta \begin{cases} \varepsilon & \text{if } [x, y]^* \in \delta_l\Lambda \cup \delta_r\Lambda \\ 1 & \text{otherwise} \end{cases} \qquad (6.6)$$

$$J_\varepsilon^\sqcap(V; x, y) = \beta \begin{cases} \varepsilon & \text{if } [x, y]^* \in \delta\Lambda \backslash \delta_b\Lambda \\ 1 & \text{otherwise} \end{cases}$$

We are now in a position to state the three main equilibrium estimates that are crucial for our analysis of the spectral gap of the generator with plus or free boundary conditions. The corresponding proofs can be found in [CGMS].

Proposition 6.1. *Let $\beta > \beta_c$ and let $\varepsilon, \alpha \in (0, 1]$. Given a positive integer L we set $M = \lfloor \varepsilon L \rfloor$, $k = \lfloor \varepsilon L/10 \rfloor$. Let $\Lambda = R_{L,M} + (0, h)$ be a vertical translate of $R_{L,M}$ contained in B_L and let $\bar{\Lambda}$ be a vertical translate of $R_{L,N}$ (with $M \leq N < L$) such that the bottom sides of Λ and $\bar{\Lambda}$ coincide. We also let*

$$\Lambda_{\text{bot}} = (x = (x_1, x_2) \in \Lambda : x_2 \leq M - 3k + h)$$

Take $J_\alpha = J_\alpha^\square(B_L)$. Then there exists $L_0 = L_0(\beta, \alpha, \varepsilon)$ and $m = m(\beta, \alpha, \varepsilon)$ such that if $L \geq L_0$

(i) If the horizontal sides of Λ do not touch the horizontal sides of B_L, then

$$\sup_\eta |\mu_\Lambda^{J_\alpha, +}(A) - \mu_\Lambda^{J_\alpha, [0]\eta}(A)| \leq e^{-mL} \qquad \forall A \in \mathcal{F}_{\Lambda_{\text{bot}}} \qquad (6.7)$$

(ii)

$$|\mu_\Lambda^{J_\alpha, +}(A) - \mu_{\bar{\Lambda}}^{J_\alpha, +}(A)| \leq e^{-mL} \qquad \forall A \in \mathcal{F}_{\Lambda_{\text{bot}}} \qquad (6.8)$$

To state the next result we have to introduce the notion of *plus *-chain* for a given configuration σ.

We say that $\{x^1, \ldots, x^n\}$ is a *plus *-chain* for σ if $d(x^{i+1}, x^i) = 1$ and $\sigma(x^i) = +1 \quad \forall i$.

Proposition 6.2. *In the same assumptions as in Proposition 6.1, let*

$$\Lambda_{\text{top}} = \Lambda \backslash \Lambda_{\text{bot}}$$
$$\Lambda_{\text{middle}} = \{x = (x_1, x_2) \in \Lambda : |x_2 - h| \leq 2k\}$$

and consider the event

$$A = \{\sigma \in \Omega_\Lambda : \exists \text{ a plus *-chain } \{x^1, \ldots, x^n\} \subset \Lambda_{\text{top}} \text{ such that } x_1^1 = -L, \; x_1^n = L\}$$

There exists $L_0 = L_0(\beta, \alpha, \varepsilon)$ such that if $L \geq L_0$ then

$$\inf_\eta \mu_\Lambda^{J_\alpha^\sqcup, [0]\eta}(A) \geq e^{-\beta(\tau_\beta + 14\varepsilon)(2L+1)} \qquad (6.9)$$

Moreover, if $\nu_{\Lambda}^{J^{\sqcup}_{\alpha},[0]_{\eta}}$ denotes the conditional measure $\mu_{\Lambda}^{J^{\sqcup}_{\alpha},[0]_{\eta}}(\cdot \,|A)$ then there exists $m = m(\beta, \alpha, \varepsilon)$ such that, for any $L \geq L_0$

$$\sup_{\eta} |\nu_{\Lambda}^{J^{\sqcup}_{\alpha},[0]_{\eta}}(B) - \mu_{\Lambda}^{J^{\sqcup}_{\alpha},+}(B)| \leq e^{-mL} \qquad \forall\, B \in \mathcal{F}_{\Lambda_{\text{bot}}}$$

$$\sup_{\eta} |\nu_{\Lambda}^{J^{\sqcup}_{\alpha},[0]_{\eta}}(B) - \mu_{\Lambda}^{J^{\parallel}_{\alpha},[0]_{\eta}}(B)| \leq e^{-mL} \qquad \forall\, B \in \mathcal{F}_{\Lambda_{\text{middle}}}$$

(6.10)

In order to formulate our last equilibrium bound we define $m_{\Lambda}(\sigma) = \frac{1}{|\Lambda|} \sum_{x \in \Lambda} \sigma(x)$. Then we have (see [CGMS])

Proposition 6.3. Let $\beta > \beta_c$ and $m \in (-m^*(\beta), m^*(\beta))$. Then

$$\liminf_{L \to \infty} -\frac{1}{\beta(2L+1)} \log\left[\mu_{Q_L}^{\beta,\emptyset}\{ m_{Q_L}(\sigma) = m \}\right] \geq \varphi(m)$$

where

$$\varphi(m) = \frac{1}{2} w \sqrt{\frac{m^*(\beta) - (|m| \vee m_1)}{2m^*(\beta)}}$$

and the constant w is the value of the Wulff functional W_τ on the Wulff curve (see e.g. [Pf] or [DKS]). The singularity point m_1 satisfies the equation

$$\frac{1}{2} w \sqrt{\frac{m^*(\beta) - m_1}{2m^*(\beta)}} = \tau_\beta$$

6.2 A Geometric Bound on the Spectral Gap

In this part, following [M], we establish via some geometric ideas (see e.g. [SC] for a review) a basic estimate on the spectral gap in a rectangle $R(l_1, l_2)$ which, besides being of independent interest, will play an important role in the determination of the exact asymptotics in the thermodynamic limit of the spectral gap gap($L_{Q_L}^{\beta,\emptyset}$). Although we believe that our bound can be derived by other means we think that our derivation is a good illustration of how these geometric techniques work in our context. In what follows we will omit for simplicity in all the notation the boundary condition τ, the interaction J and the set Λ.

We start by introducing the set of *canonical paths* in Ω_{Λ} between configurations σ and η with $\sigma \neq \eta$. Let us first order the sites in Λ as follows

$$x < y \quad \text{if} \quad x_2 < y_2 \quad \text{or} \quad x_1 < y_1 \quad \text{and} \quad x_2 = y_2$$

Given now $\sigma, \eta \in \Omega_V$ we define the path $\gamma(\sigma, \eta)$ as the sequence of configurations obtained from σ by adjusting one by one, in increasing order, the values of its spins to those of of η. More precisely, if $x^1, \ldots, x^n$ are the sites in Λ ordered as above and such that $\sigma(x^i) \neq \eta(x^i)$, then $\gamma(\sigma, \eta) = (\sigma^0 \ldots \sigma^n)$ where $\sigma^0 = \sigma$ and for $i = 1, \ldots, n$,

$$\sigma^i(x) = \begin{cases} \eta(x) & \text{if } x \leq x^i \\ \sigma(x) & \text{if } x > x^i \end{cases}$$

(6.11)

Next, for any allowed transition of the Glauber dynamics $\sigma \to \sigma^x$ and any f we set

$$e = (\sigma, \sigma^x), \qquad Q(e) = \mu(\sigma)c(x, \sigma), \qquad f(e) = f(\sigma^x) - f(\sigma) \qquad (6.12)$$

and we say that the transition e belongs to the canonical path γ if, for some i, $e = (\sigma^i, \sigma^{i+1})$. Finally we define the constant ρ_Λ as

$$\rho_\Lambda = \sup_e \sum_{\substack{\sigma, \eta: \\ e \in \gamma(\sigma, \eta)}} \frac{\mu(\sigma)\mu(\eta)}{Q(e)} \qquad (6.13)$$

Then we have the following inequality

$$\text{gap}(L_\Lambda) \geq \frac{1}{|\Lambda|} \frac{1}{\rho_\Lambda} \qquad (6.14)$$

Although the proof of (6.14) can be found in [SC], we reproduce it below because of its simplicity. Write

$$\begin{aligned}
\text{Var}(f) &= \frac{1}{2} \sum_{\sigma, \eta} \mu(\sigma)\mu(\eta)[f(\sigma) - f(\eta)]^2 \\
&= \frac{1}{2} \sum_{\sigma, \eta} \mu(\sigma)\mu(\eta)[\sum_{e \in \gamma(\sigma, \eta)} f(e)]^2 \\
&\leq \frac{1}{2} |\Lambda| \rho_\Lambda \sum_e Q(e) f(e)^2 \\
&= |\Lambda| \rho_\Lambda \mathcal{E}(f, f)
\end{aligned}$$

and (6.14) follows from the variational characterization of the spectral gap (3.12).

With the above notation our result for the the two dimensional Ising model with Hamiltonian (6.3) reads as follows

Theorem 6.4. Let $\Lambda = R_{L,M}$ with $L \leq M$ and assume that for any $[x, y] \in \bar{\mathcal{E}}_\Lambda$ $J(x, y) \leq \beta$. Then, uniformly in the boundary conditions τ,

$$\rho_\Lambda \leq c_m^{-1} e^{4\beta(2L+1)}$$

$$\inf_\tau \text{gap}(L_\Lambda^{J,\tau}) \geq \frac{1}{|\Lambda|} c_m \, e^{-4\beta(2L+1)}$$

where the constant c_m has been defined in (3.3).

Remark. The above estimate on the spectral gap is a very bad one for temperatures above the critical one, *i.e.* $\beta < \beta_c$, since in this case strong mixing holds (see the discussion at the end of section 2.3) and theorem 4.5 applies. At low temperature, instead, when the infinite volume dynamics is not ergodic, it gives the right dependence on the size of the set Λ, namely a negative exponential of the surface and not of the volume. However the constant appearing in the exponential is wrong by a factor 2 even in the limit $\beta \to \infty$. A more precise bound will be discussed in section 6.4.

Proof. The proof is based on a nice idea (see [JS1]) to bound the number of canonical paths that use a given transition e. Given a transition $e = (\xi, \xi^x)$ we define an *injective* mapping Φ from the set of all the canonical paths that use the transition e, $\Gamma(e)$, to Ω_Λ as follows:

$$\Phi(\gamma)(y) = \sigma(y) \quad \text{if} \quad y < x$$
$$\Phi(\gamma)(y) = \eta(y) \quad \text{if} \quad y \geq x$$

where σ and η denote the starting and end point of γ. It is clear that Φ is iniective. In fact the knowledege of the transition $e = (\xi, \xi^x)$ together with $\Phi(\gamma)$ allow us to reconstruct completely the initial and final configurations σ and η and thus the path itself, simply because

$$\sigma(y) = \xi(y) \qquad \forall \, y \geq x$$
$$\sigma(y) = \Phi(\gamma)(y) \qquad \forall \, y < x$$

and similarly for η.

Let now c_0 be the smallest constant such that for any transition e and any canonical path $\gamma \in \Gamma(e)$, $\gamma \equiv \gamma(\sigma, \eta)$,

$$\mu(\Phi(\gamma)) \, Q(e) \geq \frac{1}{c_0} \mu(\sigma)\mu(\eta) \tag{6.15}$$

Then we have

$$\rho_\Lambda \leq c_0 \tag{6.16}$$

Using (6.15) we can in fact estimate the r.h.s. of (6.13) by:

$$c_0 \sup_e \sum_{\gamma \in \Gamma(e)} \mu(\Phi(\gamma))$$

Since the map Φ is injective and μ is a probability measure, the above sum is not greater than one and (6.16) follows.

In order to estimate the constant c_0, let $e = (\xi, \xi^x)$ and $\gamma = \gamma(\sigma, \eta) \in \Gamma(e)$ be given. Then, by direct inspection

$$|H(\sigma) + H(\eta) - H(\xi) - H(\Phi(\gamma))| \leq \sup_{[x,y] \in \mathcal{E}_\Lambda} J(x,y) \, 4(2L+1) \leq 4\beta(2L+1)$$

so that

$$\frac{\mu(\sigma)\mu(\eta)}{\mu(\Phi(\gamma))Q(e)} = \frac{\mu(\sigma)\mu(\eta)}{\mu(\Phi(\gamma))\mu(\xi)c(\xi,x)} \leq c_m^{-1} e^{4\beta(2L+1)}$$

Thus the constant c_0 can be taken equal to

$$c_0 = c_m^{-1} e^{4\beta(2L+1)}$$

and the theorem follows. $\square$

Remark. It is amusing to observe that, if one applies the above construction to the one dimensional case, one gets $\sup_\Lambda \rho_\Lambda < \infty$ even if the energy $H(\sigma)$ is replaced by a more general expression like

$$H(\sigma) = -\frac{1}{2} \sum_{x,y \in \Lambda} J(|x-y|)\sigma(x)\sigma(y)$$

provided that $\sum_{\substack{x \leq 0 \\ y > 0}} |J(|x-y|)| < \infty$. Therefore in this case the spectral gap in a segment of length L in $\mathbf{Z}$ has a lower bound which is only proportional to L^{-1} without any negative exponential of L.

On the other hand it is known that a long range potential $J(|x-y|)$ satisfying the above condition is not able to induce any phase transition, the reason being that the energy between two semi-infinite lines is finite uniformly in the spin configuration. Thus, in some sense, the geometric approach is able to capture, at least at the level of the exponential, some (but certainly not all) of the physical aspects of the presence (absence) of a phase transition in the Ising model.

6.3 A Lower Bound on the Spectral Gap with + B.C.

Here we prove the bound (6.1) for any $\beta > \beta_c$ in a slightly more general case than the one discussed at the beginning of the section, namely when the b.c. are plus and the boundary coupling is weak or zero on three sides of the square B_L and strong on one. More precisely

Theorem 6.5. *Let* $\beta > \beta_c$, $\delta \in [0,1]$ *and let* $J_\delta \equiv J_\delta^\Gamma(B_L)$. *Then*

$$\lim_{L \to \infty} -\frac{1}{L} \log(\mathrm{gap}(L_\Lambda^{J_\delta,+})) = 0$$

Proof. Given $\varepsilon \in (0,1)$ and $\delta \in [0,1]$ let $\alpha = \varepsilon \vee \delta$. It easily follows from a rough estimate on the relative density of the two Gibbs measures $\mu_\Lambda^{J_\delta,+}$ and $\mu_\Lambda^{J_\alpha,+}$ that

$$\mathrm{gap}(L_\Lambda^{J_\delta,+}) \geq \frac{c_m}{c_M} e^{-C\varepsilon L} \, \mathrm{gap}(L_\Lambda^{J_\alpha,+})$$

for a suitable constant C.

Let now $\mathcal{D}_\varepsilon$ be the covering of the square $\Lambda \equiv B_L$ with thin horizontal rectangles $R_1, \ldots, R_n$, where

$$R_1 = \{ x \in \Lambda; \; -L \leq x_2 \leq -L + \lfloor \varepsilon L \rfloor \}$$

and each R^i, $i = 2 \ldots n$ is a vertical translate of the previous one by an amount $\frac{\lfloor \varepsilon L \rfloor}{2}$. Thanks to (3.30) and theorem 6.4, we have

$$\mathrm{gap}(L_\Lambda^{J_\alpha,+}) \geq \frac{1}{2} \inf_{i,\tau} \mathrm{gap}(L_{R_i}^{J_\alpha,\tau}) \, \mathrm{gap}(L_{\mathcal{D}_\varepsilon}^{J_\alpha,+})$$

$$\geq \frac{1}{4} L^{-2} c_m \, e^{-4\beta\varepsilon L} \, \mathrm{gap}(L_{\mathcal{D}_\varepsilon}^{J_\alpha,+})$$

and therefore, due to the arbitrariness of ε, in order to prove (6.1), it suffices the following key result

Lemma 6.6. *For any $\varepsilon, \alpha \in (0, 1]$ and any $\beta > \beta_c$*

$$\inf_L \operatorname{gap}(L_{\mathcal{D}_\varepsilon}^{J_\alpha, +}) > 0 \qquad (6.17)$$

Proof. The lemma follows immediately if we can show that, in the above range of parameters, there exists a finite time t_0 and a number $r \in (0, 1)$ such that for any large enough L the semigroup generated by $L_{\mathcal{D}_\varepsilon}^{J_\alpha, +}$ at time $t = t_0$ is a contraction in the sup norm, with norm less than $1 - r$. In more probabilistic terms if

$$\sup_\eta |\mathbb{E}^{\mathcal{D}_\varepsilon} f(\sigma_{t=t_0}^\eta)| \leq (1 - r)\|f\|_\infty \qquad (6.18)$$

for any f such that $\mu_\Lambda^{J_\alpha, +}(f) = 0$. Here $\mathbb{E}^{\mathcal{D}_\varepsilon}(f(\sigma_t^\eta))$ denotes the average of f over the block dynamics at time t starting from η.

In the present case, namely full plus b.c., we choose $t_0 = 1$. Let then $\{t_i\}$ be the random times at which the initial configuration σ is updated. Since the number n of rectangles depends only on ε and not on L, (6.18) follows if we show that there exists a number $\delta \in (0, 1)$ such that, for any L large enough:

$$\sup_\eta |\mathbb{E}^{\mathcal{D}_\varepsilon}(f(\sigma_{t=1}^\eta)|t_n \leq 1 < t_{n+1})| \leq (1 - \delta)\|f\|_\infty \qquad (6.19)$$

for any f such that $\mu_\Lambda^{J_\alpha, +}(f) = 0$. We will now concentrate on the proof of (6.19). Notice that, because of the definition of the block-dynamics, the following "multiple integral" formula holds

$$\mathbb{E}^{\mathcal{D}_\varepsilon}(f(\sigma_{t=1}^\eta)|t_n \leq 1 < t_{n+1}) =$$

$$\sum_{i_1 \ldots i_n \in \{1 \ldots n\}} \frac{1}{n^n} \int d\mu_{R_{i_1}}^{J_\alpha, \eta}(\sigma_1) \int d\mu_{R_{i_2}}^{J_\alpha, \sigma_1}(\sigma_2) \ldots \int d\mu_{R_{i_n}}^{J_\alpha, \sigma_{n-1}}(\sigma_n) f(\sigma_n) \qquad (6.20)$$

where the factor $\frac{1}{n^n}$ stands for the probability that during the first n updatings the rectangles $R_{i_1} \ldots R_{i_n}$ are chosen in the given order. Notice that in the above formula we did not specify explicitly that each rectangle R_i has plus boundary conditions on those sites that are not in Λ. In other words each configuration σ_i is identically equal to $+1$ outside Λ.

In turn, in order to prove (6.19), it is sufficient to show that there exist $\bar\delta$ such that, for all large enough L,

$$\sup_\eta |\int d\mu_{R_1}^{J_\alpha, \eta}(\sigma_1) \int d\mu_{R_2}^{J_\alpha, \sigma_1}(\sigma_2) \ldots \int d\mu_{R_{i_n}}^{J_\alpha, \sigma_{n-1}} f(\sigma_n)| \leq (1 - \bar\delta)\|f\|_\infty \qquad (6.21)$$

Notice that in the above formula each configuration σ_i, $i = 1 \ldots n$, is equal to the initial condition η above the top side of R_i.

For later purposes it is useful at this point to write down the expression of the constant r of (6.18) in terms of $\bar\delta$. A simple computation gives

$$r = \mathbb{P}(t_n \leq 1 < t_{n+1})n^{-n}\bar\delta \qquad (6.22)$$

Let us finally prove (6.21). Define recursively

$$g_j(\sigma) = \int d\mu_{R_{j+1}}^{J_\alpha,\sigma}(\sigma_{j+1}) \ldots \int d\mu_{R_n}^{J_\alpha,\sigma_{n-1}} f(\sigma_n) \qquad j = 1 \ldots n-1$$

and observe that

i) $\|g_j\|_\infty \leq \|f\|_\infty$.

ii) g_j depends only on the spins $\sigma(x)$ with $x \in \left(\cup_{k=1}^j R_k \right) \setminus R_{j+1}$

Because of i), ii) above, we can write, using (6.7) and (6.8) (remember that on the bottom and lateral sides of R_1 the b.c. are $+$)

$$\sup_\eta \left| \int d\mu_{R_1}^{J_\alpha,[0]\eta}(\sigma_1) \int d\mu_{R_2}^{J_\alpha,\sigma_1}(\sigma_2) \ldots \int d\mu_{R_n}^{J_\alpha,\sigma_{n-1}} f(\sigma_n) \right| =$$

$$= \sup_\eta |\mu_{R_1}^{J_\alpha,[0]\eta}(g_1)|$$

$$\leq \sup_\eta \{ |\mu_{R_1}^{J_\alpha,[0]\eta}(g_1) - \mu_{R_1 \cup R_2}^{J_\alpha,[0]\eta}(g_1)| + |\mu_{R_1 \cup R_2}^{J_\alpha,[0]\eta}(g_1)| \} \qquad (6.23)$$

$$\leq \|f\|_\infty e^{-m\varepsilon L} + \sup_\eta |\mu_{R_1 \cup R_2}^{J_\alpha,[0]\eta}(g_2)|$$

for a suitable $m = m(\varepsilon)$ and any L large enough. In the last step we used the DLR equation (2.2) to write

$$\mu_{R_1 \cup R_2}^{J_\alpha,[0]\eta}(g_1) = \mu_{R_1 \cup R_2}^{J_\alpha,[0]\eta}(g_2)$$

If we iterate $n-1$ times the above reasoning and use the fact that $\mu_\Lambda^{J_\alpha,+}(f) = 0$, we get

$$\sup_\eta \left| \int d\mu_{R_1}^{J_\alpha,\eta}(\sigma_1) \int d\mu_{R_2}^{J_\alpha,\sigma_1}(\sigma_2) \ldots \int d\mu_{R_n}^{J_\alpha,\sigma_{n-1}} f(\sigma_n) \right| \leq n e^{-m\varepsilon L} \|f\|_\infty$$

Thus (6.21) follows with $\bar\delta = 1 - n e^{-m\varepsilon L}$ and the lemma is proved. $\square$

Remark. It is interesting to notice that our lower bound on the spectral gap with plus boundary conditions on a square $\Lambda = Q_L$ allows one to estimate from below the probability of events like $|m_\Lambda(\sigma) - m| \leq 2/|\Lambda|$, $m < m^*(\beta)$, in terms of the probability of the event $m_\Lambda(s) \leq m$, with a negligible error in the leading exponential behaviour.

Let us in fact define $f(\sigma) = \chi(m_\Lambda(\sigma) > m)$. Then we have

$$\text{Var}_\Lambda^+(f) = \mu_\Lambda^+(m_\Lambda(\sigma) > m)\, \mu_\Lambda^+(m_\Lambda(\sigma) \leq m)$$

$$\mathcal{E}_\Lambda^+(f,f) \leq c_M |\Lambda| \mu_\Lambda^+(|m_\Lambda(\sigma) - m| \leq \frac{2}{|\Lambda|})$$

and therefore, by the Poincaré inequality $\mathcal{E}_\Lambda^+(f,f) \geq \text{gap}(L_\Lambda^+)\,\text{Var}_\Lambda^+(f)$, we get

$$\mu_\Lambda^+(|m_\Lambda(\sigma) - m| \leq \frac{2}{|\Lambda|}) \geq \frac{1}{c_M |\Lambda|} \text{gap}(L_\Lambda^+)\mu_\Lambda^+(m_\Lambda(\sigma) > m)\mu_\Lambda^+(m_\Lambda(\sigma) \leq m)$$

If we finally use our lower bound on $\text{gap}(L_\Lambda^+)$ together with the observation that $\lim_{L\to\infty}\mu_\Lambda^+(m_\Lambda(\sigma) > m) = 1$ for any $m < m^*(\beta)$ and any $\beta > \beta_c$ (see e.g. [Io2]), we get that for any $\delta > 0$ and any L large enough

$$\mu_\Lambda^+(\,|\,m_\Lambda(\sigma) - m\,| \le \frac{2}{|\Lambda|}) \ge e^{-\delta L}\mu_\Lambda^+(m_\Lambda(\sigma) \le m)$$

6.4 A Lower Bound on the Spectral Gap with Free B.C.

Here we show that, for any $\beta > \beta_c$,

$$\lim_{L\to\infty} -\frac{1}{\beta(2L+1)}\log\big(\text{gap}(L_{B_L}^{\beta,\emptyset})\big) \le \tau_\beta \tag{6.24}$$

To prove (6.24) we first replace the free b.c. with weak plus b.c. More precisely we fix a small number $\varepsilon \in (0,1)$, that will be put equal to zero *after* the limit $L \to \infty$, and bound from below $\text{gap}(L_{B_L}^{\beta,\emptyset})$ by

$$\text{gap}(L_{B_L}^{\beta,\emptyset}) \ge \frac{c_m}{c_M}e^{-C\varepsilon L}\,\text{gap}(L_{B_L}^{J_\varepsilon^\square,+})$$

where C is a numerical constant. Such a bound follows in a straightforward way from the very definition of the spectral gap. Then we proceed exactly as in the proof of the lower bound with full plus b.c. and reduce the problem to bound from below the gap of the block dynamics associated with the covering $\mathcal{D}_\varepsilon$ and coupling J_ε. Notice that, for simplicity, the small constant α giving the boundary couplings is the same as that fixing the height of the rectangles of the covering $\mathcal{D}_\varepsilon$. As in the case of full plus b.c. this amount to choose a finite time t_0 and bound from above the contraction constant r in (6.18). However, contrary to the case of plus b.c., if we choose t_0 independent of L here we cannot expect that the constant r stays bounded away from zero uniformly in the side L, since we know (see section 6.5 below) that $\text{gap}(L_{B_L}^{\beta,\emptyset})$ shrinks to zero as $L \to \infty$ much more rapidly than $e^{-C\varepsilon L}$, at least if ε is small. Thus we have to proceed slightly more carefully than before.
First of all we fix $t_0 = 2$ and observe that

$$\text{gap}(L^{\mathcal{D}_\varepsilon}) \ge -\frac{1}{2}\log(1-r) \ge \frac{1}{2}r$$

Next we define $\mathcal{S}$ to be the event that up to time $t = 1$ only the first rectangle R_1 has been updated and, for any function f, we set

$$f_1(\eta) \equiv p\,\mathbb{E}^{\mathcal{D}_\varepsilon}\big(f(\sigma_{t=1}^\eta)|\mathcal{S}\big) = p\mu_{R_1}^{J_\varepsilon,[0]\eta}(f)$$

where $p = \frac{1}{n}\mathbb{P}(t_1 < 1 \le t_2)$ and notice that
  i) f_1 does not depend on the spins in the first rectangle R_1
 ii) $\|f_1\|_\infty \le p\|f\|_\infty$
iii) $\mu_\Lambda^{J_\varepsilon,+}(f_1) = 0$ iff $\mu_\Lambda^{J_\varepsilon,+}(f) = 0$
Moreover

$$\sup_\eta |\mathbb{E}^{\mathcal{D}_\epsilon} f(\sigma_{t=2}^\eta)| \leq (1-p)\|f\|_\infty + \sup_\eta |\mathbb{E}^{\mathcal{D}_\epsilon} f_1(\sigma_{t=1}^\eta)|$$

$$\leq (1-pr_1)\|f\|_\infty$$

where r_1 is the smallest constant such that

$$\sup_\eta |\mathbb{E}^{\mathcal{D}_\epsilon} f(\sigma_{t=1}^\eta)| \leq (1-r_1)\|f\|_\infty \tag{6.25}$$

for all f that do not depend on the spins in R_1 and such that $\mu_\Lambda^{J_\epsilon,+}(f) = 0$. Thus $r \geq p\,r_1$ and it is enough to prove

$$\lim_{L\to\infty} -\frac{1}{\beta(2L+1)} \log(r_1) \leq \tau_\beta \tag{6.26}$$

In order to establish (6.26) we proceed exactly as in the case of full plus b.c. and we bound from below the constant $\bar\delta$ defined in (6.21). Using the same notation of the proof of (6.21) and remembering that on the bottom and lateral sides of the first rectangle we have plus b.c., we write, thanks to proposition 6.1, 6.2

$$\sup_\eta |\mu_{R_1}^{J_\epsilon^\sqcup,[0]\eta}(g_1)|$$

$$\leq \sup_\eta \left\{ \mu_{R_1}^{J_\epsilon^\sqcup,[0]\eta}(A)|\mu_{R_1}^{J_\epsilon^\sqcup,[0]\eta}(g_1\,|A)| + \mu_{R_1}^{J_\epsilon^\sqcup,[0]\eta}(A^c)\|f\|_\infty \right\}$$

$$\leq \sup_\eta \left\{ \mu_{R_1}^{J_\epsilon^\sqcup,[0]\eta}(A)\left[|\mu_{R_1}^{J_\epsilon^\sqcup,[0]\eta}(g_1\,|A) - \mu_{R_1\cup R_2}^{J_\epsilon^\|,[0]\eta}(g_1)| + |\mu_{R_1\cup R_2}^{J_\epsilon^\|,[0]\eta}(g_1)|\right] \right\}$$

$$+ \sup_\eta \mu_{R_1}^{J_\epsilon^\sqcup,[0]\eta}(A^c)\|f\|_\infty$$

$$\leq \sup_\eta \left\{ \mu_{R_1}^{J_\epsilon^\sqcup,[0]\eta}(A)\left[e^{-m\epsilon L}\|f\|_\infty + |\mu_{R_1\cup R_2}^{J_\epsilon^\|,[0]\eta}(g_1)|\right] + \mu_{R_1}^{J_\epsilon^\sqcup,[0]\eta}(A^c)\|f\|_\infty \right\} \tag{6.27}$$

where A is the event appearing in proposition 6.2. It is important to observe that we have been able to use proposition 6.2 to bound from above the error term $|\mu_{R_1}^{J_\epsilon^\sqcup,[0]}(g_1\,|A) - \mu_{R_1\cup R_2}^{J_\epsilon^\|,[0]}(g_1)|$ because, by construction, the function g_1 does depend only on the spins $\sigma_1(x)$ with $x \in (R_1)_{\text{middle}}$

It is important to understand the reasoning that led us to introduce the event A. Let us suppose that the initial configuration η is identically eqaul to -1. In this case the typical configurations of the Gibbs measure $\mu_{R_1}^{J_\epsilon^\sqcup,[0]\eta}$ consists of a sea of minus spins with a thin layer of plus spins attached to the lateral and bottom boundary and small islands of pluses in the bulk. Thus the first updating will not modify substantially the initial configuration and similarly for the successive updatings. One possibility to drastically change the initial configuration and make a transition to configurations where now the majority of the spins are $+1$ is via a large deviation in the first updating. The event A not only does the job but, apparently, is also an efficient way to make the transition in the sense that the lower bound one gets on the spectral gap is of the same order of the upper bound of the next section.

Let us now bound the remaining terms in the r.h.s. of (6.27). Notice also that the Gibbs measure $\mu_{R_1\cup R_2}^{J_\epsilon^\|,[0]\eta}$ has η b.c. on his top side, plus b.c. with boundary coupling β on its bottom side and weak plus b.c. on its lateral sides. Thus we can write

$$\mu_{R_1 \cup R_2}^{J_\epsilon^\parallel,[0]_\eta}(g_1) = \mu_{R_1 \cup R_2}^{J_\epsilon^\parallel,[0]_\eta}(g_2)$$

To estimate the remaining term $\mu_{R_1 \cup R_2}^{J_\epsilon^\parallel,[0]_\eta}(g_2)$ we can at this point proceed exactly as in the case with full b.c. since we have them on the bottom side. The final result is

$$|\mu_{R_1 \cup R_2}^{J_\epsilon^\parallel,[0]_\eta}(g_2)| \le n e^{-m\epsilon L}\|f\|_\infty + |\mu_\Lambda^{J_\epsilon^\sqcap,+}(f) - \mu_\Lambda^{J_\epsilon^\square,+}(f)| \tag{6.28}$$

To estimate the last term in the r.h.s. of (6.28) we observe that, since f does not depend on the spins in R_1, we can apply once more proposition 6.1 and get

$$|\mu_\Lambda^{J_\epsilon^\sqcap,+}(f) - \mu_\Lambda^{J_\epsilon^\square,+}(f)| \le e^{-m\epsilon L}\|f\|_\infty \tag{6.29}$$

In conclusion we have obtained that

$$\sup_\eta |\mu_{R_1}^{J_\epsilon^\sqcup,[0]_\eta}(g_1)| \le \sup_\eta \Big\{ \mu_{R_1}^{J_\epsilon^\sqcup,[0]_\eta}(A)\, 2^n\, e^{-m\epsilon L}\|f\|_\infty + \mu_{R_1}^{J_\epsilon^\sqcup,[0]_\eta}(A^c)\|f\|_\infty \Big\}$$

$$\le \big(1 - e^{-\beta\tau_\beta(1+14\epsilon)(2L+1)}(1 - 2^n e^{-m(\epsilon)\epsilon L})\big)\|f\|_\infty$$

for any L large enough, because of proposition 6.2. Thus the constant $\bar\delta$ in this new case can be bounded from below by $\frac{1}{2}e^{-\beta\tau_\beta(1+14\epsilon)(2L+1)}$. In turn, using (6.22) and the arbitrariness of the constant ϵ, the above lower bound on $\bar\delta$ immediately implies (6.26). $\square$

6.5. Upper Bound on the Spectral Gap with Free B.C.

Here we show that, for the two dimensional Ising model

$$\lim_{L\to\infty} -\frac{1}{\beta(2L+1)} \log\big(\text{gap}(L_{B_L}^{\beta,\emptyset})\big) \ge \tau_\beta \tag{6.30}$$

At the end of this paragraph we will also discuss an extension to higher dimensions. Following [M], let $\Lambda = B_L$ and let $f_\Lambda(\sigma)$ be the trial function

$$f_\Lambda(\sigma) = \chi\{m_\Lambda(\sigma) > 0\} - \chi\{m_\Lambda(\sigma) < 0\} \tag{6.31}$$

If we plug f_Λ in the variational characterization of the gap (3.12) we get

$$\text{gap}(L_\Lambda^{\beta,\emptyset}) \le 4\,c_M\,|\Lambda|\, \frac{\mu_\Lambda^{\beta,\emptyset}\{|m_\Lambda(\sigma)| \le \frac{2}{|\Lambda|}\}}{1 - \mu_\Lambda^{\beta,\emptyset}\{m_\Lambda(\sigma) = 0\}} \tag{6.32}$$

Therefore

$$\lim_{L\to\infty} -\frac{1}{\beta(2L+1)} \log\big(\text{gap}(L_\Lambda^{\beta,\emptyset})\big) \ge \lim_{L\to\infty} -\frac{1}{\beta(2L+1)} \log\Big(\mu_\Lambda^{\beta,\emptyset}\{|m_\Lambda(\sigma)| \le \frac{2}{|\Lambda|}\}\Big)$$

$$= \tau_\beta$$

because of proposition 6.3. $\square$

Remark Notice that up to (6.32) our argument was completely dimension indepen-dent. Unfortunately, in dimensions higher than two there is yet no precise bound of the probability that the magnetization is close to zero. However there is a bound (see [P]) that says that, for any $d \geq 2$, there exists $\beta_0 < +\infty$ such that for any $\beta > \beta_0$ the r.h.s of (6.32) is exponentially small in L^{d-1}, *i.e.* the surface of the cube Λ. Moreover there are good chances that β_0 coincides with the critical temperature. Alternatively one could use the result of [Th]. If we combine such bounds with the very general lower bound of theorem 3.8 we get that for any $d \geq 2$ and low enough temperature

$$\lim_{L \to \infty} \frac{1}{\log L} \log\big(-\log(\text{gap}(L_{Q_L}^{\beta,\emptyset}))\big) = d - 1$$

6.6 Mixed B.C.

We would like to conclude our analysis of the spectral gap of the two dimensional stochastic Ising model at low temperature by discussing the case of "mixed" b.c., *i.e.* b.c. that do not have a definite sign.

Remember that, when the b.c. where free *i.e.* absent, the spectral gap was very small because of the presence of a very narrow "bottleneck" in the phase space, while for plus b.c. the spectral gap was much larger because the bottleneck is either absent or much wider. This phenomenon can be associated to the fact that in the case of free b.c. the two phases are equally likely while in the case of plus b.c. the plus phase is chosen by the b.c.

It is then natural to ask whether it is possible to have intermediate situations in which a bottleneck for the Glauber dynamics is still present but nevertheless the b.c. are able to choose the phase. The main result in this direction is the following one (see [HiYo]).

Given the square Q_L and a configuration $\tau \in \Omega$, define

$$v_L(\tau) = \max\{|\sum_{x \in A} \tau(x)|; \ A \text{ is a connected subset of } \partial_{\text{ex}} Q_L\}$$

where $\partial_{\text{ex}} Q_L = \{x \in Q_L^c; \ d_2(x, Q_L) = 1\}$. Then we have

Theorem 6.7. *Let* $d = 2$ *and define a class* $\Omega_0(\vartheta)$ $(0 \leq \vartheta \leq 1)$ *of boundary conditions by*

$$\Omega_0(\vartheta) = \{\tau \in \Omega; \limsup_{L \to \infty} \frac{V_L(\tau)}{4L} \leq \vartheta\}$$

Then, for every $\vartheta < \bar{\vartheta} < \frac{1}{8}$ *and* $\alpha \in]0, 1 - 8\bar{\vartheta}[$, *there exist* $\beta_0 > 0$ *such that for every* $\tau \in \Omega_0(\vartheta)$ *and* $\beta \geq \beta_0$

$$\text{gap}(L_{Q_L}^{\beta,\tau}) \leq c(\beta)e^{-\beta\alpha L}$$

for a suitable constant $c(\beta)$ *independent of* τ.

6.7 Applications

Here we discuss some additional results and applications that follow from the spec-tral gap estimates established so far.

Let us first consider the standard two dimensional Ising model at inverse temperature $\beta > \beta_c$ and zero external field h. We denote by μ_+^β denote the so called "plus phase", *i.e.* the infinite volume Gibbs measure obtained as thermodynamic limit of the finite volume Ising Gibbs measure with plus b.c., and by $\| \ \|_{2,\mu_+^\beta}$ the L^2–norm w.r.t μ_+^β. Let also $T^\beta(t)$ be the Markov semigroup of an attractive Glauber dynamics (e.g. the heat-bath dynamics) for the above model. Then we have

Theorem 6.8. *Let $\alpha \in [0, \infty)$ be given. Then there exists $c < +\infty$ such that for any local function f*

$$\|T^\beta(t)f - \mu_+^\beta(f)\|_{2,\mu_+^\beta} \leq \frac{c}{t^\alpha}\|\|f\|\| \quad \forall \, t \geq 1$$

Proof. Using the reversibility of the semigroup $T^\beta(t)$ w.r.t the Gibbs measure μ_+^β we write

$$\|T^\beta(t)f - \mu_+^\beta(f)\|_{2,\mu_+^\beta}^2$$
$$= \int d\mu_+^\beta(\eta)\left\{ \int d\mu_+^\beta(\eta')[T^\beta(t)f(\eta) - T^\beta(t)f(\eta')] \right\}^2$$
$$\leq \int d\mu_+^\beta(\eta) \int d\mu_+^\beta(\eta')[T^\beta(t)f(\eta) - T^\beta(t)f(\eta')]^2$$
$$= \int d\mu_+^\beta(\eta) \int d\mu_+^\beta(\eta')[\mathbb{E}(f(\sigma_t^{\beta,\eta}) - f(\sigma_t^{\beta,\eta'}))]^2$$
$$\leq \|\|f\|\| \sum_x \|\nabla_x f\|_\infty \int d\mu_+^\beta(\eta) \int d\mu_+^\beta(\eta') \, \mathbb{P}(\sigma_t^{\beta,\eta}(x) \neq \sigma_t^{\beta,\eta'}(x))$$
$$= \|\|f\|\|^2 \int d\mu_+^\beta(\eta) \int d\mu_+^\beta(\eta') \, \mathbb{P}(\sigma_t^{\beta,\eta}(0) \neq \sigma_t^{\beta,\eta'}(0))$$
$$\leq 2\|\|f\|\|^2 \int d\mu_+^\beta(\eta) \, \mathbb{P}(\sigma_t^{\beta,\eta}(0) \neq \sigma_t^{\beta,+}(0))$$

$$(6.33)$$

where we have used the standard inequality

$$|f(\sigma) - f(\eta)| \leq \sum_x \|\nabla_x f\|_\infty \chi(\sigma(x) \neq \eta(x))$$

together with translation invariance.

Let us now bound the factor $\int d\mu_+^\beta(\eta) \, \mathbb{P}(\sigma_t^{\beta,\eta}(0) \neq \sigma_t^{\beta,+}(0))$ in the r.h.s of (6.33). Using monotonicity together with reversibility and denoting by $m^*(\beta)$ the spontaneous magnetization, $m^*(\beta) = \mu_+^\beta(\sigma(0))$, we have

$$\int d\mu_+^\beta(\eta)\, \mathbf{P}(\sigma_t^{\beta,\eta}(0) \neq \sigma_t^{\beta,+}(0)) =$$

$$= \int d\mu_+^\beta(\eta)\Big[\mathbf{P}(\sigma_t^{\beta,+}(0) = +1) - \mathbf{P}(\sigma_t^{\beta,\eta}(0) = +1)\Big]$$

$$\leq \mathbf{P}(\sigma_t^{B_L,\beta,+,+}(0) = +1) - \frac{1}{2}(m^*(\beta)+1) \tag{6.34}$$

$$= \frac{1}{2}\Big\{\mathbf{E}(\sigma_t^{B_L,\beta,+,+}(0)) - m^*(\beta)\Big\}$$

where $\sigma_t^{B_L,\beta,+,+}$ denotes the dynamics in B_L with plus boundary conditions and starting from the configuration identically equal to plus one. Thanks to (3.15),

$$\mathbf{E}(\sigma_t^{B_L,\beta,+,+}(0)) \leq \mu_{B_L}^{\beta,+}(\eta(0)) + e^{2\beta L^2 - t\,\mathrm{gap}(L_{B_L}^{\beta,+})} \tag{6.35}$$

If we plug (6.35) into (6.34) we obtain that the r.h.s. of (6.34) is bounded from above by:

$$\frac{1}{2}\big(\mu_{B_L}^{\beta,+}(\eta(0)) - m^*(\beta)\big) + e^{2\beta L^2 - t\,\mathrm{gap}(L_{B_L}^{\beta,+})} \leq \frac{1}{2}C_1 e^{-mL} + e^{2\beta L^2 - t\,\mathrm{gap}(L_{B_L}^{\beta,+})} \tag{6.36}$$

because of the well known bound

$$0 \leq \mu_{B_L}^{\beta,+}(\eta(0)) - m^*(\beta) \leq C_1 \exp(-mL) \tag{6.37}$$

valid for any $\beta > \beta_c$ and suitable constants C_1, m. We now choose the size L depending on t as:

$$L = 2\frac{\alpha \log(t)}{m} \tag{6.38}$$

and apply (6.1) to (6.36). We get that, for any large enough t, the r.h.s of (6.36) is bounded from above by $3C_1 t^{-2\alpha}$. If we finally plug this bound in the r.h.s of (6.33) we get the sought polynomial decay. $\square$

The above result as an interesting consequence in terms of coercive inequality for the Glauber dynamics in the plus phase. Denote by $\mathcal{E}_+^\beta(f,f)$ and $\mathrm{Var}_+^\beta(f)$ the Dirichlet form and variance of a local function f w.r.t the Gibbs measure μ_+^β. Then

Corollary 6.9. *In the same hypotheses of theorem 6.8, for any $\delta > 0$ there exists a finite constant c such that for any local function f*

$$\mathrm{Var}_+^\beta(f)^{1+\delta} \leq c\,\mathcal{E}_+^\beta(f,f)\,\|f\|^{2\delta}$$

Proof. Given a local function f with $\mu_+^\beta(f) = 0$, let $d\rho_f(\lambda)$ be its spectral measure w.r.t the selfadjoint operator $-L^\beta$ in $L^2(\Omega, \mu_+^\beta)$. Then, for any $\delta > 0$ and $t > 0$, we have

$$\mathrm{Var}^\beta_+(f) = \int_0^\infty d\rho_f(\lambda)$$

$$\leq e \int_0^\infty d\rho_f(\lambda)e^{-\lambda t} + t \int_0^\infty d\rho_f(\lambda)\lambda$$

$$= e\|T^\beta(t)f\|^2_{2,\mu^\beta_+} + t\mathcal{E}^\beta_+(f,f) \tag{6.39}$$

$$\leq c(\delta)\frac{\|\|f\|\|^2}{t^{1/\delta}} + t\mathcal{E}^\beta_+(f,f)$$

because of theorem 6.8. It suffices now to optimize the choice of t to get the final result. □

As a next topic we consider in more details the dynamics in a finite square Q_L with free b.c. and large β. For notation convenience, in what follows σ^η_t will always denote $\sigma^{Q_L,\beta,\eta,\emptyset}_t$.

As we said at the beginning of the section, the slow relaxation process to equilibrium is due to the presence of a bottleneck in the configuration space that somehow separates configuration with positive magnetization from those with negative magnetization. It is however clear that if the flip rates are invariant under a global spin flip $\sigma \to -\sigma$ and f is function that is even w.r.t such a sign inversion, then the relaxational behaviour of $f(\sigma^\eta_t)$ should not be affected by the presence of the bottleneck. In particular one would expect that functions that are *even* under global spin flip, i.e. they do not distinguish between the two phases, will relax to their equilibrium value in a time much shorter than the global relaxation time T_{spin} given by the inverse of the spectral gap. More precisely, if we denote by T_{contour} the inverse of the spectral gap in the spectrum of the generator $L^{\beta,\emptyset}_{Q_L}$ restricted to the invariant subspace of functions that are even with respect to a global spin flip, we expect that $T_{\mathrm{contour}} << T_{\mathrm{spin}}$. Since even functions can be thougth of as functions of the broken lines in the dual lattice separating regions of plus spins from regions of minus spins (Peierls contours), one could say that T_{contour} characterizes the relaxation process as $t \to \infty$ of the probability distribution of the Peierls contours generated by the dynamics at time t.

The main result of [MM] concerning this new time scale goes as follows

Theorem 6.10. *There exists a positive constant β_o such that for any $\beta \geq \beta_o$*

$$\lim_{L\to\infty} \frac{1}{L} \log(\frac{T_{\mathrm{spin}}}{T_{\mathrm{contour}}}) > 0$$

Using the above result one can draw some interesting consequences that make the picture found in [M] more precise. The first one says that, under the dynamics, any initial configuration relaxes to one of the two phases within a time scale of the order of $L^2 T_{\mathrm{contour}}$ much shorter than T_{spin}. More precisely we have (see theorem 3.1 in [MM])

Theorem 6.11. Let $t_o \equiv 10\beta L^2 T_{\text{contour}}$. There exist positive constants β_o, L_o such that for any $\beta \geq \beta_o$ and any $L \geq L_o$

$$\sup_\eta \mathbb{P}(\sigma_t^+ \neq \sigma_t^\eta \neq \sigma_t^-) \leq e^{-[t/t_o]} \quad \forall t \geq t_o$$

The second one (see theorem 3.2 in [MM]) says that, once the system decides to jump from one phase to the opposite one, then, with large probability, it does it on a time scale not larger than $L^3 T_{\text{contour}}$, again much shorter than the average time one has to wait in order to see the jump. One could say that in our case the Glauber dynamics has a behavior similar, in some sense, to that of a finite dimensional reversible Markov processes with invariant measure having a symmetric double well structure in the low noise regime.

In order to formulate the result, let us define recursively, for a fixed small δ, the following sequence of stopping times:

$$s_0 \equiv 0$$
$$t_i \equiv \inf\{t > s_{i-1}; \, \big|\, |m(\sigma_t^\eta)| - m^*(\beta)\,\big| \geq 2\delta\}$$
$$s_i \equiv \inf\{t > t_i; \, \big|\, |m(\sigma_t^\eta)| - m^*(\beta)\,\big| < \delta\}$$

where $m(\sigma)$ denotes the (normalized) magnetization of σ, i.e. $m(\sigma) = \frac{1}{L^2}\sum_{x\in Q_L} \sigma(x)$. We also define the random variable $\nu(\eta)$ as

$$\nu(\eta) \equiv \min\{i; \, |m(\sigma_{s_i}^\eta) + m^*(\beta)| < \delta\}$$

Then we have:

Theorem 6.12. Let t_0 be as in theorem 6.11. Then there exist positive constants β_o, L_o and c such that for any $\beta \geq \beta_o$ and any $L \geq L_o$

i)

$$\sup_\eta \mathbb{P}\big(s_{\nu(\eta)} - t_{\nu(\eta)} \geq Lt_0\big) \leq e^{-cL}$$

ii)

$$\lim_{L\to\infty} \frac{1}{L} \log\Big(\frac{\mathbb{E}(s_{\nu(+)})}{T_{\text{spin}}}\Big) = 0$$

Remark. If η is such that $m(\eta) > m^*(\beta) - 2\delta$, then we may call $s_{\nu(\eta)} - t_{\nu(\eta)}$ and $s_{\nu(\eta)}$ the time scale of the last excursion before leaving the set $\{\sigma; m(\sigma) \geq -m^*(\beta) + \delta\}$ and the transition time for η respectively. Thus we may conclude that the last excursion occurs on a time scale much shorter than the average transition time, at least if the dynamics starts from the plus configuration.

We conclude this part by stating two last results on the time–dependent magnetization $m(\sigma_t^\eta)$ (see [M]). The first one concerns the large deviations of $m(\sigma_t^\eta)$ and makes a precise connection with analogous results for the equilibrium magnetization obtained in the framework of a rigorous justification on the Wulff construction (see [Sh], [Io1], [Io2], [Pf] and [CGMS]).

Let $\rho \in (-m^*(\beta), m^*(\beta))$ and let ρ_L be a sequence of integers such that

$$\lim_{L \to \infty} \frac{\rho_L}{L^2} = \rho$$

$$\rho_L - L^2 = 0 \mod 2$$

Let also $\tau_{\rho_L}(\eta)$ be the stopping time

$$\tau_{\rho_L}(\eta) = \inf \{t \geq 0; \quad m(\sigma_t^\eta) \leq \rho_L\}$$

Then we have (see theorem 5.1 in [M])

Theorem 6.13. *For any β large enough and any ρ_L as above:*

$$\lim_{L \to \infty} \frac{1}{\beta L} \log \mathbb{E}(\tau_{\rho_L}(+)) = \lim_{L \to \infty} -\frac{1}{\beta L} \log \mu_{Q_L}^{\beta,\emptyset}(\rho \vee 0) \equiv \psi(\rho \vee 0)$$

Remark The rate function $\psi(\rho)$ can be computed easily in terms of the Wulff functional.

The second result makes more precise the picture of the Glauber dynamics as a symmetric double well Markov chain jumping from one phase to the opposite one in a time scale T_{spin}. Let M be the two state space $\{-m^*(\beta), m^*(\beta)\}$ and let Y_t be a continuous time Markov chain on M with unitary jump rate for both states. Then (see theorem 6.1 in [M]) we have

Theorem 6.14. *For any β large enough and for any $\varepsilon \in (0, \frac{m^*(\beta)}{2})$ there exists $\alpha(L) \equiv \alpha(L, \beta, \varepsilon)$ such that*
i) for any choice of times $t_1 < t_2 < \ldots < t_k$ and numbers $m_i \in M, i = 1 \ldots k$

$$\lim_{L \to \infty} \mathbb{P}^\mu \left(|m(\sigma_{\alpha(L)t_1}^\eta) - m_1| < \varepsilon, \ldots, |m(\sigma_{\alpha(L)t_k}^\eta) - m_k| < \varepsilon \right)$$

$$= \mathbb{P}^\nu (Y_{t_1} = m_1, \ldots, Y(t_k) = m_k)$$

where $\mathbb{P}^\mu$ ($\mathbb{P}^\nu$) denotes the probability over the Glauber dynamics (the chain Y_t) with initial distribution the invariant measure $\mu_{Q_L}^{\beta,\emptyset}$ (ν).
ii)

$$\lim_{L \to \infty} \frac{1}{\beta L} \log(\alpha(L)) = \tau_\beta$$

7. Glauber Dynamics for Random Systems

In this final section we discuss what happens to the relaxational properties of the Glauber dynamics when we remove the assumption of *translation invariance* of the interaction, see definition 2.1, and consider in particular *random interactions*. The simplest example of such a system is the (bond) dilute Ising ferromagnet. In this case the couplings between nearest neighbor spins, that in (2.4) were assumed to be constant, become a collection of i.i.d random variables $\{J_{xy}\}$ that take only two values, $J_{xy} = 0$ and $J_{xy} = \beta$ with probability $1 - p$ and p respectively, independently for each pair of nearest neighbors $x, y \in \mathbf{Z}^d$. In a more pictorial form one starts from the standard Ising model and removes, independently for each bond $[x, y]$, the coupling J_{xy} with probability $1 - p$.

Since the $\{J_{xy}\}$ are uniformly bounded, at sufficiently high temperatures (*i.e.* sufficiently small values of β) Dobrushin's uniqueness theory applies and detailed information about the unique Gibbs measure and the relaxation to equilibrium of an associated Glauber dynamics are available using the concept of complete analyticity [DS], [SZ], [MO1] and [MO2]. This regime is usually referred to as the *paramagnetic* phase and, at least for the two dimensional dilute Ising model, it is known to cover the whole interval $\beta < \beta_c$ where β_c is the critical value for the "pure" Ising system.

There is then a range of temperatures, below the paramagnetic phase, where, even if the Gibbs state is unique, certain characteristics of the paramegnetic phase like the analyticity of the free energy as a function of the external field disappear. This is the so called *Griffiths' regime* [G] (see also [F] for additional discussion on this and many other related topics). This "anomalous behavior" is caused by the presence of arbitrarily large clusters of bonds associated with "strong" couplings J_{xy}, which can produce a long–range order inside the cluster. Even above the percolation threshold, *i.e.* when one of such clusters is infinite with probability one, there may be a Griffiths phase for values of $\beta \in (\beta_c, \beta_c(p))$, where β_c is the critical value for the Ising model on $\mathbf{Z}^d$ and $\beta_c(p)$ the critical value of the dilute model above which there is a phase transition (see [F]). What happens is that for almost all realizations of the disorder J and for all site x there is a finite length scale $l(J, x)$, such that correlations between $\sigma(x)$ and $\sigma(y)$ start decaying exponentially at distances greater than $l(J, x)$.

In [BD] an "elementary" approach was given to the problem of uniqueness of the equilibrium state of disordered systems in the Griffiths regime (see also [FI]). In another paper [D] Dobrushin prepared the mathematical background for the study of (arbitrary order) truncated correlation functions for spin glasses. Bounds on $T_c(p)$ have been obtained in [ACCN] and [OPG]. More recent references where, at least for the statics, the situation has been considerably cleared up are [DKP], [GM1], [GM2], [GM3] and [Be]. In particular, under suitable conditions on the couplings distribution, one proves that the infinite volume Gibbs state is unique with probability one and the static correlation functions decay exponentially fast uniformly in the size of the system and its boundary conditions.

The effect of the Griffiths' singularities on the dynamical properties are much more serious since, as we will see, the long time behaviour of any associated Glauber dynamics is dominated by the islands of strongly coupled spins produced by large statistical fluctuations in the disorder (see [B1], [B2] for a non rigorous treatment and [GZ1], [GZ2], [CMM1], [CMM2]). In what follows we will analyze in some detail the simple case of the dilute Ising model. Although at first sight such a model is a very special one, it is important to say that most of the results, particularily those concerning the dynamical behaviour inside the Griffiths phase, apply with minor changes to a much wider class of models (see [CMM1], [CMM2]).

7.1 The dynamics in the paramagnetic phase

In this first part we assume $\beta < \beta_c$, where β_c is the critical value for the "pure" Ising model (2.4), while the parameter p is an arbitrary value in the interval $[0, 1]$. It is not difficult to see that in this case, for *any* couplings configuration $\{J_{xy}\}$ there exists a unique infinite volume Gibbs measure with exponentially decaying correlations functions. Let us in fact consider, for an arbitrary configuration $J = \{J_{xy}\}$, the quantity

$$\mu_{B_L}^{J,+}(\sigma(0)) - \mu_{B_L}^{J,-}(\sigma(0)) \tag{7.1}$$

It is a well known fact that (7.1) is an increasing function of each coupling J_{xy}. Therefore, since $J_{xy} \leq \beta$, (7.1) is bounded from above by the analogous quantity computed for the pure Ising model with constant coupling β. Thus, since $\beta < \beta_c$, (7.1) decays exponentially fast in L uniformly in the configuration J. In particular weak mixing holds (see (4.1)) uniformly in J. Uniqueness of the infinite volume Gibbs measure then follows from a simple argument (see for details in a more general case [Be]).

Let us now consider an attractive Glauber dynamics for the dilute Ising model like the heat bath dynamics or Metropolis. All the general results of section 3 apply without change also to this random case. Moreover, using the fact that weak mixing holds uniformly in J, we could repeat word by word the proof of theorem 4.1 and get exponential ergodicity in the $\| \cdot \|_\infty$ norm uniformly in J.

We can conclude that, as long as $\beta < \beta_c$, the dynamical dilute Ising model does not behave differently from the corresponding pure model at inverse temperature β.

7.2 The Dynamics in the Griffiths Phase: $P < P_c$

Here we consider the more interesting case of $\beta > \beta_c$ but we assume that p is below the percolation threshold p_c for the d-dimensional independent bond percolation on $\mathbb{Z}^d$ (see e.g. [Gri]). It is not difficult to convince oneself that in this case there still exists with probability one a unique infinite volume Gibbs measure. Let us in fact denote by $\{W_i\}$ the connected components of the set

$$W = \{x \in \mathbb{Z}^d : \exists y \text{ such that } d_2(x, y) = 1 \text{ and } J_{xy} = \beta \} \tag{7.2}$$

and let us call "regular" any site x that does not belong to the set W. It is well known that if $p < p_c$ all W_i are finite with probability one. Therefore, with probability one, the infinite volume Gibbs measure is simply the product of the Ising Gibbs measure at inverse temperature β and free boundary conditions for each component W_i and the Bernoulli measure of parameter $\frac{1}{2}$ for each site $x \in W^c$.

Although the situation might appear, and indeed is, quite simple, it is nevertheless very instructive to analyze the relaxational properties of the dynamics in this case. We will see in fact later on that many (but not all) of the features of the case $p < p_c$ remain true also for values of p slightly above the percolation threshold p_c.

Let us first observe that, with probability one, the Glauber dynamics in our case is a product dynamics for each of the clusters $\{W_i\}$ and for each of the "regular" sites $x \in W^c$. Thus, if we consider a local function f that for simplicity we can take as the spin at the origin, $\pi_0(\sigma) = \sigma(0)$, we get that

$$\|T^J(t)\pi_0\|_{L^2(\mu^J)} \le e^{-\lambda_0 t} \tag{7.3}$$

where, for any $x \in \mathbf{Z}^d$,

$$\lambda_x = \begin{cases} 1 & \text{if } x \text{ is regular} \\ \text{gap}(L_{W_i}^{J,\theta}) & \text{if } x \in W_i \end{cases}$$

Notice that in (7.3) we used the obvious property $\mu^J(\pi_0) = 0$ valid because of the simmetry $\sigma \to -\sigma$. Since the clusters W_i are finite with probability one, we can immediately conclude that $\|T^J(t)\pi_0\|_{L^2(\mu^J)}$ converges exponentially fast to its equilibrium value but with an exponential rate, λ_0 in our case, that *depends* on the chosen local function, π_0, through its support. It is important to outline here two important features of the dynamics in the present case

i) in $d = 2$ for any $\beta > \beta_c$ or in $d \ge 3$ and any β large enough, $\inf_x \lambda_x = 0$ with probability one. By ergodicity we have in fact that, with probability one, for any $L \ge 1$ we can find $x(L)$ such that $Q_L(x(L)) = W_i$ for some i. Thanks to the results of section 6 the spectral gap of the cluster W_i is thus exponentially small in L^{d-1}. In particular the spectral gap of the infinite volume dynamics is zero. We can say that such non–uniformity of the rates λ_x is a first signal of the Griffiths phase.

ii) the fact that local functions relax exponentially fast although with a non–uniform rate is a specific feature of the dilute model and it does not extend to more general systems in which the interaction between clusters of strongly interacting spins is weak but non zero (see next section for more details)

Although the analysis of the relaxation to equilibrium for a fixed realization of the random couplings $\{J_{xy}\}$ is certainly interesting, much more relevant from the physical point of view is the same analysis when one takes average over the disorder. It is here that the differences between the dynamics in the paramagnetic phase and in the Griffiths phase appear more pronounced.

Let us denote by $\mathbf{E}(\cdot)$ the average w.r.t the disorder and let us compute upper and lower bounds on $\mathbf{E}(\|T^J(t)\pi_0\|_{L^2(\mu^J)})$. For this purpose, for any $x \in \mathbf{Z}^d$, it is convenient to denote with C_x the set

$$C_x = \begin{cases} x & \text{if } x \text{ is regular} \\ W_i & \text{if } x \in W_i \end{cases} \tag{7.4}$$

Then, remembering theorem 3.8, we can write

$$\mathbb{E}\big(\|T^J(t)\pi_0\|_{L^2(\mu^J)}\big) \leq \sum_{V \subset\subset \mathbb{Z}^d} \mathbb{P}(C_0 = V)e^{-\operatorname{gap}(L_V^{\beta,\theta})t}$$

$$\leq \sum_{V \subset\subset \mathbb{Z}^d} \mathbb{P}(C_0 = V)e^{-t\,e^{-c(\beta)|V|^{\frac{d-1}{d}}}} \tag{7.5}$$

$$\leq \sum_{n \geq 1} C(p)e^{-m(p)n - t e^{-c(\beta)n^{\frac{d-1}{d}}}}$$

$$\leq C_1(p,\beta)e^{-C_2(p,\beta)\log(t)^{\frac{d}{d-1}}}$$

for suitable constants C_1, C_2. In the third inequality above we used the well known fact (see e.g. [Gri]) that

$$\mathbb{P}(|C_0| = n) \leq C(p)\,e^{-m(p)n} \qquad \forall\, p < p_c$$

Notice that the same computation carried out at $\beta < \beta_c$ would have led to a pure exponential decay since (see theorem 4.1 a) of [SchYo])

$$\inf_{V \subset \mathbb{Z}^d} \operatorname{gap}(L_V^{\beta,\theta}) > 0 \qquad \forall\, \beta < \beta_c$$

Let us now turn to the proof of a lower bound on the same quantity. We assume $\beta > \beta_c$ if $d = 2$ or β large enough if $d \geq 3$ but, contrary to the proof of the upper bound, we don't need to restrict ourselves to values of p below the percolation threshold; we just need $p < 1$.

Let $\Lambda = Q_L$ and let $\bar{\Theta}$ be the set of all interactions $J \in \Theta$ such that

(a) $J_{xy} = \beta$ for all $\{x,y\}$ such that $\{x,y\} \subset \Lambda$
(b) $J_{xy} = 0$ for all $\{x,y\}$ which intersect both Λ and Λ^c (the boundary edges)

If we denote by $m_\Lambda = |\Lambda|^{-1}\sum_{x \in \Lambda}\sigma(x)$ the normalized magnetization in Λ, we can write (remember that $\mu^J(\pi_0) = 0$)

$$\mathbb{E}\|T^J(t)\pi_0\|_{L^2(\mu^J)} \geq \mathbb{E}\|T^J(t)m_\Lambda\|_{L^2(\mu^J)} \geq \mathbb{P}(\bar{\Theta})\inf_{J\in\bar\Theta}\|T^J(t)m_\Lambda\|_{L^2(\mu^J)} \tag{7.6}$$

Choose $J \in \bar{\Theta}$ and let

$$F_\Lambda = \{\sigma \in \Omega : m_\Lambda(\sigma) > \tfrac{1}{2}\}$$

Then we have

$$\|T^J(t)m_\Lambda\|_{L^2(\mu^J)} \geq \sqrt{\mu(F_\Lambda)}\,\|T^J(t)m_\Lambda\|_{L^2(\mu^J(\cdot\,|\,F_\Lambda))}$$

and

$$\|T^J(t)m_\Lambda\|_{L^2(\mu^J(\cdot\,|\,F_\Lambda))} \geq \|T^J(t)m_\Lambda\|_{L^1(\mu^J(\cdot\,|\,F_\Lambda))} \geq \mu^J(T^J(t)m_\Lambda\,|\,F_\Lambda) \tag{7.7}$$

For $\sigma \in \Omega$, let $\{\eta_t^\sigma\}_{t\geq 0}$ be the process associated with $T^J(t)$ with initial condition $\eta_0^\sigma = \sigma$, and let $\{\eta_t^\mu\}_{t\geq 0}$ be the stationary process (the one with initial distribution

μ^J). Consider the events

$$G^\sigma_{\Lambda,t} \equiv \{ \exists s \in [0,t] : |m_\Lambda(\eta^\sigma_s) - 1/2| \le 1/(100) \} \qquad \sigma \in \Omega$$

For each $\sigma \in F_\Lambda$, if $|\Lambda| > 100$, we have

$$m_\Lambda(\eta^\sigma_t) \ge \frac{1}{2} \mathbf{1}_{(G^\sigma_{\Lambda,t})^c} - \mathbf{1}_{G^\sigma_{\Lambda,t}} = \frac{1}{2} - \frac{3}{2} \mathbf{1}_{G^\sigma_{\Lambda,t}}$$

which implies

$$\mu^J(T^J(t)\, m_\Lambda \mid F_\Lambda) \ge \frac{1}{2} - \frac{3}{2} \int_\Omega \mu^J(d\sigma \mid F_\Lambda) \operatorname{Prob}(G^\sigma_{\Lambda,t}) \ge \frac{1}{2} - \frac{3}{2} \mu^J(F_\Lambda)^{-1} \operatorname{Prob}(G^\mu_{\Lambda,t})$$
(7.8)

If $t_1, t_2, \ldots$ are the (random) times at which the stationary process η^μ_t is updated inside Λ and n_t is the number of updates up to time t, we have, for all $j \in \mathbf{Z}_+$,

$$P(G^\mu_{\Lambda,t}) \le j\mu^J\{|m_\Lambda(\sigma) - 1/2| \le 1/(100)\} + \operatorname{Prob}\{n_t > j\} \qquad (7.9)$$

which, taking $j = k|\Lambda|t$ with $k = 2c_M$, can be bounded by

$$k\,|\Lambda|\, t\, \mu^J\{|m_\Lambda(\sigma) - 1/2| \le 1/(100)\} + e^{-k'|\Lambda|t}$$

$$\le k\,|\Lambda|\, t\, e^{-c(\beta)L^{d-1}} + e^{-k'|\Lambda|t} \qquad (7.10)$$

for suitable positive constants $C(\beta)$, k' thanks to the results of section 6.5. Moreover, because of the same results,

$$\mu^J(F_\Lambda) \ge \frac{1}{3}$$

for any L large enough.

Take now $L = L_t$ as the smallest integer for which $C(\beta)L^{d-1} \ge 4\log t$. In this way we find

$$\mu(T^J(t)\, m_\Lambda \mid F_\Lambda) \ge \frac{1}{3} \qquad (7.11)$$

for all t large enough. From (7.6) ... (7.11) it follows

$$\mathbf{E}\, \|T^J(t)\pi_0\|_{L^2(\mu^J)} \ge \frac{1}{3} \mathbf{P}(\bar\Theta) \ge \exp\left[-k'' (\log t)^{\frac{d}{d-1}} \right]$$

for a suitable positive constant k''.

Thus we can conclude that, for any $\beta > \beta_c$ in $d = 2$ or for any β large enough in higher dimensions there exists two positive constants C_1, C_2 such that, for any large enough time t

$$e^{-C_1 \log(t)^{\frac{d}{d-1}}} \le \mathbf{E}\, \|T^J(t)\pi_0\|_{L^2(\mu^J)} \le e^{-C_2 \log(t)^{\frac{d}{d-1}}}$$

We would like to conclude this part with a short discussion of the almost sure scaling law of $\operatorname{gap}(L^{J,\tau}_{Q_L})$ as $L \to \infty$. Such a discussion is instructive since it represents a main guideline for the coming analysis of the dynamical dilute Ising model *above* the percolation threshold.

The starting observation is the following. For any $L \geq 1$ and any J let

$$v(J, L) \equiv \max_{x \in Q_L} |C_x|$$

Since the probability distribution of the volume of the cluster C_0 containing the origin has an exponential tail for any $p < p_c$, it follows from standard arguments that there exists a positive constant $k = k(p)$ and a set $\bar{\Theta}$ of measure one such that for any $J \in \bar{\Theta}$ there exists $L_0(J)$ such that for any $L \geq L_0(J)$

i) $v(J, L) \leq k \log(L)$

ii) there exists $x \equiv x(L, J)$, with $|x| \leq L/2$, such that $C_x = Q_l(x)$ and *all* couplings inside the cube $Q_l(x)$, $l = (k^{-1} \log L)^{1/d}$, are equal to β and *all* couplings connecting a point inside $Q_l(x)$ with one of its nearest neighbors outside it are zero. By construction $Q_l(x) \subset Q_L$ if L is large enough.

It follows immediately from i) above, theorem 3.8, proposition 3.9 and the simple observation that

$$\begin{aligned}
\mathrm{gap}(L_{Q_L}^{J,\tau}) &= \min_{x \in Q_L} \mathrm{gap}(L_{C_x}^{J,\tau}) \\
c_s(L_{Q_L}^{J,\tau}) &= \max_{x \in Q_L} c_s(L_{C_x}^{J,\tau})
\end{aligned} \tag{7.12}$$

since the couplings among disjoint clusters are zero, that for any $J \in \bar{\Theta}$ and any $L \geq L_0(J)$

$$\begin{aligned}
\mathrm{gap}(L_{Q_L}^{J,\tau}) &\geq e^{-kv(J,L)^{\frac{d-1}{d}}} \geq e^{-k' \log(L)^{\frac{d-1}{d}}} \\
c_s(L_{Q_L}^{J,\tau}) &\leq kv(J,L)e^{kv(J,L)^{\frac{d-1}{d}}} \leq e^{k'' \log(L)^{\frac{d-1}{d}}}
\end{aligned} \tag{7.13}$$

for suitable constants k, k', k''. Again, using the results of section 6.5 and ii) above, one also obtain an upper bound on the spectral gap of the same order but with a different constant $\bar{k}$ in the usual range of inverse temperature β. Similarily for the logarithmic Sobolev constant. We can thus conclude that, almost surely, the spectral gap in the box Q_L shrinks, as $L \to \infty$, roughly as $e^{-k \log(L)^{\frac{d-1}{d}}}$. We can actually go further and estimate also the probability of a large deviation from the "typical" behaviour. Using (7.13) we can write for any $\varepsilon \in (0, 1]$

$$\begin{aligned}
\mathbb{P}\left(c_s(L_{Q_L}^{J,\tau}) \geq L^\varepsilon\right) &\leq \mathbb{P}\left(kv(J,L)e^{kv(J,L)^{\frac{d-1}{d}}} \geq L^\varepsilon\right) \\
&\leq L^d \mathbb{P}\left(|C_0| \geq k_\varepsilon' \log(L)^{\frac{d}{d-1}}\right) \\
&\leq e^{-k_\varepsilon'' \log(L)^{\frac{d}{d-1}}}
\end{aligned} \tag{7.14}$$

for suitable constants $k_\varepsilon', k_\varepsilon''$ and any L large enough.

In the next final section we will see how to extend the above simple analysis to a much more complicate situation in which one connected component of the set of non–regular sites W is infinite.

7.3 The Dynamics in the Griffiths Phase: $P \geq P_c$

Here we analyze the more complex case when $\beta > \beta_c$ and $p \geq p_c$ but there still exists a unique infinite volume Gibbs state. More precisely, following [CMM1], we make the following assumption (recall definition 2.6 in section 2) that easily implies (see [Be]) almost sure uniqueness of the infinite volume Gibbs measure and exponential decay of the average (over the disorder) of the absolute value of covariances.

(H) There exist $L_0 \in \mathbf{Z}_+$, $\alpha > 0$, $\vartheta > 0$ such that for all $L \geq L_0$

$$\mathbf{P}\{\, SMT(Q_L, L/4, \alpha)\,\} \geq 1 - e^{-\vartheta L}$$

Notice that the above assumption is quite general and flexible in that it does not assume anything specific on the couplings (e.g. that strong couplings do not percolate).

It is rather elementary to show that hypothesis (H) holds as long as $p < p_c$. Much more interesting and less trivial is the problem of showing that it holds also for some value of p above the percolation threshold. The following result (see [CMM2]) provides a partial answer in two dimensions. For an extension to higher dimensions we refer the reader to [ACCMM]. Rember that in two dimensions $p_c = \frac{1}{2}$.

Theorem 7.1. *If*

$$\beta < \begin{cases} \frac{\beta_c}{p} \vee \log(\frac{p}{2p-1}) & \text{if } p > 1/2 \\ \infty & \text{if } p \leq 1/2 \end{cases}$$

then hypothesis (H) holds.

The picture below illustrates the region described in theorem 7.1.

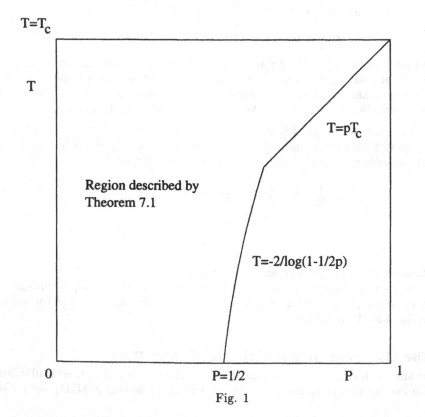

Fig. 1

Using hypothesis (H) we will now state the main results (see [CMM1]) concerning the relaxational properties of the dynamics, almost surely and in average, and the growth of the logarithmic Sobolev constant in the cube Q_L as $L \to \infty$.

Theorem 7.2. *Assume (H). Then*
(a) *If $d \geq 1$ there exists a set $\bar{\Theta} \subset \Theta$ of full measure such that for each $J \in \bar{\Theta}$ there exists a unique infinite volume Gibbs measure μ^J. Moreover there exists a constant k and, for each $J \in \bar{\Theta}$ and for any local function f there exists $0 < t_0(J, f) < \infty$ such that for all $t \geq t_0$*

$$\|T^J(t)f - \mu^J(f)\|_\infty \leq \exp\left[-t \exp\left[-k \left(\log t\right)^{\frac{d-1}{d}} \left(\log \log t\right)^{d-1}\right]\right] \qquad (7.15)$$

(b) *Assume $d \geq 2$. Then there exists a constant k and for any local function f there exists $0 < t_0(f) < \infty$ such that, if $t \geq t_0(f)$ then*

$$\mathbb{E}\|T^J(t)f - \mu^J(f)\|_\infty \leq \exp\left[-k \left(\log t\right)^{\frac{d}{d-1}} \left(\log \log t\right)^{-d}\right] \qquad (7.16)$$

Remark. The almost sure speed of relaxation to equilibrium is faster than any stretched exponential, *i.e.* a decay like e^{-Ct^δ} with $\delta < 1$, and , as the next theorem shows, it cannot be improved in general. Also the bound (b) on the relaxation for the *averaged* dynamics in the bounded case is, apart from the technical factor $(\log \log t)^d$, optimal (see theorem 7.3 below).

Recall now that $\pi_0(\sigma) = \sigma(0)$ and that, by simmetry, $\mu^J(\pi_0) = 0$. Then we have

Theorem 7.3. *For each $d \geq 2$ there is $\beta_1(d) > 0$ such that the following holds: if $\beta > \beta_1(d)$ is such that for almost all J there exists a unique Gibbs measure μ^J, then*
(a) *for all large enough t,*

$$\mathbb{E}\|T^J(t)\pi_0\|_{L^2(\mu^J)} \geq \exp\left[-k \left(\log t\right)^{\frac{d}{d-1}}\right] \qquad (7.17)$$

for some k which depends on d, p_1 and p_2.
(b) *Assume $p > p_c$ and choose the transition rates of the heat-bath dynamics given in (3.5). Then there exists $k > 0$ and a set Θ_0 of positive probability such that for all $J \in \Theta_0$ there exists $0 < t_0(J) < \infty$ such that for all $t \geq t_0$ we have*

$$\|T^J(t)\pi_0\|_\infty \geq \exp\left[-t \exp\left[-k \left(\log t\right)^{\frac{d-1}{d}}\right]\right] \qquad (7.18)$$

Theorem 7.4.
(i) *Assume (H). Then there exist C_1, C_2 and L_1 depending on d, α and ϑ such that for all $L \geq L_1$,*

$$\mathbb{P}\left\{ c_s(L_{Q_L}^J) > C_1 \exp\left[C_2 \beta \left(\log \log L\right)^{d-1} \left(\log L\right)^{\frac{d-1}{d}} \right] \right\} < L^{-2d} \qquad (7.19)$$

(ii) Assume (H) and $d \geq 2$. Then for any $\varepsilon \in (0,1]$ there exist positive constants C_3 and L_2 depending on d, r, α, ϑ, J_0 and ε such that for all $L \geq L_2$

$$\mathbb{P}\{\, c_s(L^J_{Q_L}) > L^\varepsilon \,\} < \exp\big[-C_3 \,(\log\log L)^{-d}(\log L)^{\frac{d}{d-1}}\big] \qquad (7.20)$$

Remark. Using (i) together with the Borel–Cantelli lemma, it follows that, with probability one, $c_s(L^J_{Q_L})$ does not grow faster than the exponential appearing in (7.19). Such a growth is, apart from the $\log\log L$ factor, optimal if one compares it with the almost sure growth obtained for $p < p_c$. Similar comment apply also to the bound on the probability of a large deviations of $c_s(L^J_{Q_L})$.

7.4 A Coarse Graining Description Above P_c

This paragraph represents really the core of the analysis of the dynamical dilute Ising model in the Griffiths phase above the percolation threshold. We give a deterministic upper bound on the logarithmic Sobolev constant $c_s(\mu^{J,\tau}_\Lambda)$ in the cube $\Lambda \equiv Q_L$.

For this purpose, recall first the result at the end of section 7.2. There we proved that if $\sup_i |W_i \cap Q_L| \leq v$ where $\{W_i\}$ are the connected components of the set of non–regular sites W, then

$$c_s(\mu^{J,\tau}_\Lambda) \leq C_1 v \, e^{C_2 v^{\frac{d-1}{d}}} \qquad (7.21)$$

for suitable constants C_1 and C_2 independent of L and v. A key ingredient for the proof of the above bound was the observation that the connected components of W are non–interacting since they are separated from each other by a "safety belt" of regular sites, *i.e.* sites completely decoupled from their complement. Unfortunately, when p is larger than p_c, the same approach becomes useless since the non–regular set W has exactly one infinite connected component (the percolating cluster) with a positive density so that v will typically be of the same order of magnitude as the original volume $|Q_L|$.

It is however very important to observe that, thanks to some of the results of section 4.2, the above conclusions in the non percolating regime remains true, modulo some irrelevant constant factors, even if the value $J_{xy} = 0$ is replaced by a very small number J_{min}, provided that $|J_{min}||W \cap Q_L| \ll 1$.

This remark suggests how to transpose to the dilute model above the percolation threshold the previous ideas. In a certain sense, if the coupling J are such that truncated correlations functions decay exponentially fast on a length scale $l_0 \ll L$, then our original model behaves, after a suitable "coarse-graining", quite closely to the same dilute model *below* the percolation threshold. Let us in fact make a coarse–grained description of the model on a new scale $l_0 \ll L$, by replacing sites with disjoint cubes C_i of side l_0 and declare "regular" those cubes C_i in which truncated correlations decay exponentially fast with rate $\alpha > 0$. In this way, if B is a collection of "non–regular" cubes C_i surrounded by a safety–belt of regular cubes, then the effective interaction of B with any other region outside the safety–belt will be not larger than $|B| \exp(-\alpha l_0)$. Thus, if l_0 is chosen so large that the effective interaction among the connected components of the set W_{l_0} of non–regular cubes C_i is much smaller than one, e.g. if $|W_{l_0}| \exp(-\alpha l_0) \ll 1$, then our system, on scale l_0, will behave like a diluted Ising model even if, strictly speaking, the regular cubes are connected by non–zero couplings. In particular we will be able to apply the results of Section 4.2 and, as a consequence, we will get the bound (7.13) on the spectral gap, with v equal to the volume of the largest connected component of the set W_{l_0}. Moreover, if the probability p_{l_0} that a cube C_i of side l_0 is non–regular is very small, *i.e.* if assumption (H) holds, then the net effect of the coarse–graining will be that of replacing the original parameter $p > p_c$ with a new, renormalized one, $p_{l_0} \ll p_c$.

Although the above reasoning looks quite appealing from a physical point of view, it is still unsatisfactory for the following reason. In a typical configuration of J, the volume of the set $W_{l_0} \cap Q_L$ is roughly $p(l_0)L^d$, where $p(l_0)$ is the probability that a cube C_i is not regular. Using our basic assumption (H), $p(l_0) \approx \exp(-\vartheta l_0)$ so that the minimal scale l_0 satisfying $|W_{l_0}| \exp(-\alpha l_0) \ll 1$ becomes of order $\log L$. This unfortunately is a too large scale: since v is at least l_0^d, the corresponding bound (7.21) on the logarithmic Sobolev constant, becomes at least of the order of a power of L.

In order to overcome this difficulty, we appeal to Proposition 4.10. More precisely we introduce an intermediate length scale $l_1 \ll L$ and we assume that the J in Λ are such that the hypotheses of Proposition 4.10 apply for l_1. If this is the case, then Proposition 4.10 basically allows us to replace the initial cube $\Lambda = Q_L$ with a smaller cube $Q_{l_1}(x)$, for a suitable $x \in \Lambda$. Once we have reduced the initial scale L to the new scale l_1, we make the coarse–grained analysis on scale $l_0 \ll l_1$ on the new cube $Q_{l_1}(x)$ and proceed as explained before.

The advantage of the above two–scale analysis is twofold. First of all the shortest scale l_0 is now at most of the order of $\log l_1$ instead of $\log L$. If we assume (H) it follows from the Borel–Cantelli lemma that in a typical configuration the intermediate scale l_1 can be taken already of the order of $\log L$. Thus we see that, with probability one, the smallest scale becomes $l_0 \approx \log \log L$ with an enormous gain in precision. We conclude this short heuristic discussion by observing that it is precisely the coarse–grained analysis on scale $\log \log L$ that is responsible for the various $\log \log L$ factors in Theorem 7.4.

We are now ready for a precise formulation of our results. Remember that $\mathcal{F}_l$ is the class of subsets of $\mathbf{Z}^d$ which are multiple of Q_l.

Definition 7.5. *A cube $C = Q_l(x)$ is said to be α–regular if, letting $n(l) = \lfloor l/(2\gamma_1) \rfloor$, $SMT(Q_n(y), n/2, \alpha)$ holds for all $y \in Q_l(x)$ where the constants γ_1 and $\widehat{m}(\alpha)$ are those appearing in Proposition 2.9.*

We immediately observe that, thanks to proposition 2.9 and the second remark after its proof, if $V \in \mathcal{F}_{n(l)}$ (*i.e.* if l is a multiple of the integer $2\gamma_1$) is the union of α–regular cubes of side length l then, if l is large enough depending on β and d, $SMT(V, l/2, \widehat{m}(\alpha))$ holds.

Definition 7.6. *Let $l \in \mathbf{Z}_+$, let $\alpha > 0$, let Λ be a multiple of Q_l and write $\Lambda = \cup_{i=1}^{n} Q_l(x_i)$ for some $n \in \mathbf{Z}_+$ and $x_i \in l\mathbf{Z}^d$. Let K be the set of all $i \in \{1, \ldots, n\}$ such that $Q_l(x_i)$ is not α–regular and define*

$$W(\Lambda, l, \alpha) = \left\{ x \in \Lambda : d_2\big(x, \cup_{i \in K} Q_l(x_i)\big) \leq 2l \right\}$$

$v(\Lambda, l, \alpha) =$ *the cardinality of the largest connected component of $W(\Lambda, l, \alpha)$*

Proposition 7.7. *Choose the transition rates c_J as in (3.6). Then, for each $\alpha > 0$ there exists $\bar{l}(d, \alpha, \beta)$ and $k(d, \alpha)$ such that the following holds for all integers $l_0 \geq \bar{l}$ which are multiples of $2\gamma_1$: let V be a multiple of Q_{l_0} and let $v = v(V, l_0, \alpha)$ (see Definition 7.6) and let $m = \hat{m}(d, \alpha)$ be the decay rate given by Proposition 2.9. Assume that $ml_0 \geq \max\{200d\beta, C, 10(1 + \log|W(V, l_0, \alpha)|)\}$ where C is the constant appearing in Corollary 3.6. Then,*

$$\inf_{\tau \in \Omega} \mathrm{gap}(L_V^\tau) \geq \exp\left[-k\left(2d\beta v^{\frac{d-1}{d}} + l_0^{d-1}\right)\right] \tag{7.22}$$

Proof. Write $V = \bigcup_{i=1}^s C_i$, where $C_i = Q_{l_0}(y_i)$ for some $y_i \in l_0 \mathbb{Z}^d$. Let $B = W(V, l_0, \alpha)$ and let A be the union of all those (α–regular) cubes C_i such that $d(C_i, C_j) > l_0$ for all C_j which are not α–regular. Let also $A_0 = A \cap \partial_{l_0}^+ B$ and $B_0 = B \cap \partial_r^+ A$, $r = 1$. By Proposition 3.4 we have

$$\inf_{\tau \in \Omega} \mathrm{gap}(L_V^{J,\tau}) \geq \frac{1}{2}\left[\inf_{\tau \in \Omega} \inf_{D \in \{A,B\}} \mathrm{gap}(L_D^{J,\tau})\right] \inf_{\tau \in \Omega} \mathrm{gap}(L_{\{A,B\}}^{J,\tau}) \tag{7.23}$$

The proof of the Proposition can then be organized in the following steps:

(a) We can use Corollary 3.6 to show that the gap of the block dynamics generator $L_{\{A,B\}}^{J,\tau}$ is at least $1/2$. In order to show that 3.6 does indeed apply to our case, we first notice that $d(A_0, B_0) \geq l_0$, which, together with the trivial inequality $|B_0| \leq |W(V, l_0, \alpha)|$, implies the hypothesis (i) of 3.6. Then we observe that $A \backslash A_0$ can be expressed as a union of α–regular cubes C_i. So, by Proposition 2.9, the property $SMT(A \backslash A_0, l_0/2, m)$ holds.

(b) Since the set A is a union of α–regular cubes of side lenght l_0, property $SMT(V, l_0/2, \hat{m}(\alpha))$ holds for any subset V of A which is a multiple of Q_{l_0}. Thus, using Proposition 4.10, one proves that $\mathrm{gap}(L_A^{J,\tau})$ is bounded from below by a quantity which does not depend on the size of A:

$$\mathrm{gap}(L_A^{J,\tau}) \geq \frac{1}{2} \inf_\tau \inf_{C_i \subset A} \mathrm{gap}(L_{C_i}^{J,\tau}) \geq C_1 e^{-C_2 l_0^{d-1}} \tag{7.24}$$

(c) For what concerns the gap of $L_B^{J,\tau}$, we write B as the disjoint union of its connected components $\tilde{B}_1, \ldots, \tilde{B}_n$. Since $L_{\tilde{B}_i}^{J,\tau}$ commutes with $L_{\tilde{B}_j}^{J,\tau}$ for all $i \neq j$, it follows that

$$\mathrm{gap}(L_B^{J,\tau}) = \inf_{i \in \{1,\ldots,n\}} \mathrm{gap}(L_{\tilde{B}_i}^{J,\tau})$$

From (7.23), (a), (b) and (c), together with Theorem 3.8 (for the dynamics (3.6) we can take $c_m = 1/2$ in (3.3)) and the fact that trivially $\|J^{(\lambda)}\| \leq 2d\beta$, we get

$$\inf_{\tau \in \Omega} \mathrm{gap}(L_V^{J,\tau}) \geq \frac{1}{4} \min\left\{\frac{1}{2} \inf_i e^{-\left(k\, 2d\beta |\tilde{B}_i|^{\frac{d-1}{d}}\right)}, C_1 e^{-C_2 l_0^{d-1}}\right\}$$

In order to obtain (7.22) we now observe that by definition of v, we have $|\tilde{B}_i| \leq v$, and that the minimum of the two quantities in braces is greater than their product if l_0 is such that both terms are less than 1. $\square$

Theorem 7.8. *If the transition rates are given by (3.6), then for each $\alpha > 0$ there exist $\bar{l}$, C_1 and C_2 depending on d, β and α such that the following holds for all positive integers $l_0 \geq \bar{l}$ multiples of $2\gamma_1$: let l_1 be a multiple of l_0 and let Λ be a multiple of Q_{l_1} so that we can write*

$$\Lambda = \bigcup_{i=1}^{n} B_i = \bigcup_{i=1}^{s} C_i \tag{7.25}$$

where $B_i = Q_{l_1}(x_i)$ and $C_i = Q_{l_0}(y_i)$ for some $x_i \in l_1 \mathbb{Z}^d$ and $y_i \in l_0 \mathbb{Z}^d$ Let $v = v(\Lambda, l_0, \alpha)$ (see Definition 7.6), and let $m = \hat{m}(d, r, \alpha)$. Assume that

(i) *For each $i \in \{1, \dots n\}$ the cube B_i is α−regular*

(ii) *$kv^{\frac{d-1}{d}} \leq m\delta l_1$, where $k(d, \alpha, \beta)$ and δ are the constants appearing in Theorem 3.8 and Proposition 4.10 respectively*

(iii) *$30d \log l_1 \leq m l_0 \leq (l_1)^{1/(2d)}$*

Then we have

$$\sup_{\tau \in \Omega} c_s(\mu_\Lambda^{J,\tau}) \leq C_1 \exp\left[C_2 \left(v^{\frac{d-1}{d}} + l_0^{d-1} \right) \right]$$

Proof. Since Λ is the union of α−regular cubes B_i, for any $I \subset \{1, \dots, n\}$ property $SMT(V, l_1/2, \hat{m}(\alpha))$ holds for $V = \bigcup_{i \in I} B_i$. Therefore, thanks hypothesis (ii) and to proposition 4.10

$$\sup_{\tau \in \Omega} c_s(L_\Lambda^{J,\tau}) \leq 2 \sup_{\tau \in \Omega} \sup_{B_i \in \Lambda} c_s(L_{B_i}^{J,\tau})$$

$$\leq C' \left[4 + 4l_1^d (2d\beta + 2\log 2) \right] \sup_{B_i \in \Lambda} \text{gap}(L_{B_i}^{J,\tau})^{-1}$$

We now observe that hypothesis iii) allow us to apply proposition 7.7 above to any cube B_i in Λ. Thus

$$\sup_{B_i \in \Lambda} \text{gap}(L_{B_i}^{J,\tau})^{-1} \leq \exp\left[+k\left(2d\beta v^{\frac{d-1}{d}} + l_0^{d-1} \right) \right]$$

and the result follows. $\square$

7.5 Proof of the Main Results Above P_c

In this final part we prove our main results above the percolation threshold, namely theorem 7.2, 7.3 and 7.4. First we need the following simple estimate.

Lemma 7.9. *Assume (H). Then there exist $L_0 \in \mathbb{Z}_+$, $\vartheta' > 0$ such that for all $L \geq L_0$,*

$$\mathbb{P}\{ Q_L \text{ is } \alpha\text{−regular} \} \geq 1 - e^{-\vartheta' L}$$

Proof. The probability that Q_L is not α−regular is bounded by

$$\mathbb{P}\{SMT(Q_n(y), n/4, \alpha) \text{ does not hold for some } y \in Q_L\} \leq$$
$$\leq L^d e^{-\vartheta L/(3n)}$$
$$\leq e^{-\vartheta' L}$$

if L is greater than some L_0. $\square$

Proof of Theorem 7.4. We give the proof in the special case of L which is a power of 2, which is enough to prove Theorem 7.2. We also assume, without loss of generality, that the constant γ_1 of proposition 2.9 is also a power of 2. A proof which works for all L requires a modification of Theorem 7.8 where one considers more general coverings of Λ with cubes and cuboids with slightly different sidelengths. This generalization is straightforward.

Part (i). We are going to use the key deterministic estimate of $c_s(\mu_{Q_L}^{J,\tau})$ given in Theorem 7.8. The idea is to prove that with probability greater than $1 - 3L^{-2}$, it is possible to choose the two parameters in Theorem 7.8, l_0, l_1 in such a way that the deterministic upper bound on $c_s(\mu_{Q_L}^{J,\tau})$ given in that proposition is not greater than

$$\exp\left[C(\log\log L)^d(\log L)\right]$$

for a suitable constant C. More precisely we define l_0 and l_1 as those powers of 2 (they are uniquely defined) such that

$$\frac{60d}{m}\log\log L \leq l_0 < \frac{120d}{m}\log\log L \qquad \frac{3d}{\vartheta'}\log L \leq l_1 < \frac{6d}{\vartheta'}\log L \qquad (7.26)$$

where $m \equiv \hat{m}(\alpha)$ (see Proposition 2.9) and ϑ' is given in lemma 7.9. We then take

$$v_* = l_0^d \log L \qquad (7.27)$$

Since l_0 divides l_1 and l_1 divides L, we can write Q_L as in (7.25). We now observe that, if L is large enough, the hypotheses (i) – (iii) of Theorem 7.8 are satisfied for all $J \in \tilde{\Theta} \equiv \cap_{i=1}^2 \Theta_i$, where

$\Theta_1 = \{J : \text{each } B_i \text{ is } \alpha-\text{regular}\}$
$\Theta_2 = \{J : v(Q_L, l_0, \alpha) \leq v_*\}$

and $v(Q_L, l_0, \alpha)$ has been defined in 7.6. By Theorem 7.8, (7.26) and (7.27), for any $J \in \tilde{\Theta}$, we have

$$c_s(\mu_{Q_L}^{J,\tau}) \leq \exp\left[C(\log\log L)^d \log L\right]$$

for a suitable constant C. In order to prove the theorem it is therefore sufficient to estimate from above $\mathbb{P}(\tilde{\Theta}^c)$.

From Proposition 7.9, it follows

$$\mathbb{P}(\Theta_1^c) \leq L^d e^{-\vartheta' l_1} \leq L^{-2d} \qquad (7.28)$$

for all L large enough. Let $p(l)$ be the probability that a cube Q_l is not $\alpha-$regular. Then $p(l)$ goes to zero as $l \to \infty$, and a standard estimate for $2-$dependent site percolation implies

$$\mathbb{P}(\Theta_2^c) \leq L^d \left(k_1 \, p(l_0)\right)^{k_2 v_* l_0^{-d}} \leq L^{-3d} \qquad (7.29)$$

for L large enough, where k_1 and k_2 are two suitable geometrical constants. This completes the proof.

Proof of part (ii). The proof is the same as in part (i), with a different choice of the three basic parameters l_0, l_1 and v_*. More precisely we define l_0 and l_1 as those powers of 2 (they are uniquely defined) such that (let again $m = \hat{m}(\alpha)$)

$$\frac{60d^2}{(d-1)m} \log \log L \le l_0 < \frac{120d^2}{(d-1)m} \log \log L \qquad (\log L)^{\frac{d}{d-1}} \le l_1 < 2(\log L)^{\frac{d}{d-1}}$$

Given $\varepsilon \in (0,1)$ we then let

$$v_* = \left(\frac{\varepsilon \log L}{2 J_0 C_2}\right)^{\frac{d}{d-1}} \tag{7.30}$$

where C_2 appears in Theorem 7.8. Write Q_L as in (7.25) and define the events $\tilde{\Theta}$, Θ_i as in the proof of part (i). Thanks to Theorem 7.8, we get

$$c_s(\mu_{Q_L}^{J,\tau}) \le C_1 L^{\frac{s}{2}} \exp(C_2 l_0^d) \le L^\varepsilon \qquad \forall J \in \tilde{\Theta} \tag{7.31}$$

for all L sufficiently large. In order to prove (7.20) it is therefore sufficient to bound from above $\mathbb{P}(\tilde{\Theta}^c)$. As before, we find

$$\mathbb{P}(\Theta_1^c) \le L^d e^{-\vartheta' l_1} = L^d \exp\left[-\vartheta' (\log L)^{\frac{d}{d-1}}\right] \tag{7.32}$$

and

$$\mathbb{P}(\Theta_2^c) \le L^d (k_1 p(l_0))^{k_2 v_* l_0^{-d}} \le \exp\left[-C_3 (\log \log L)^{-d} (\log L)^{\frac{d}{d-1}}\right] \tag{7.33}$$

for a suitable constant C_3 and all L large enough. Clearly (7.32) and (7.33) complete the proof of (ii). $\square$

Proof of Theorem 7.2. The proof of the almost sure bounds (part (a)) is a simple consequence of Theorem 7.4. Let $\bar{\Theta}$ be the set of interactions J such that for each $J \in \bar{\Theta}$ there exists $L_1(J)$ such that for all $L \ge L_1(J)$ (C is given in Proposition 7.4)

(i) $c_s(L_{B_L}^J) < \exp\left[C (\log L)(\log \log L)^d\right]$
(ii) $SMT(B_L, \gamma_1(2L+1), \alpha)$ holds

Using Theorem 7.4, (H) and the Borel–Cantelli lemma, one can check that $\mathbb{P}(\bar{\Theta}) = 1$. Moreover, thanks to (ii) and (iii), for all $J \in \bar{\Theta}$ there exists a unique infinite volume Gibbs measure that in the sequel will be denoted by μ^J. Let, in fact, f be any local function on Ω, and take L large enough such that $B_L \supset \Lambda_f$. Then, given two arbitrary boundary conditions τ and η, and using a telescopic interpolation between them, we get

$$\sup_{\tau,\eta \in \Omega} \left| \mu_{B_L}^{J,\tau}(f) - \mu_{B_L}^{J,\eta}(f) \right| \le |\partial_r^+ B_L| \sup_{x \in \partial_r^+ B_L} \|\nabla_x[\mu_{B_L}^J(f)]\|_\infty =$$

$$= |\partial_r^+ B_L| \sup_{x \in \partial_r^+ B_L} \sup_{\tau \in \Omega} \left| \frac{\mu_{B_L}^{J,\tau}(h_x, f)}{\mu_{B_L}^{J,\tau}(h_x)} \right| \tag{7.34}$$

where $h_x \equiv \exp[-\nabla_x H_{B_L}]$.

Therefore, if L is larger than $L_1(J)$ and if $d(\Lambda_f, (B_L)^c) > \gamma_1(2L+1)+1$, we can use $SMT(B_L, \gamma_1(2L+1), \alpha)$, and write

$$\sup_{\tau, \eta \in \Omega} \left| \mu_{B_L}^{J,\tau}(f) - \mu_{B_L}^{J,\eta}(f) \right| \le kL^{d+4} |\Lambda_f| \|f\|_\infty e^{-\alpha d(\Lambda_{h_x}, \Lambda_f)} \tag{7.35}$$

for a suitable constant k, and the uniqueness follows. At this point, in order to prove inequality (7.15), we simply appeal to (3.23) .

Proof of part (b). Let $L_t = \lfloor k_1 t \rfloor$ for some $k_1 > k_0$ such that L_t is a power of 2 (k_0 is given in Lemma 3.2) and, for simplicity, let $\Lambda_t = B_{L_t}$. For any $\varepsilon \in (0,1)$, let $\Theta(t,\varepsilon)$ be the set of interactions J such that

(i) $c_s(L_{\Lambda_t}^J) \le L_t^\varepsilon$

(ii) $SMT(\Lambda_t, (\log L_t)^{\frac{d}{d-1}}, \hat{m}(\alpha))$.

We can write, for any $\tau \in \Omega$,

$$\mathbb{E}\|T^J(t)f - \mu^J(f)\|_\infty \le \|f\| \, \mathbb{P}(\Theta(t,\varepsilon)^c) + \sup_{J \in \Theta(t,\varepsilon)} \|T_{\Lambda_t}^{J,\tau}(t)f - \mu_{\Lambda_t}^{J,\tau}(f)\|_\infty +$$
$$+ \|T^J(t)f - T_{\Lambda_t}^{J,\tau}(t)f\|_\infty + \sup_{J \in \Theta(t,\varepsilon)} |\mu_{\Lambda_t}^{J,\tau}(f) - \mu^J(f)| \tag{7.36}$$

We denote by X_1, X_2, X_3 and X_4 the four terms on the RHS of (7.36). The last two terms, using lemma 3.2 and property $SMT(\Lambda_t, (\log L_t)^{\frac{d}{d-1}}, \hat{m}(\alpha))$ are exponentially small in t:

$$X_3 + X_4 \le C_f \, e^{-2t} \tag{7.37}$$

Furthermore, we have

$$\mathbb{P}(\Theta(t,\varepsilon)^c) \le \mathbb{P}\{ c_s(L_{\Lambda_t}^J) \ge L^\varepsilon \} +$$
$$+ \mathbb{P}\{ SMT(\Lambda_t, (\log L)^{\frac{d}{d-1}}, \hat{m}(\alpha)) \text{ does not hold } \} \tag{7.38}$$

Of the above two terms the first one is estimated via (ii) of Theorem 7.4, which implies

$$\mathbb{P}\{ c_s(L_{\Lambda_t}^J) \ge L^\varepsilon \} \le \exp\left[-C_3 (\log \log L_t)^{-d} (\log L_t)^{\frac{d}{d-1}} \right] \tag{7.39}$$

provided that t is large enough. The second term in the RHS of (7.38) can be bounded from above, using Proposition 2.9, by the probability that there exists a cube $Q_l(x)$ in Λ_t, with $l = \lceil (\log L)^{\frac{d}{d-1}} \rceil$, which is not α−regular. Using Proposition 7.9 such a probability is bounded from above by

$$L_t^d \exp[-\vartheta' (\log L_t)^{\frac{d}{d-1}}] \tag{7.40}$$

provided that t is so large that $L_t \ge L_0$. In this way we have obtained

$$X_1 \le \|f\| \left[\exp[-C_3 (\log \log L_t)^{-d} (\log L_t)^{\frac{d}{d-1}}] + L_t^d \exp[-\vartheta (\log L_t)^{\frac{d}{d-1}}] \right] \tag{7.41}$$

As for X_2, we use (3.23) and the fact that now $c_s(L_{\Lambda_t}^J) \le L_t^\varepsilon$, and we get

$$X_2 \le 2 \|f\| \exp\left[-k' t^{1-\varepsilon} \right] \tag{7.42}$$

for any t sufficiently large. From (7.37), (7.41) and (7.42) we get that for large t the dominant term in (7.36) is the first one and, by consequence (7.16) follows. $\square$

Proof of the lower bound, Theorem 7.3

It is enough to prove part b) since part a) has already been proved in section 7.2.

The main idea for the lower bound on the a.s. relaxation of the spin at the origin can be divided into two distinct parts. The first part consists in showing that, with positive probability, for any L large enough, there exists a local function f_L, with $\Lambda_{f_L} \subset B_L \cap C_0$, C_0 being the cluster of the origin, whose relaxational behaviour is not faster than $\exp\left[-t\exp\left[-k\left(\log L\right)^{\frac{d-1}{d}}\right]\right]$.

The second part amounts to proving that the influence of the slow relaxation of f_L on the spin at the origin is not smaller than a negative exponential of L. This implies a lower bound on $\|T^J(t)\pi_0\|_\infty$ of the order of $\exp\left[-mL - t\exp\left[-k\left(\log L\right)^{\frac{d-1}{d}}\right]\right]$ and the result (remember that $\mu^J(\pi_0) = 0$ by symmetry) follows by optimizing over $L \leq t$. Let us now implement these sketchy ideas.

We say that a path $\gamma = (x_0, \ldots, x_n)$ connects x to y if

$$x_0 = x, \quad x_n = y \quad d_2(x_i, x_{i+1}) = 1 \quad \text{and} \quad J_{x_i, x_{i+1}} = \beta \quad \forall i = 0, \ldots, l-1$$

We will then write $\gamma : x \to y$ and set $|\gamma| = n$. Given a configuration J of the random couplings, a local function f and a finite set $\Lambda \subset C_0$, where C_0 is the cluster containing the origin, we set

$$d^J(\Lambda) = \sup_{x \in \Lambda} \inf_{\gamma : 0 \to x} |\gamma|; \quad e^J(f) = \frac{\mathcal{E}^J(f, f)}{\operatorname{Var}^J(f)}; \quad e^J(\Lambda) = \inf_{\substack{f \in L^2(\Omega, d\mu^J) \\ \Lambda_f \subset \Lambda}} e^J(f) \quad (7.43)$$

(both the Dirichlet form and the variance are with respect to the unique infinite volume Gibbs measure). With the above definition we have the following two key results.

Lemma 7.10. *Under the same assumptions of part (b) of Theorem 7.3 there exists a set $\bar{\Theta} \subset \Theta$ of positive measure and two positive constants k and L_0 such that for each $J \in \bar{\Theta}$ and any $L \geq L_0$ there exists f such that*

$$\Lambda_f \subset C_0$$
$$d^J(\Lambda_f) \leq kL$$
$$e^J(f) \leq \exp\left[-k(\log L)^{\frac{d-1}{d}}\right]$$

Lemma 7.11. *There exists $m > 0$ such that for any $t \geq 1$ and any local function f such that $\Lambda_f \subset C_0$*

$$\|T^J(t)\pi_0\|_\infty \geq (8|\Lambda_f|)^{-1}\exp\left[-md^J(\Lambda_f) - 2e^J(f)t\right]$$

Before proving the two lemmas we complete the proof of part (b) of the theorem. For this purpose choose J in the set of positive measure given by Lemma 7.10, define $L(t) = t\exp\left[-(\log t)^{\frac{d-1}{d}}\right]$ and assume that t is so large that $L(t) \geq L_0$. If we apply

Lemma 7.11 to the function f given by lemma 7.10 for $L = L_t$ and use the upper bound on $e^J(f)$ given in Lemma 7.10 we immediately get the sought lower bound on $\|T^J(t)\pi_0\|_\infty$.

Sketch of the proof of Lemma 7.10. A detailed proof of the lemma is given in [ACCMM]. Here we only sketch the main ideas. Fix ε small enough and E large enough. Let $I(\varepsilon, E, L)$ be the event that inside the cube B_L there exists a smaller cube $Q_l(x)$, $l \geq \varepsilon \log(L)$, such that all the couplings inside $Q_l(x)$ are equal to β, all but one the boundary couplings are zero and $Q_l(x)$ is connected to the origin by a path γ of length smaller than EL. Using the fact that $p > p_c$ one can prove that the probability of $I(\varepsilon, E, L)$ is strictly bounded away from zero uniformly in L.

Let now $f(\sigma) \equiv \mathbf{I}\{m_{Q_l(x)}(\sigma) \geq 0\}$ where $m_{Q_l(x)}(\sigma)$ denotes the (normalized) magnetization in $Q_l(x)$. Notice that $\Lambda_f \subset Q_l(x) \subset B_L$. If we compute $e^J(f)$ and use the estimates of section 6.5 together with the symmetry under global spin flip, we get

$$e^J(f) \leq \frac{c_M\, |Q_l(x)|\, \mu^J\{\, |m_{Q_l(x)} - 1/2| \leq 1/100\,\}}{\mathrm{Var}^J(f)} \leq \exp\left[-k(\log L)^{\frac{d-1}{d}}\right]$$

for a suitable constant k depending on ε and any L large enough. □

Proof of Lemma 7.11. For any given J we set $F^J(x, t) \equiv T^J(t)\pi_x(\underline{1})$. Notice that, since the nearest neighbor coupling J_{xy} are non–negative, the heat–bath dynamics is attractive (see section 3.4) so that $F^J(x, t)$ is a non–increasing function of t and $\|T^J(t)\pi_0\|_\infty = F^J(0, t)$. Next we define $m = m(\beta)$ by

$$\inf_{\substack{d_2(x,y)=1 \\ J_{xy}=\beta}} \quad \inf_{\substack{\sigma,\eta\in\Omega \\ \sigma(y)=1,\eta(y)=-1 \\ \sigma(z)\geq\eta(z)\ \forall z\neq y}} T^J(1)\pi_x(\sigma) - T^J(1)\pi_x(\eta) \equiv 2e^{-m} \tag{7.44}$$

Thanks to attractivity the quantity in (7.44) is non–negative, and, in particular, it is strictly positive with our choice of the transition rates.

Fix now a local function f and a path $\gamma = (x_0, \ldots, x_l) \subset C_0$. The result of the lemma is then a direct consequence of the following three simple inequalities valid for any $t \geq 1$ (see [CMM1] for a proof).

$$F^J(0, t) \geq e^{-ml} F^J(x_l, t) \tag{7.45}$$

$$\|T^J(t)f - \mu^J(f)\|^2_{L^2(\mu^J)} \leq 4\,\mathrm{Var}^J(f) \sum_{x\in\Lambda_f} F^J(x, t) \tag{7.46}$$

$$\|T^J(t)f - \mu^J(f)\|^2_{L^2(\mu^J)} \geq \frac{\mathrm{Var}^J(f)}{2} \exp[-2e^J(f)t] \tag{7.47}$$

In fact, for any local function f such that $\Lambda_f \subset C_0$, if we sum (7.45) over $x \in \Lambda_f$ and use (7.46) and (7.47), we get

$$F^J(0, t) \geq \frac{1}{8|\Lambda_f|} \exp[-md^J(\Lambda_f) - 2e^J(f)t] \geq \frac{1}{8|\Lambda|} \exp[-md^J(\Lambda) - 2e^J(f)t] \tag{7.48}$$

□

8. Open Problems

We conclude these notes with a partial list of open problems.

Pb. 1 Prove (or disprove) the exponential ergodicity of the infinite volume dynamics for a general non–attractive system satisfying only the weak mixing condition of section 2 in dimensions $d \geq 3$.

Pb. 2 Discuss the validity of the strong mixing condition of section 2 for the three dimensional Ising model for $\beta < \beta_c$ or $\beta > \beta_c$ and $h \neq 0$.

Pb. 3 Prove the analogous of theorem 5.2 for the true three dimensional Ising model without making the SOS approximation.

Pb. 4 Analyze the relaxational properties under a Glauber dynamics for the Ising model at the critical point ($h = 0$, $\beta = \beta_c$). Such an interesting and difficult point was only marginally touched in these notes in the example at the end of section 3

Pb. 5 Consider a Glauber dynamics for the two dimensional Ising model in a box Q_L at low temperature, zero external field and plus boundary conditions. Prove that the spectral gap, as a function of L, shrinks as L^{-2}. Use then such a bound to prove (or disprove) the conjectured (see [FH] and [OG]) $e^{-\sqrt{t}}$ decay of the time autocorrelation of the spin at the origin for the infinite volume Glauber dynamics in the plus phase (see section 6.7). Discuss the same problem in three dimension. Here the conjectured decay is no longer a stretched exponential $e^{-\sqrt{t}}$ but, rather, a pure exponential. In particular the Glauber dynamics in the plus phase should have zero spectral gap in two dimensions and positive spectral gap in three dimensions. We refer the reader to [CSS] for some related results in a simplified context (zero temperature dynamics).

Pb. 6 This question has been posed to me by R.Dobrushin. Consider a Glauber dynamics for the two dimensional Ising model in the phase coexistence region and run the dynamics starting from a configuration which is all pluses except a large (e.g. spherical) bubble of minuses centered at the origin. As time goes on the bubble will shrink under the influence of the external plus spins and the question is whether, under a suitable time–dependent rescaling, it will reach asymptotically in time a limiting shape and whether the limiting shape has anything to do with the Wulff shape (see [DKS]).

Pb. 7 Improve theorem 6.10 by showing that in the phase coexistence region the spectral gap of the generator in a large box B_L with free boundary conditions restricted to the subspace of even functions shrinks not faster than an inverse power of L.

Pb. 8 Consider the two dimensional spin glass for which the random couplings take value $\pm J$ with equal probability. Extend to such model part a) of theorem 7.3 when $J > \beta_c$, β_c being the critical value of the two dimensional Ising model (see [RSP] for an heuristic approach).

References

[ACCN]   M. Aizenman, J. T. Chayes, L. Chayes and C. M. Newman: *The phase boundary in dilute and random Ising and Potts ferromagnets.* J. Phys. A: Math. Gen. **20**, L313 (1987)

[ACCMM] K.Alexander, L. Chayes, F. Cesi, C. Maes and F. Martinelli: *Relaxation of Dilute Magnets in the Griffiths' Regime* in preparation

[B1]   A. J. Bray: *Upper and lower bounds on dynamic correlations in the Griffiths phase.* J. Phys. A: Math. Gen. **22**, L81 (1989)

[B2]   A. J. Bray: *Dynamics of dilute magnets above T_c.* Phys. Rev. Lett. **60**, No 8, 720 (1988)

[BD]   L. A. Bassalygo and R. L. Dobrushin: *Uniqueness of a Gibbs field with random potential–an elementary approach.* Theory Prob. Appl. **31**, 572 (1986)

[Be]   J. van den Berg: *A constructive mixing condition for 2-D Gibbs measures with random interactions.* Preprint 1996

[BEF]   J. Bricmont, A. El Mellouki and J. Frölich: *Random surfaces in statistical mechanics: Roughening, Rounding, Wetting, . . .* J. Stat. Phys **42**, Nos 5/6, 743 (1986)

[BM]   J. van den Berg and C. Maes: *Disagreement percolation in the study of Markov fields.* Ann. Prob. **22**, 749 (1994)

[BZ]   L. Bertini and B. Zegarlinski: *Coercive inequalities for Gibbs measures* Preprint 1997

[BDZ]   E. Bolthausen, J. Deuschel and O. Zeitouni: *Entropy repulsion of the lattice free field.* Commun. Math. Phys. **170**, No 2, 417, (1995)

[CGMS]   F. Cesi, G. Guadagni, F. Martinelli and R. Schonmann: *On the 2D Stochastic Ising Model in the Phase Coexistence Region Near the Critical Point.* Journal. Stat. Phys. **85**, No 1-2, 55, (1996)

[CM1]   F. Cesi and F. Martinelli: *On the layering transition of an SOS surface interacting with a wall I. Equilibrium results.* Journ. Stat. Phys. **82**, No 3-4, 823, (1996)

[CM2]   F. Cesi and F. Martinelli: *On the layering transition of an SOS surface interacting with a wall II. The Glauber dynamics.* Commun. Math. Phys. **177**, No 1, 173, (1996)

[CMM1]   F. Cesi, C. Maes and F. Martinelli: *Relaxation of Disordered Magnets in the Griffiths' Regime* Comm. Math. Phys.

[CMM2]   F. Cesi, C. Maes and F. Martinelli: *Relaxation to equilibrium for two dimensional disordered Ising models in the Griffiths phase.* Comm. Math. Phys. to appear.

[CSS]   L. Chayes, R. H. Schonmann and G. Swindle: *Lifshitz law for the volume of a two-dimensional droplet at zero temperature.* Journal of Stat. Phys. **79**, 821 (1995).

[D]   R. L. Dobrushin: *A formula of full Semiinvariants.* In 'Cellular Automata and Cooperative Systems', N. Boccara, E. Goles, S. Martinez and P. Picco (eds.), Kluwer Acad. Publ., Dordrecht-Boston-London, pp. 135-140 (1993).

[DS]   R. L. Dobrushin and S. Shlosman: *Constructive criterion for the uniqueness of Gibbs fields.* Statistical Physics and Dynamical Systems. Fritz, Jaffe and Szász eds. Birkhauser 347 (1985)

[DeSt] J. D. Deuschel and D. W. Stroock: *Large deviations.* Academic Press, Series in Pure and Applied Mathematics, **137**, (1989)

[DM] E. I. Dinaburg and A. E. Mazel: *Layering transition in SOS model with external magnetic field.* Journal. Stat. Phys. **74** Nos 3/4, 533 (1994)

[DKP] H. Von Dreyfus, A. Klein and J. F. Perez: *Taming Griffiths singularities: infinite differentiability of quenched correlations functions.* Commun. Math. Phys. **170**, 21 (1995)

[DKS] R.Dobrushin, R.Kotecký, S.Shlosman: *Wulff Construction. A Global Shape From Local Interaction* Translation of Math. Monographs, AMS **104** (1992)

[DRS] D. Dhar, M. Randeria and J. P. Sethna: *Griffiths singularities in the dynamics of disordered Ising models.* Europhys. Lett., **5**, No 6, 485 (1988)

[EFS] A. C. .D. van Enter, R. Fernandez and A. D. Sokal: *Regularity properties and pathologies of position-space renormalization-group transformations: scope and limitations of Gibbsian theory.* Journal of Stat. Phys. **72**, Nos 5/6, 879, (1993)

[F] J. Fröhlich: *Mathematical aspects of the physics of disordered systems.* in "Critical Phenomena, Random Systems, Gauge Theories" Eds. K.Osterwalder and R. Stora, Elsevier (1986)

[FI] J. Fröhlich and J. Z .Imbrie: *Improved perturbation expansion for disordered systems: beating Griffiths singularities.* Commun. Math. Phys. **96**, 145 (1984)

[FH] D. Fisher and D. Huse: Phys.Rev.B **35**, 6841 (1987)

[FP1] J. Frölich and C. Pfister: *Semi-infinite Ising model. I.* Commun. Math. Phys. **109**, 493 (1987)

[FP2] J. Frölich and C. Pfister: *Semi-infinite Ising model. II.* Commun. Math. Phys. **112**, 51 (1987)

[G] R. Griffiths: *Non-analytic behaviour above the critical point in a random Ising ferromagnet.* Phys. Rev. Lett. **23**, 17 (1969)

[GM1] G. Gielis and C. Maes: *Percolation Techniques in Disordered Spin Flip Dynamics: Relaxation to the Unique Invariant Measure.* Commun. Math. Phys. **177**, 83 (1996).

[GM2] G. Gielis and C. Maes: *Local analyticity and bounds on the truncated correlation functions in disordered systems.* Markov Proc. Relat. Fields **1**, 459 (1995)

[GM3] G. Gielis and C. Maes: *The Uniqueness regime of Gibbs Fields with Unbounded Disorder.* Journal Stat. Phys. **81**, 829 (1995)

[Gri] G. Grimmett: *Percolation.* Springer-Verlag, (1989)

[GZ1] A. Guionnet and B. Zegarlinski: *Decay to equilibrium in random spin systems on a lattice.* Commun. Math. Phys. **181**, No 3, 703 (1996)

[GZ2] A. Guionnet and B. Zegarlinski: *Decay to equilibrium in random spin systems on a lattice II.* Journal Stat. Phys. **86**, No 3-4, 899 (1997)

[Hi] Y. Higuchi: *Coexistence of infinite (*)-clusters II:-Ising percolation in two dimension.* Prob. Th. Rel. Fields, **97**, 1, (1993)

[HiYo] Y. Higuchi, N. Yoshida: *Slow relaxation of stochastic Ising models with random and non-random boundary conditions.* New trends in stochastic analysis, ed. by K. D. Elworthy, S. Kusuoka, I. Shigekawa, World Scientific Publishing, 153-167, (1997).

[Ho1] R.Holley: *Possible Rates of Convergence in Finite Range Attractive Spin Systems.* Contemp.Math. **41**, 215-234 (1985)

[Ho2] R.Holley: *Rapid convergence to equilibrium in ferromagnetic stochastic Ising models.* Reseñas IME–USP, **1**, No 2/3, 131 (1993)

[HS] R. A. Holley and D. W. Strook: *Uniform and L^2 convergence in one dimensional stochastic Ising models.* Commun. Math. Phys. **123**, 85 (1989)

[Io1] D. Ioffe: *Large Deviation for the 2D Ising Model: a Lower Bound without Cluster Expansion.* Journ. Stat. Phys. **74**, 411, (1994)

[Io2] D. Ioffe: *Exact Large Deviation Bounds up to T_c for the Ising Model in Two Dimensions.* Probab.Theory Related Fields, **102**, No 3, 313, (1995)

[J] S. Jain: *Anomalously slow relaxation in the diluted Ising model below the percolation threshold.* Physica A, **218**, 279 (1995)

[JS1] M.Jerrum, A.Sinclair: *Approximating the Permanent* Siam Journ.Comput. **18**, 1149, (1989)

[JS2] M.Jerrum, A.Sinclair: *Polynomial-Time Approximation Algorithms for the Ising Model* SIAM Journ. Comput. **22**, 1087, (1993)

[L] T. M. Ligget: *Interacting particles systems.* Springer–Verlag (1985)

[La] E. Laroche: *Hypercontractivité pour de systèmes des spin de portée infinie* Probab.Theor.Relat.Fields **101**, 89 (1995)

[LMaz] J. Lebowitz and A. E. Mazel: *A remark on the low temperature behavior of an sos interface in half space* Journal Stat. Phys. **84**, no 3/4, 379 (1996)

[LM] C. Maes and J. Lebowitz: *The effect of an external field on an interface, entropic repulsion.* Journal Stat. Phys. **46** No 1/2, 39 (1987)

[LY] ShengLin Lu, H.T.Yau: *Spectral Gap and Logarithmic Sobolev Inequality for Kawasaki and Glauber Dynamics.* Comm.Math.Phys. **156**, 399 (1993)

[M] F. Martinelli: *On the two dimensional dynamical Ising model in the phase coexistence region.* Journal Stat. Phys. **76**, No. 5/6, 1179 (1994)

[MM] E. Marcelli and F. Martinelli: *Some new results on the 2D kinetic Ising model in the phase coexistence region.* Journal Stat. Phys. **84**, No 3/4, 655, (1996)

[MO] F. Martinelli and E. Olivieri: *Finite volume mixing conditions for lattice spin systems and exponential approach to equilibrium of Glauber Dynamics* in "Cellular Automata and Cooperative Systems" ed. N.Boccara, E.Goles, S.Martinez and P.Picco, Nato Asi series, **396**, 473, (1993)

[MO1] F. Martinelli and E. Olivieri: *Approach to equilibrium of Glauber dynamics in the one phase region I: The attractive case.* Commun. Math. Phys. **161**, 447 (1994)

[MO2] F. Martinelli and E. Olivieri: *Approach to equilibrium of Glauber dynamics in the one phase region II: The general case.* Commun. Math. Phys. **161**, 487 (1994)

[MO3] F.Martinelli, E.Olivieri: *Some remarks on pathologies of renormalization group transformations for the Ising model* Journ. Stat. Phys. **72**, 1169 (1994)

[MO4] F.Martinelli, E.Olivieri: *Instability of Renormalization Group Pathologies under Decimation* Journ. Stat. Phys. **79**, 25 (1995)

[MOS] F. Martinelli, E. Olivieri and R.H.Schonmann: *For Gibbs state of 2D lattice spin systems weak mixing implies strong mixing.* Commun. Math. Phys. **165**, 33 (1994)

[O] E. Olivieri: *On a cluster expansion for lattice spin systems and finite size condition for the convergence.* Journal Stat. Phys. **50**, 1179 (1988)

[OG] Ogielski: *Dynamics of fluctuations in the order phase of kinetic Ising model.* Phys. Rev. Lett. **36**, No 13. 7315 (1987)

[OP] E. Olivieri and P. Picco: *Cluster expansion for D–dimensional lattice systems and finite volume factorization properties.* Journal Stat. Phys. **59**, 221 (1990)

[OPG] E. Olivieri, F. Perez and S. Goulart–Rosa–Jr.: *Some rigorous results on the phase diagram of the dilute Ising model.* Phys. Lett. **94A**, No 6,7, 309 (1983)

[P] A. Pisztora: *Surface order large deviations for Ising, Potts and percolation models.* Probab. Th. Rel. Fields **104**, 427 (1996)

[Pf] C.Pfister: *Large deviations and phase separation in the two dimensional Ising model.* Elv. Phys. Acta **64**, 953 (1991)

[Po] G. Posta: *Spectral gap for an unrestricted Kawasaki type dynamics* ESAIM, Probab. Statist. **1**, 145, (1995/97)

[RSP] M. Randeria, J. P. Sethna and R. G. Palmer: *Low-frequency relaxation in Ising spin-glasses.* Phys. Rev. Lett. **54**, No. 12, 1321 (1985)

[SC] L. Saloff-Coste: *Markov Chains* preprint 1996, to appear in the proceedings of the Saint Flour summer school in probability theory 1996

[Sch] R. H. Schonmann: *Slow droplet–driven relaxation of stochastic Ising Models in the vicinity of the phase coexistence region.* Commun. Math. Phys. **170**, 453 (1995)

[SchYo] R. H. Schonmann and N. Yoshida: *Exponential relaxation of Glauber dynamics with some special boundary conditions* Comm. Math.Phys. in press

[ShSch] R. H. Schonmann and S. B. Shlosman: *Complete analiticity for 2D Ising model completed.* Comm.Math.Phys. **170**, 453 (1995)

[SZ] D. W. Stroock and B. Zegarlinski: *The logarithmic Sobolev inequality for discrete spin systems on a lattice.* Commun. Math. Phys. **149**, 175 (1992)

[Th] L.Thomas: *Bounds on the Mass Gap for the Finite Volume Stochastic Ising Models at Low Temperatures.* Comm.Math.Phys. **126**, 1 (1989)

[Yo1] N. Yoshida: *Relaxed criteria of the Dobrushin-Shlosman mixing condition* Journal Stat. Phys. **87**, No 1–2, 293 (1997)

[Yo2] N. Yoshida: *Finite volume Glauber dynamics in a small magnetic field* to appear in Journal Stat.Phys.

[Z] B. Zegarlinski: *Strong decay to equilibrium in one dimensional random spin systems.* Journal Stat. Phys. **77**, 717 (1994)

[Zha] M.Zahradník: *An alternate versione of Pirogov–Sinai theory* Comm.Math.Phys. **93**, 559 (1984)

PROBABILITY ON TREES :

AN INTRODUCTORY CLIMB

Yuval PERES

Contents

1 Preface

These notes are based on lectures delivered at the Saint Flour Summer School in July 1997. The first version of the notes was written and edited by Dimitris Gatzouras. The notes were then expanded and revised by David Levin and myself. I hope that they are useful to probabilists and graduate students as an introduction to the subject; a more complete account is in the forthcoming book co-authored with Russell Lyons.

The first 10 chapters are devoted to basic facts about percolation on trees, branching processes and electrical networks, with an emphasis on several key techniques: moment estimates, the use of percolation to determine dimension, and the "method of random paths" to construct flows of finite energy. These 10 chapters are the "introductory climb" alluded to in the title.

More advanced topics start in Chapter 11, where the method of random paths is refined in order to establish the Grimmett-Kesten-Zhang Theorem: *Simple random walk on the infinite percolation cluster in $\mathbf{Z}^d$, $d \geq 3$ is transient.*

Chapters 12 and 13 contain a regularity property of subperiodic trees, and its application to random walks on groups. In Chapter 14 we discuss capacity estimates for hitting probabilities; these are used in Chapter 15 to derive intersection-equivalence of fractal percolation and Brownian paths.

In Chapter 16 we analyze the phase transition in a broadcasting model considered by computer scientists: A random bit is propagated, with errors, from the root of a tree to its boundary, and the goal is to reconstruct the original bit from the boundary values. Remarkably, the same model arose independently in genetics, as a mutation model, and in mathematical physics, where it is equivalent to the Ising model on a tree. In Chapter 17, the Ising model on a tree is used to construct a nearest-neighbor process on $\mathbf{Z}$ that is "less predictable" than simple random walk.

In Chapters 18 and 19, we study the speed and recurrence properties of tree-indexed processes; in particular, we relate three natural notions of speed (cloud speed, burst speed, and sustainable speed) to three well-known dimension indices (Minkowski dimension, packing dimension, and Hausdorff dimension). In Chapter 20 we consider a dynamical variant of percolation, where edges open and close according to independent Poisson processes. At any fixed time, the random configuration is a sample of Bernoulli percolation, but we focus on exceptional random times when the number of infinite open clusters is atypical. There are striking parallels between the study of these exceptional times for dynamical percolation, and the study of multiple points for Brownian motion. We conclude in Chapter 21 by describing some results on stochastic domination between randomly labeled trees, and stating some open problems for other graphs.

I was first drawn to thinking about general trees in a lecture of I. Benjamini in 1989, when H. Furstenberg noted that certain trees that appeared in the lecture could be interpreted (via b-adic expansions) as Cantor sets with different Hausdorff and Minkowski dimensions. I. Benjamini and I proceeded to examine relations between properties of trees and properties of the corresponding compact sets; these connections had unexpected uses later (see Chapter 15). For example, consider a subset Λ of the unit square in the plane and the corresponding tree $T(\Lambda, b)$ in base b. Then Λ is hit by planar Brownian motion (i.e., it has positive logarithmic capacity) iff simple random walk on $T(\Lambda, b)$ is transient.

We then learned that a year earlier, R. Lyons (building on works of Furstenberg, Shepp, Kahane and Fan) had established some remarkably precise connections between random walks, percolation and capacity on trees. R. Lyons and R. Pemantle had already used these ideas to determine the *sustainable speed* of first-passage percolation on trees.

The point of view of these lectures was largely developed in the ensuing collaboration with Itai Benjamini, Russell Lyons and Robin Pemantle, whose influence pervades these notes. Other coauthors whose insights and ideas are represented here include Chris Bishop (see Chapter 15), Will Evans, Claire Kenyon, and Leonard Schulman (see Chapter 16), Olle Häggström and Jeff Steif (see Chapter 20).

In fact, probability on trees is a rich and fast-growing subject, so the account presented in these notes is necessarily incomplete. Natural complements are the two conference proceedings volumes: *Trees*, edited by B. Chauvin, S. Cohen and A. Rouault (Birkhäuser 1996) and *Classical and Modern Branching Processes*, edited by K. B. Athreya and P. Jagers (Springer 1996). Continuum random trees are fascinating objects studied in several papers by David Aldous; Tom Liggett is writing a detailed account of the contact process on trees. Superprocesses, which can be obtained as scaling limits of branching random walks, have been studied by numerous authors. I apologize to the many researchers whose results involving probability on trees are not described here.

Acknowledgements I am grateful to the participants in the St. Flour summer school for their comments and to the organizer, Pierre Bernard, for his warm hospitality.

I am greatly indebted to Dimitris Gatzouras and David Levin for their help in preparing these notes. I thank Itai Benjamini, Dayue Chen, Amir Dembo, Nina Gantert, David Grabiner, Olle Häggström, Davar Khoshnevisan, Elon Lindenstrauss, Elhanan Mossel, Oded Schramm and Balint Virag for their comments on the manuscript.

Yuval Peres
Jerusalem, December 1998

2   Basic Definitions and a Few Highlights

A **tree** is a connected graph containing no cycles. All trees considered in these notes are locally finite: the degree $\deg(v)$ is finite for each vertex v, although $\deg(v)$ may be unbounded as a function of v.

Why study general trees?

1. More can be done on trees than on general graphs. Percolation problems, for example, are easier to analyze on trees. The insight and techniques developed for trees can sometimes be extended to more general models later.

2. Trees occur naturally. Some examples are:

   (a) **Galton-Watson trees.** Let L be a non-negative integer-valued random variable and set $Z_0 \equiv 1$, $Z_1 = L$, and $Z_{n+1} = \sum_{i=1}^{Z_n} L_i^{(n+1)}$, where the $L_i^{(n)}$ are i.i.d. copies of L. Then Z_n is the number of individuals in generation n of a Galton-Watson branching process, a population which starts with one individual and in which each individual independently produces a random number of offspring with the same distribution as L. The collection of all individuals form the vertices of a tree, with edges connecting parents to their children.

   (b) **Random spanning trees in networks.** A *spanning tree* of a graph G is a tree which is a subgraph of G including all the vertices of G. There are several interesting algorithms for generating random spanning trees of finite graphs.

3. Trees describe well the complicated structure of certain compact sets in $\mathbf{R}^d$. Examples include Cantor sets on intervals and fractal percolation, a collection of nested random subsets of the unit cube described below.

Example 2.1 Fractal Percolation is a recursive construction generating random subsets $\{A_n\}$ of the unit cube $[0,1]^d$. Tile $A_0 = [0,1]^d$ by b^d similar subcubes with side-length b^{-1}. Generate A_1 by taking a union of some of these subcubes, including each independently with probability p. In general, A_n will be a union of b-adic cubes of order n (cubes with side-length b^{-n} and vertices with coordinates of the form kb^{-n}). A_{n+1} is obtained by tiling each such cube contained in A_n by b^d b-adic subcubes of order $n+1$, and taking a union which includes each subcube independently with probability p. The limit set of this construction $\bigcap_{n=0}^{\infty} A_n$ is denoted by $Q_d(p)$.

There is a tree associated with each realization of fractal percolation. The vertices at level n correspond to b-adic cubes of order n which are contained in A_n, and a vertex v at level n is the parent of a vertex w at level $n+1$ if the cube corresponding to v contains the cube corresponding to w. $\triangle$

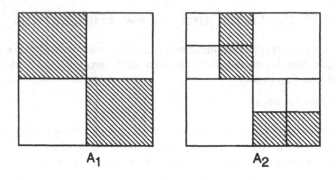

Figure 1: **A realization of A_1 and A_2 for $d = 2$, $b = 2$.**

Let $Q_3(\frac{1}{2}) \subset [0,1]^3$ denote the limit set of fractal percolation with $b = 2$, $d = 3$, and $p = \frac{1}{2}$. In Chapter 15, we will see that the random set $Q_3(\frac{1}{2})$ is **intersection-equivalent** in the cube to the Brownian motion path started uniformly in the cube. By this we mean the following: if $[B]$ denotes the range $\{B(t) : t \geq 0\}$ of a three-dimensional Brownian motion started uniformly in $[0,1]^3$, then for some constants $C_0, C_1 > 0$ and all closed sets $\Lambda \subset [0,1]^3$,

$$C_0 \, \mathbf{P}(Q_3(1/2) \cap \Lambda \neq \emptyset) \; \leq \; \mathbf{P}([B] \cap \Lambda \neq \emptyset) \; \leq \; C_1 \, \mathbf{P}(Q_3(1/2) \cap \Lambda \neq \emptyset) \,.$$

Consequently, hitting probabilities for Brownian motion can be related to hitting probabilities of $Q_3(\frac{1}{2})$. This gives a new perspective on the classical study of intersections and multiple points of Brownian paths.

For example, consider two independent copies $Q_3(\frac{1}{2})$ and $Q_3'(\frac{1}{2})$. Then the intersection $Q_3(\frac{1}{2}) \cap Q_3'(\frac{1}{2})$ has the same distribution as $Q_3(\frac{1}{4})$. Since the tree corresponding to $Q_3(\frac{1}{4})$ is a Galton-Watson tree with mean offspring 2, it survives with positive probability. Hence $Q_3(\frac{1}{4}) \neq \emptyset$ with positive probability, and intersection-equivalence shows that two independent Brownian paths in $\mathbf{R}^3$ intersect with positive probability, a result first proved in [21].

It also follows that three Brownian paths in space do not intersect (as first proved in [22]). By intersection-equivalence, it is enough to show that the intersection of the limit sets of three independent fractal percolations, which has the same distribution as $Q_3(\frac{1}{8})$, is empty a.s. But the tree corresponding to $Q_3(\frac{1}{8})$ is a critical Galton-Watson process and hence dies out, see Chapter 3.

Infinite family trees arising from supercritical Galton-Watson Branching processes, (Galton-Watson trees in short) play a prominent role in these notes.

Question 2.2 *In what ways are Galton-Watson trees like regular trees?*

First we establish a simple property of regular trees.

Example 2.3 Simple random walk $\{X_n\}_{n\geq 0}$ on a graph is a Markov chain on the vertices, with transition probabilities

$$\mathbf{P}(X_{n+1} = w | X_n = v) = \begin{cases} \frac{1}{\deg(v)} & \text{if } w \sim v, \\ 0 & \text{otherwise}. \end{cases}$$

The notation $u \sim v$ means that the vertices u and v are connected by an edge. Now suppose the graph is a tree, and let $|v|$ stand for the distance of a vertex v from the root ρ, *i.e.*, $|v|$ is the number of edges on the unique path from ρ to v. On the b-ary tree,

$$\mathbf{E}\Big[|X_{n+1}| - |X_n| \,\Big|\, X_n\Big] \geq \frac{b}{b+1}(+1) + \frac{1}{b+1}(-1) = \frac{b-1}{b+1}.$$

(We have an inequality here because X_n may be at the root.) Hence the distance of the random walk on the tree from the root stochastically dominates an upwardly biased random walk on $\mathbf{Z}$. It is therefore transient and will visit 0 only finitely many times. After the last visit of the random walk to the root,

$$\mathbf{E}\Big[|X_{n+1}| - |X_n| \,\Big|\, X_n\Big] = \frac{b-1}{b+1},$$

and the strong law of large numbers for martingale differences implies that, almost surely, $n^{-1}|X_n| \to \frac{b-1}{b+1}$. $\triangle$

One specific case of Question 2.2 is

Question 2.4 *On a Galton-Watson (GW) tree with mean $m = \sum_k k p_k > 1$, is simple random walk transient on survival of the GW process?*

We will see later that the answer is positive; this was first proved by Grimmett and Kesten (1984).

For a tree Γ, denote $\Gamma_n = \{v : |v| = n\}$. Define the **lower growth** and **upper growth** of Γ as $\underline{\mathrm{gr}}(T) := \liminf |\Gamma_n|^{1/n}$ and $\overline{\mathrm{gr}}(T) = \limsup |\Gamma_n|^{1/n}$ respectively. If $\underline{\mathrm{gr}}(\Gamma) = \overline{\mathrm{gr}}(\Gamma)$, we speak of the **growth** of the tree Γ and denote it by $\mathrm{gr}(\Gamma)$.

Question 2.5 *Is $\underline{\mathrm{gr}}(\Gamma) > 1$ sufficient for transience of simple random walk on Γ? Is it necessary?*

The answer to both questions is negative. An analogous situation holds for Brownian motion on manifolds, where exponential volume growth is not sufficient and not necessary for transience.

Example 2.6 (3–1 tree) The 3-1 tree Γ has $\mathrm{gr}(\Gamma) = 2$ (actually $|\Gamma_n| = 2^n$), but simple random walk is recurrent on it. Γ can be embedded in the upper half-plane, with its root ρ at the origin. The root has two offspring, and for $n \geq 1$, each level Γ_n

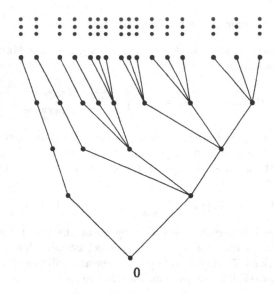

Figure 2: **The 3-1 Tree.**

has 2^n vertices which can be ordered from left to right as $v_1^n, \ldots, v_{2^n}^n$. For $k \leq 2^{n-1}$, each v_k^n has only one child, while for $2^{n-1} < k \leq 2^n$, each v_k^n has three children. Observe that for any vertex not on the right-most path to infinity, the subtree above it will eventually have no more branching (because "powers of 3 beat powers of 2"). The random walk on Γ will have excursions on left-hand branches, but must always return to the right-most branch (because of recurrence of simple random walk on the line). If these excursions are ignored, then we have a simple random walk on the right-most path, i.e., on $\mathbf{Z}^+$, which is recurrent. $\triangle$

It is even easier to construct transient trees of polynomial growth: *E.g.*, replace every edge at level k of the ternary tree by a path consisting of 2^k edges. Simple random walk on the resulting tree, considered just when it visits branch points, dominates an upward biased random walk on the integers, whence it is transient.

On the other hand, positive speed implies exponential growth:

Theorem 2.7 *Define the* **speed** *of a random walk as* $\lim_n n^{-1}|X_n|$*, when this limit exists. If the speed of simple random walk on a tree Γ exists and is positive, then Γ has exponential growth, i.e.,* $\underline{\mathrm{gr}}(\Gamma) > 1$*.*

This follows from Theorem 5.4 below.

Example 2.6 suggests that $\mathrm{gr}(\Gamma)$ does not give much information on the behavior of a random walk on Γ. The growth $\mathrm{gr}(\Gamma)$ barely takes into account the structure of Γ, and a more refined notion is required.

A **cutset** Π is a set of vertices such that any infinite self-avoiding path from emanating the root ρ must pass through some vertex in Π. The **branching number** of a tree Γ is defined as

$$\mathrm{br}(\Gamma) = \sup\left\{ \lambda \geq 1 : \inf_{\Pi \text{ cutset}} \sum_{v \in \Pi} \lambda^{-|v|} > 0 \right\}. \tag{1}$$

The function $\inf\left\{ \sum_{v \in \Pi} \lambda^{-|v|} : \Pi \text{ a cutset} \right\}$ is decreasing in λ and positive at $\lambda = 1$.

The **boundary of a tree** Γ, denoted $\partial\Gamma$, is the set of all infinite self-avoiding paths (rays) emanating from the root ρ of Γ. A natural metric on the boundary $\partial\Gamma$ is $d(\xi, \eta) = e^{-n}$, where n is the number of edges shared by ξ and η. $\dim_H(\partial\Gamma)$ will denote the Hausdorff dimension of $\partial\Gamma$ with respect to this metric d. Because an open cover of $\partial\Gamma$ corresponds to a cutset of Γ, and vice-versa, the Hausdorff dimension of $\partial\Gamma$ is related to the branching number of Γ by

$$\log \mathrm{br}(\Gamma) = \dim_H(\partial\Gamma).$$

Similarly, $\mathrm{gr}(\Gamma)$ is related to the Minkowski dimension $\dim_M(\partial\Gamma)$ by

$$\log \mathrm{gr}(\Gamma) = \dim_M(\partial\Gamma).$$

Generally, $\mathrm{br}(\Gamma) \leq \underline{\mathrm{gr}}(\Gamma)$, since for $\lambda > \underline{\mathrm{gr}}(\Gamma)$ we must have

$$\inf_n |\Gamma_n| \lambda^{-n} = \inf_n \sum_{v \in \Gamma_n} \lambda^{-|v|} = 0;$$

using the fact that Γ_n is itself a cutset yields the inequality. If $\partial\Gamma$ is countable, then $\mathrm{br}(\Gamma) = 1$, because $\dim_H A = 0$ for countable sets A. For the 3-1 tree in Example 2.6, $\partial\Gamma$ is countable, and consequently $\mathrm{br}(\Gamma) = 1$.

As an indication that the branching number $\mathrm{br}(\Gamma)$ contains more information about the tree than the growth $\mathrm{gr}(\Gamma)$, we mention two results that we shall prove later, in Chapters 7 and 13.

Bernoulli(p) percolation on a tree Γ is the random subgraph of Γ obtained by independently including each original edge of Γ with probability p, and discarding each with probability $1 - p$. The retained edges are called **open**, and $\mathbf{P}_p$ is the probability corresponding to this process (see Chapter 4 for the formal definition of the probability space.) The first quantity of interest in percolation is

$$p_c(\Gamma) = \inf\{p \in [0, 1]: \mathbf{P}_p(\rho \leftrightarrow \infty) > 0\}, \tag{2}$$

where $\{\rho \leftrightarrow \infty\}$ denotes the event that the root ρ is connected to ∞, i.e., that there is an infinite self-avoiding path emanating from ρ, that consists of open edges.

Theorem 2.8 (R. Lyons 1990) *For an infinite and locally finite tree* Γ,

$$p_c(\Gamma) = \frac{1}{\mathrm{br}(\Gamma)}. \tag{3}$$

Theorem 2.9 (R. Lyons 1990) *If* $\mathrm{br}(\Gamma) > 1$, *then simple random walk on* Γ *is transient.*

We close with an equivalent description of the branching number $\mathrm{br}(\Gamma)$ of a tree Γ. If u, v are vertices in Γ so that v is a child of u, denote by uv the edge connecting them. A **flow** θ on Γ from the root ρ to ∞ is an edge function obeying $\theta(uv) = \sum \theta(vw)$, where the sum is over all children w of v. This property is known as **Kirchhoff's node law**. Imagine the tree as a network of pipes through which water can flow entering at the root. However much water enters a pipe must leave through the other end, splitting up among the outgoing pipes (edges). Define $\theta(v)$, for a vertex $v \neq \rho$, to be the amount of flow that reaches v, i.e., $\theta(v) := \theta(uv)$ for u the parent of v. The **strength** of a flow θ, denoted $\|\theta\|$, is the amount flowing from the root, $\sum_{v:v\sim\rho} \theta(v)$. When $\|\theta\| = 1$, we call θ a **unit flow**.

Lemma 2.10 *For a tree* Γ,

$$\mathrm{br}(\Gamma) = \sup\{\lambda \geq 1 : \exists \text{ a nonzero flow } \theta \text{ from } \rho \text{ to } \infty : \forall v, \ \theta(v) \leq \lambda^{-|v|}\}. \tag{4}$$

Proof. This follows directly from the Min-cut/Max-flow Theorem, which in our setting says that

$$\sup\{\|\theta\| : \theta(v) \leq \lambda^{-|v|} \ \forall v\} = \inf_{\Pi \text{ cutset}} \sum_{v\in\Pi} \lambda^{-|v|}. \tag{5}$$

For details, see Lyons and Peres (1999). □

Remark: As mentioned above, $\mathrm{br}(\Gamma) \leq \underline{\mathrm{gr}}(\Gamma) = \liminf_n |\Gamma_n|^{1/n}$. In general, to get an upper bound for $\mathrm{br}(\Gamma)$ one can seek explicit 'good' cutsets. To get lower bounds use either

(i) Theorem 2.8, which in particular says that $\mathrm{br}(\Gamma) \geq 1/p_c(\Gamma)$, or

(ii) find a good flow θ on Γ such that $\theta(v) \leq \lambda^{-|v|}$ for all v; then $\mathrm{br}(\Gamma) \geq \lambda$. (Recall that $\theta(v)$ denotes the flow from the unique parent of v to v.)

A flow θ on Γ induces a measure μ on $\partial\Gamma$: for cylinder sets $[v] = \{\xi \in \partial\Gamma : \xi \text{ passes through } v\}$, define $\mu([v])$ as $\theta(v)$. If $[v_1], \ldots, [v_n]$ are disjoint cylinders (which means that no v_i is an ancestor of another), and $[v] = \bigcup_{i=1}^n [v_i]$ (i.e., the $\{v_i\}$ form a cutset for the subtree Γ^v rooted at v), then Kirchhoff's node law implies (by induction on n) that $\mu([v]) = \sum_{i=1}^n \mu([v_i])$. Countable additivity can be proven using the compactness of $\partial\Gamma$: Cylinders form a basis consisting of open sets and are also closed in the natural topology on $\partial\Gamma$. Thus countable additivity follows from finite additivity.

3 Galton-Watson Trees

Let L be a non-negative integer-valued random variable and let $p_k = \mathbf{P}(L = k)$ for $k = 0, 1, 2, \ldots$. To avoid trivial cases, we assume throughout that $p_1 < 1$. Let $\{L_i^{(n)}\}_{i, n \in \mathbf{N}}$ be independent and identically distributed copies of L, set $Z_0 = 1$, and define

$$Z_{n+1} = \begin{cases} \sum_{i=1}^{Z_n} L_i^{(n+1)} & \text{if } Z_n > 0, \\ 0 & \text{if } Z_n = 0. \end{cases}$$

The variables Z_n are the population sizes of a Galton-Watson branching process. The tree associated with a realization of this process has Z_n vertices at level n, and for $i \le Z_n$, the i'th vertex in level n has $L_i^{(n+1)}$ children in level $n + 1$.

Generating functions are an indispensable tool in the analysis of Galton-Watson processes. Set $f(s) = \mathbf{E}[s^L]$ and define inductively

$$f_0(s) = s, \qquad f_1(s) = f(s), \qquad f_{n+1}(s) = f \circ f_n(s), \qquad 0 \le s \le 1.$$

It can be verified by induction that $f_n(s) = \mathbf{E}[s^{Z_n}]$ for all n, that is, f_n is the generating function of Z_n. Note that $f(s) = \sum_{k=0}^{\infty} p_k s^k$ and $f'(1) = \mathbf{E}[L] = m$. We always have $f''(s) \ge 0$ for $s \ge 0$, so f is convex on $\mathbf{R}^+$.

Define q to be the smallest fixed point of f in $[0, 1]$. Note that if $p_0 = 0$, then $q = 0$. Observe that $\lim_n \mathbf{P}(Z_n = 0) = \lim_n f_n(0) \le q$, and since $\lim_n f_n(0)$ must be a fixed point of f, it follows that $q = \lim_n \mathbf{P}(Z_n = 0)$. So

$$q = \mathbf{P}(Z_n \to 0) = \text{probability of extinction}.$$

Since f is convex, if $1 \ge m = f'(1)$, then $q = 1$. If instead $1 < m = f'(1)$, then $q < 1$. Thus, a Galton-Watson process dies out a.s. if and only if $m \le 1$.

A property of trees A is **inherited** if all finite trees have property A, and all the immediate descendant subtrees $\Gamma^{(i)}$ of Γ have A when Γ has A. (The immediate descendant subtrees $\Gamma^{(i)}$ of Γ are the subtrees of Γ rooted at the children of the root ρ.)

Example 3.1 The following are all inherited properties:

1. $\{\Gamma : \sup_n |\Gamma_n| < \infty\}$.

2. $\{\Gamma : |\Gamma_n| \text{ grows polynomially in } n\}$.

3. $\{\Gamma : \Gamma \text{ finite or } \mathrm{br}(\Gamma) \le c\}$. $\qquad\qquad\qquad\qquad\qquad\qquad\qquad\qquad\qquad \triangle$

Proposition 3.2 (0-1 Law) *Let $\mathbf{P}$ be the probability measure on trees corresponding to a GW process with $m > 1$. If A is inherited, then*

$$\mathbf{P}(A \mid \text{non-extinction}) \in \{0, 1\}.$$

Proof. We have

$$\mathbf{P}(\Gamma \in A | Z_1 = k) \leq \mathbf{P}\left(\bigcap_{i=1}^{k}\{\Gamma^{(i)} \in A\} \mid Z_1 = k\right) = \mathbf{P}(\Gamma \in A)^k.$$

Thus,

$$\mathbf{P}(\Gamma \in A) = \sum_k p_k \mathbf{P}(\Gamma \in A | Z_1 = k) \leq f(\mathbf{P}(\Gamma \in A)).$$

Convexity of f implies that the only numbers $x \in [0,1]$ satisfying $x \leq f(x)$ are $x = 1$ and all $x \in [0,q]$. Since A holds for all finite trees, $\mathbf{P}(\Gamma \in A) \geq q$. So $\mathbf{P}(\Gamma \in A) \in \{q, 1\}$. □

Observe that $m^{-n} Z_n$ is a non-negative martingale and hence converges to some finite random variable $W < \infty$. If $m \leq 1$, then $Z_n = 0$ eventually, so a.s. $W = 0$. The case $m > 1$ is treated by the following theorem.

Theorem 3.3 (Kesten and Stigum (1966a)) *When $m > 1$,*

$$P(W > 0 \mid non\text{-}extinction) = 1 \quad if\ and\ only\ if \quad \mathbf{E}[L \log^+ L] < \infty.$$

A conceptual proof of Theorem 3.3 appears in Lyons, Pemantle, and Peres (1995).

Hawkes (1981), under the assumption that $\mathbf{E}[L \log^2 L] < \infty$, proved that for Galton-Watson trees Γ,

$$\mathbf{P}(\dim_H(\partial\Gamma) = \log m \mid \text{non-extinction}) = 1.$$

This is equivalent to

$$\mathbf{P}(\text{br}(\Gamma) = m \mid \text{non-extinction}) = 1. \tag{6}$$

R. Lyons discovered a simpler proof without the assumption $\mathbf{E}[L \log^2 L] < \infty$, which is given below in Corollary 5.2. Because a.s. $m^{-n} Z_n \to W$, where $0 \leq W < \infty$, it follows that a.s. $\overline{\text{gr}}(\Gamma) \leq m$. This, together with the general inequality $\text{br}(\Gamma) \leq \underline{\text{gr}}(\Gamma)$ and (6), implies that a.s. given non-extinction,

$$m = \text{br}(\Gamma) \leq \underline{\text{gr}}(\Gamma) \leq \overline{\text{gr}}(\Gamma) \leq m.$$

4 General percolation on a connected graph

General (bond) percolation on a connected graph G is a random subgraph $G(\omega)$ of G such that, for any edge e in G, the event that e is an edge of $G(\omega)$ is measurable. **Independent** $\{p_e\}$ **percolation** is the percolation obtained when each edge e is retained (or declared open) with probability p_e, independently of other edges (and removed or declared closed otherwise). We already discussed in Chapter 2 the special case of **Bernoulli**(p) **percolation** where all probabilities p_e are the same, $p_e \equiv p$.

Formally, the sample space for a general bond percolation is $\Omega = \{0,1\}^E$, where E is the edge set of the graph G. The σ-field $\mathcal{F}$ on Ω is generated by the finite-dimensional cylinders, sets of the form $\{\omega \in \Omega : \omega(e_1) = x_1, \ldots, \omega(e_m) = x_m\}$ for $x_i \in \{0,1\}$. The probability measures $\mathbf{P}_{\{p_e\}}$ and $\mathbf{P}_p$, corresponding to independent $\{p_e\}$ percolation and Bernoulli(p) percolation respectively, are product measures on $(\Omega, \mathcal{F})$.

We write the event that vertex sets A and B are connected by a path in $G(\omega)$ by $\{A \leftrightarrow B\}$; when G is an infinite tree Γ, we write $\{\rho \leftrightarrow \partial\Gamma\}$ for the event that there is an infinite path emanating from ρ with all edges open.

The connected components of open edges in percolation are called **clusters**, and the cluster containing v is denoted by $\mathcal{C}(v)$. Define

$$\mathcal{C} := \{\exists v \in G \text{ with } |\mathcal{C}(v)| = \infty\};$$

$\mathcal{C}$ is the event that there is an infinite cluster somewhere in the percolation on G. We write $\mathcal{C}_G$ when there is a possibility of ambiguity.

For Bernoulli(p) percolation, at any fixed vertex v,

$$\mathbf{P}_p(|\mathcal{C}(v)| = \infty) > 0 \text{ if and only if } \mathbf{P}_p(\mathcal{C}) = 1. \tag{7}$$

One implication in (7) follows immediately from Kolmogorov's zero-one law: $\mathcal{C}$ does not depend on the status of any finite number of edges, hence $\mathbf{P}_p(\mathcal{C}) \in \{0,1\}$. To see the other implication, assume $\mathbf{P}_p(\mathcal{C}) = 1$ and take a ball $B_n(v)$ large enough so that

$$\mathbf{P}_p(\text{there exists an infinite path intersecting } B_n(v)) > 0.$$

Then clearly

$$\mathbf{P}_p(\partial B_n(v) \leftrightarrow \infty) > 0.$$

Because $B_n(v)$ is finite, the event that all edges in $B_n(v)$ are open has positive probability. By independence of disjoint edge sets,

$$\begin{aligned}
\mathbf{P}_p(|\mathcal{C}(v)| = \infty) &\geq \mathbf{P}_p(\text{all edges in } B_n(v) \text{ are open and } \partial B_n(v) \leftrightarrow \infty) \\
&= \mathbf{P}_p(\text{all edges in } B_n(v) \text{ are open})\mathbf{P}_p(\partial B_n(v) \leftrightarrow \infty) \\
&> 0.
\end{aligned}$$

Alternatively, one can use the FKG inequality for the events $A = \{$all edges in $B_n(v)$ are open$\}$ and $B = \{$there exists an infinite path connecting $B_n(v)$ to $\infty\}$, as both these events are increasing. See Grimmett (1989) for details.

For Bernoulli(p) percolation on an arbitrary graph G, the **critical probability** (already mentioned in the case of trees) is

$$p_c(G) = \inf\{\, p : \mathbf{P}_p(\mathcal{C}) = 1 \,\}.$$

For this definition to make sense, $p \mapsto \mathbf{P}_p(\mathcal{C})$ must be non-decreasing. This can be seen by by coupling the measures $\mathbf{P}_p$ for all p together, see Grimmett (1989).

5 The First-Moment Method

The *first moment method* is straightforward but useful. For general percolation on a tree Γ with root ρ, it asserts that

$$\mathbf{P}(\rho \leftrightarrow \infty) \leq \sum_{v \in \Pi} \mathbf{P}(\rho \leftrightarrow v) \tag{8}$$

for any cutset Π. For Bernoulli(p) percolation on the tree, the inequality becomes

$$\mathbf{P}_p(\rho \leftrightarrow \infty) \leq \sum_{v \in \Pi} p^{|v|}.$$

When $p < 1/\mathrm{br}(\Gamma)$, this can be made arbitrarily small for appropriate choice of cutset. This proves

Proposition 5.1 *For any locally finite Γ,*

$$p_c(\Gamma) \geq \frac{1}{\mathrm{br}(\Gamma)}. \tag{9}$$

In general there is equality here, as advertised previously in Theorem 2.8. The proof of equality is in §7.

Corollary 5.2 *Let T be a GW tree with mean $m > 1$. Almost surely on non-extinction, $\mathrm{br}(T) = m$ and $p_c(T) = 1/m$.*

Proof. Let $\mathbf{P}_{GW}$ be the distribution of T on the space of rooted trees $\mathcal{T}$, and let $Z_n = |T_n|$ be the size of level n of T. Given $t \in \mathcal{T}$, let $\mathbf{P}_{p,t}$ be Bernoulli(p) percolation on t.

Observe that

$$m \geq \overline{\mathrm{gr}}(T) \geq \underline{\mathrm{gr}}(T) \geq \mathrm{br}(T) \geq \frac{1}{p_c(T)}. \tag{10}$$

The first inequality follows since Z_n/m^n converges to a finite random variable, the middle inequalities hold in general, and the right-most is the content of Proposition 5.1. Thus it is enough to show that for $p > m^{-1}$,

$$\mathbf{P}_{GW}\left(t \,:\, \mathbf{P}_{p,t}(|\mathcal{C}(\rho)| = \infty) > 0 \mid \text{non-extinction}\right) = 1. \tag{11}$$

Combine the measures $\mathbf{P}_{GW}$ and $\mathbf{P}_{p,t}$: Given the Galton-Watson tree T, perform Bernoulli(p) percolation on T and let T' be the component of ρ in the percolation. T' is itself a Galton-Watson tree, where the number of individuals in the first generation is $Z_1' = \sum_{i=1}^{Z_1} Y_i$, where $\{Y_i\}$ are i.i.d. Bernoulli(p) random variables. Because $\mathbf{E}[Z_1'] = mp > 1$, with positive probability T' is infinite:

$$\mathbf{P}(|T'| = \infty) = \int \mathbf{P}_{p,t}(|\mathcal{C}(\rho)| = \infty\}) d\mathbf{P}_{GW}(t) > 0.$$

We conclude that the integrand must be positive with positive $\mathbf{P}_{GW}$-probability:

$$\mathbf{P}_{GW}\left(t : \mathbf{P}_{p,t}(|\mathcal{C}(\rho)| = \infty) > 0\right) > 0.$$

Since the set

$$\{t : \mathbf{P}_{p,t}(|\mathcal{C}(\rho)| = \infty) = 0\}$$

defines an inherited property, Proposition 3.2 implies that (11) holds. This proves that a.s. on survival, $p_c(T) = m^{-1}$, whence (10) yields that $\mathrm{br}(T) = m$. $\qquad\square$

Kahane and Peyrière (1976) calculated the dimension of the limit set of fractal percolation; their methods were different. The proof above is due to R. Lyons.

Question 5.3 (Häggström) *Suppose simple random walk $\{X_n\}_{n\geq 0}$ on Γ has positive lower speed, i.e., for some positive number s*

$$\mathbf{P}\left(\liminf_n \frac{|X_n|}{n} > s\right) > 0. \tag{12}$$

Is it necessarily true that $\mathrm{br}(\Gamma) > 1$?

The answer is positive, and the proof relies on the first-moment method again.

Theorem 5.4 *If (12) holds, then $\mathrm{br}(\Gamma) \geq e^{I(s)/s}$, where*

$$I(s) = \frac{1}{2}[(1 + s)\log(1 + s) + (1 - s)\log(1 - s)].$$

Proof. By (12) above, there exists L such that

$$\mathbf{P}\left(|X_n| > ns \text{ for all } n \geq L\right) > 0.$$

Define a general percolation on Γ by

$$\Gamma(\omega) = \left\{v \in \Gamma : |v| \leq L \text{ or } X_n = v \text{ for some } n < |v|s^{-1}\right\}.$$

More precisely, if $e(v)$ denotes the edge from the parent of v to v, we retain $e(v)$ if $|v| \leq L$ or if $X_n = v$ for some $n < |v|s^{-1}$. By the definition of this percolation,

$$\mathbf{P}(\rho \leftrightarrow \infty) \geq \mathbf{P}\left(|X_n| > ns \text{ for all } n \geq L\right) > 0. \tag{13}$$

On the other hand, we claim that if S_n is simple symmetric random walk on $\mathbf{Z}$, then for $|v| > L$,

$$\mathbf{P}(\rho \leftrightarrow v) = \mathbf{P}(X_n = v \text{ for some } n < |v|s^{-1}) \leq \mathbf{P}\left(\max_{n < |v|s^{-1}} |S_n| \geq |v|\right). \tag{14}$$

Consider a particle on Γ which moves with X when X moves along the unique path from ρ to v, but remains stationary during excursions (possibly infinite) of X from this path. This particle performs a simple random walk on the path with (possibly infinite) holding times between moves. The probability on the left in (14) is the chance that this particle reaches v before time $|v|s^{-1}$, which is at most the chance that simple random walk on $\mathbf{Z}$ travels distance $|v|$ from the origin in the same time. This proves (14).

By the reflection principle,

$$\mathbf{P}\left(\max_{n\leq N}|S_n|\geq sN\right)\leq 2\,\mathbf{P}\left(\max_{n\leq N}S_n\geq sN\right)\leq 4\,\mathbf{P}(S_N\geq sN)\leq 4\,e^{-NI(s)}\,,$$

where $I(s)$ is the large deviations rate function for simple random walk on $\mathbf{Z}$ (see, e.g., Durrett 1996). Thus for $|v|>L$ we have

$$\mathbf{P}(\rho\leftrightarrow v)\leq 4\exp\left(-|v|\frac{I(s)}{s}\right)\,.$$

Combine this with (13) and (8) to conclude that if $\lambda=e^{I(s)/s}$, then

$$0<\mathbf{P}(\rho\leftrightarrow\infty)\leq\sum_{v\in\Pi}\mathbf{P}(\rho\leftrightarrow v)\leq 4\sum_{v\in\Pi}\lambda^{-|v|}$$

for any cutset Π at distance more than L from the root. Hence $\mathrm{br}(\Gamma)\geq e^{I(s)/s}$. $\quad\square$

Conjecture 1 *Under the assumptions of Question 5.3 above*

$$s\leq\frac{\mathrm{br}(\Gamma)-1}{\mathrm{br}(\Gamma)+1}\,,\qquad i.e.,\qquad \mathrm{br}(\Gamma)\geq\frac{1+s}{1-s}.$$

Remark. Very recently, this conjecture was proved by B. Virag (1998).

Recall that for simple random walk on the b-ary tree, the speed a.s. equals $\frac{b-1}{b+1}$.

Example 5.5 Take a binary tree and a ternary tree rooted together. The simple random walk on this tree does not have an a.s. constant speed. $\quad\triangle$

The **Fibonacci tree** Γ_{fib} is a subtree of the binary tree. We label vertices as (L) and (R) (for "left" and "right"). The root is labeled (L). Every vertex labeled (L) has two offspring, one labeled (L) and one labeled (R). Every vertex labeled (R) has one offspring, which is labeled (L).

Exercise 5.6 *Justify the name Fibonacci tree. Also, show that*

$$\mathrm{br}(\Gamma_{\mathrm{fib}})=\mathrm{gr}(\Gamma_{\mathrm{fib}})=(1+\sqrt{5})/2.$$

Hint: Use a two state Markov chain to define a 'good' flow.

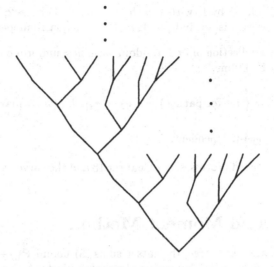

Figure 3: **The Fibonacci tree.**

6   Quasi-independent Percolation

Consider Bernoulli(p) percolation on a tree Γ. If v and w are vertices in Γ, then

$$\mathbf{P}(\rho \leftrightarrow u \ \text{ and } \ \rho \leftrightarrow w) = \frac{p^{|v|}p^{|w|}}{p^{|v \wedge w|}} = \frac{\mathbf{P}(\rho \leftrightarrow u)\mathbf{P}(\rho \leftrightarrow w)}{\mathbf{P}(\rho \leftrightarrow u \wedge w)},$$

where $v \wedge w$ is the vertex at which the paths from the root ρ to v and w separate. This turns out to be a key property of independent percolation, and we therefore make the following definition.

A **quasi-independent percolation** on a tree Γ is any general percolation so that for some $M < \infty$ and any vertices $u, v \in \Gamma$,

$$\mathbf{P}(\rho \leftrightarrow v \ \text{and} \ \rho \leftrightarrow w) \le M \frac{\mathbf{P}(\rho \leftrightarrow u)\mathbf{P}(\rho \leftrightarrow w)}{\mathbf{P}(\rho \leftrightarrow v \wedge w)}. \tag{15}$$

Example 6.1 *Percolation induced by i.i.d. labels.*

1. Let E be the edge set of a tree Γ, and let $\{X_e\}_{e \in E}$ be i.i.d. $\{-1, 1\}$-valued random variables with $\mathbf{P}(X_e = 1) = 1/2$. Write $\mathbf{path}(v)$ for the unique path in Γ from the root to v. A **tree-indexed random walk** $\{S_v\}$ is defined for vertices v of Γ by

$$S_v = \sum_{e \in \mathbf{path}(v)} X_e.$$

Define $\Gamma(\omega) = \{v \colon S_v(\omega) \in [0, b)\}$. For $b = 2$, this is equivalent to Bernoulli($1/2$) percolation: the only infinite paths in $\Gamma(\omega)$ are those for which each 1 is followed by -1, and each -1 by 1 (with 1 in the first step). For $b > 2$, the corresponding percolation process is not independent, but it is quasi-independent.

2. Let $\{U_e\}$ be a collection of i.i.d. random variables, uniform on $[0,1)$, indexed by the edges of Γ. Define

$$\Gamma(\omega) = \left\{ v : \text{ for } \mathbf{path}(v) = e_1 e_2 \cdots e_{|v|}, \ U_{e_1}(\omega) = \max_{k \le |v|} U_{e_k}(\omega) \right\}.$$

This is *not* quasi-independent.

For more on tree-indexed processes, see Chapter 18 and the survey article by Pemantle (1995). △

7 The Second Moment Method

For general percolation on a tree, the cutset sums (8) bound $\mathbf{P}(\rho \leftrightarrow \partial\Gamma)$ from above. We get lower bounds by using the *second moment method*, which we describe next.

By our standing assumption about local finiteness of trees,

$$\{\rho \leftrightarrow \partial\Gamma\} = \bigcap_n \{\rho \leftrightarrow \Gamma_n\}.$$

We extend the definition of the boundary $\partial\Gamma$ to finite trees by

$$\partial\Gamma = \begin{cases} \text{leaves of } \Gamma, \text{ i.e., vertices with no offspring} & \text{if } \Gamma \text{ is finite,} \\ \text{infinite paths starting at } \rho & \text{if } \Gamma \text{ is infinite.} \end{cases}$$

Consider the case Γ finite first. Let μ be a probability measure on $\partial\Gamma$ and set

$$Y = \sum_{x \in \partial\Gamma} \mu(x) \mathbf{1}_{\{\rho \leftrightarrow x\}} \frac{1}{\mathbf{P}(\rho \leftrightarrow x)} .$$

Then $\mathbf{E}[Y] = \sum_{x \in \partial\Gamma} \mu(x) = 1$, and

$$\begin{aligned}
\mathbf{E}[Y^2] &= \mathbf{E}\left[\sum_{x \in \partial\Gamma} \sum_{y \in \partial\Gamma} \mu(x)\mu(y) \frac{\mathbf{1}_{\{\rho \leftrightarrow x\} \cap \{\rho \leftrightarrow y\}}}{\mathbf{P}(\rho \leftrightarrow x)\mathbf{P}(\rho \leftrightarrow y)} \right] \\
&= \sum_{x \in \partial\Gamma} \sum_{y \in \partial\Gamma} \mu(x)\mu(y) \frac{\mathbf{P}(\rho \leftrightarrow x \text{ and } \rho \leftrightarrow y)}{\mathbf{P}(\rho \leftrightarrow x)\mathbf{P}(\rho \leftrightarrow y)} .
\end{aligned}$$

$$(16)$$

Thus, in the case of quasi-independent percolation,

$$\mathbf{E}[Y^2] \le M \sum_{x,y \in \partial\Gamma} \mu(x)\mu(y) \frac{1}{\mathbf{P}(\rho \leftrightarrow x \wedge y)} . \tag{17}$$

In the case of independent percolation, there is an equality with $M = 1$ in (17).
Define the **energy** of the measure μ in the kernel K as

$$\mathcal{E}_K(\mu) = \sum_{x,y \in \partial \Gamma} K(x,y)\mu(x)\mu(y) = \int_{\partial \Gamma} \int_{\partial \Gamma} K(x,y)\mu(dx)\mu(dy).$$

When the kernel is

$$K(x,y) = \frac{1}{\mathbf{P}(\rho \leftrightarrow x \wedge y)} \qquad \text{for } x, y \in \partial \Gamma,$$

(17) can be rewritten as

$$\mathbf{E}[Y^2] \le M\mathcal{E}_K(\mu).$$

By the Cauchy-Schwarz inequality,

$$(\mathbf{E}[Y])^2 = (\mathbf{E}[Y 1_{\{Y>0\}}])^2 \le \mathbf{E}[Y^2]\mathbf{P}(Y > 0),$$

and consequently

$$\mathbf{P}(Y > 0) \ge \frac{(\mathbf{E}[Y])^2}{\mathbf{E}[Y^2]} \ge \frac{1}{M} \frac{1}{\mathcal{E}_K(\mu)}.$$

Since $\mathbf{P}(\rho \leftrightarrow \partial \Gamma) \ge \mathbf{P}(Y > 0)$,

$$\mathbf{P}(\rho \leftrightarrow \partial \Gamma) \ge \frac{1}{M} \frac{1}{\mathcal{E}_K(\mu)}.$$

The left-hand side does not depend on μ, so optimizing the right-hand side with respect to μ yields

$$\mathbf{P}(\rho \leftrightarrow \partial \Gamma) \ge \frac{1}{M} \sup_{\mu : \mu(\partial \Gamma) = 1} \frac{1}{\mathcal{E}_K(\mu)} = \frac{1}{M} \mathrm{Cap}_K(\partial \Gamma), \qquad (18)$$

where we define the **capacity** of $\partial \Gamma$ in the kernel K to be

$$\mathrm{Cap}_K(\partial \Gamma) = \sup_{\mu : \mu(\partial \Gamma) = 1} \frac{1}{\mathcal{E}_K(\mu)}.$$

For Γ infinite, let μ be any probability measure on $\partial \Gamma$. μ induces a probability measure on Γ_n : for a vertex $x \in \Gamma_n$, set

$$\mu(x) = \mu(\text{infinite paths through } x).$$

By the finite case considered above,

$$\mathbf{P}(\rho \leftrightarrow \Gamma_n) \ge \frac{1}{M} \frac{1}{\sum\limits_{x,y \in \Gamma_n} K(x,y)\mu(x)\mu(y)}.$$

Each path ξ from the root ρ to ∞ must pass through some vertex x in Γ_n ; write $x \in \xi$ if the path ξ goes through vertex x. If $x \in \xi$ and $y \in \eta$, then $\xi \wedge \eta$ is a descendant of $x \wedge y$. This implies that $K(x, y) \leq K(\xi, \eta)$ for $x \in \xi$ and $y \in \eta$. Therefore,

$$
\begin{aligned}
\int_{\partial \Gamma} \int_{\partial \Gamma} K(\xi, \eta) d\mu(\xi) d\mu(\eta) &= \sum_{x, y \in \Gamma_n} \int_{x \in \xi} \int_{y \in \eta} K(\xi, \eta) d\mu(\xi) d\mu(\eta) \\
&\geq \sum_{x, y \in \Gamma_n} K(x, y) \mu(x) \mu(y) \\
&\geq \frac{1}{M} \frac{1}{\mathbf{P}(\rho \leftrightarrow \Gamma_n)} .
\end{aligned}
$$

Hence

$$
\mathbf{P}(\rho \leftrightarrow \Gamma_n) \geq \frac{1}{M} \frac{1}{\mathcal{E}_K(\mu)}
$$

for any probability measure μ on $\partial \Gamma$. Optimizing over μ and passing to the limit as $n \to \infty$, we get

$$
\mathbf{P}(\rho \leftrightarrow \partial \Gamma) \geq \frac{1}{M} \mathrm{Cap}_K(\partial \Gamma). \tag{19}
$$

To summarize, we have established the following proposition.

Proposition 7.1 *Let Γ be finite or infinite, $\mathbf{P}$ the probability measure corresponding to a quasi-independent percolation on Γ, and K the kernel on $\partial \Gamma$ defined by $K(x, y) = \mathbf{P}(\rho \leftrightarrow x \wedge y)^{-1}$. Then*

$$
\mathbf{P}(\rho \leftrightarrow \partial \Gamma) \geq \frac{1}{M} \mathrm{Cap}_K(\partial \Gamma), \tag{20}
$$

where $M = 1$ in the case of independent percolation.

For Bernoulli percolation, we have already proven that $p_c(\Gamma) \geq 1/\mathrm{br}(\Gamma)$ in Proposition 5.1, using the first-moment method. We will now prove the reverse inequality, thus showing equality. For convenience, we restate the result.

Theorem 2.8 (R. Lyons 1990) *For Bernoulli(p) percolation on a tree Γ,*

$$
p_c(\Gamma) = 1/\mathrm{br}(\Gamma).
$$

Proof. Take $p > 1/\mathrm{br}(\Gamma)$ and $1/p < \lambda < \mathrm{br}(\Gamma)$. By Lemma 2.10, there exists a unit flow μ from ρ to the boundary satisfying $\mu(v) \leq C\lambda^{-|v|}$ for each vertex $v \in \Gamma$. We may identify μ with a probability measure on $\partial \Gamma$ (see the discussion following Lemma 2.10).

Consider the kernel

$$
K(\xi, \eta) = \frac{1}{\mathbf{P}(\rho \leftrightarrow \xi \wedge \eta)} = p^{-|\xi \wedge \eta|}.
$$

The energy $\mathcal{E}_K(\mu)$ of μ in the kernel K is given by

$$\int_{\partial\Gamma}\int_{\partial\Gamma}p^{-|\xi\wedge\eta|}d\mu(\xi)d\mu(\eta) = \sum_v p^{-|v|}\int\int_{\xi\wedge\eta=v}d\mu(\xi)d\mu(\eta).$$

Since the set of pairs (ξ,η) with $\xi\wedge\eta=v$ is contained in the set of pairs (ξ,η) with $v\in\xi, v\in\eta$, the right-hand side above is not larger than

$$\begin{aligned}
\sum_v p^{-|v|}[\mu(v)]^2 &= \sum_{n=0}^{\infty}p^{-n}\sum_{|v|=n}[\mu(v)]^2 \\
&\leq \sum_{n=0}^{\infty}p^{-n}\sum_{|v|=n}C\lambda^{-|v|}\mu(v) \\
&= C\sum_{n=0}^{\infty}(p\lambda)^{-n}\mu(\Gamma_n).
\end{aligned}$$

The last sum is finite since $\lambda p > 1$. Applying Proposition 7.1 yields

$$\mathbf{P}_p(\rho\leftrightarrow\partial\Gamma) \geq C^{-1}(1-1/\lambda p) > 0.$$

$\square$

8 Electrical Networks

The basic reference for the material in this chapter is Doyle and Snell (1984). Here we will not restrict ourselves to trees, but will discuss general graphs.

While electrical networks are only a different language for reversible Markov chains, the electrical point of view is useful because of the insight gained from the familiar physical laws of electrical networks.

A **network** is a finite connected graph G, endowed with non-negative numbers $\{c_e\}$, called **conductances**, that are associated to the edges of G. The reciprocal $r_e = 1/c_e$ is the **resistance** of the edge e. A network will be denoted by the pair $\langle G, \{c_e\}\rangle$. Vertices of G are often called **nodes**. A real-valued function h defined on the vertices of G is **harmonic** at a vertex x of G if

$$\sum_{y\sim x}\frac{c_{xy}}{\pi_x}h(y) = h(x), \qquad \text{where } \pi_x = \sum_{y\sim x}c_{xy}. \tag{21}$$

(Recall that the notation $y\sim x$ means y is a neighbor of x.)

We distinguish two nodes, $\{a,z\}$, which are called the *source* and the *sink* of the network. A function V which is harmonic on $G\setminus\{a,z\}$ will be called a **voltage**. A voltage is completely determined by its boundary values, V_a, V_z. In particular, the following result is derived from the maximum principle.

Proposition 8.1 *Let h be a function on a network G which is harmonic on $G\setminus\{a,z\}$ and such that $h(a) = h(z) = 0$. Then h must vanish everywhere on G.*

Proof. We will first show that $h \leq 0$. Suppose this is not the case. Then $h(x_*) := \max_G h > 0$. By harmonicity on $G \setminus \{a, z\}$, if $x \notin \{a, z\}$ belongs to the set $A = \{x : h(x) = \max_G h\}$ and $y \sim x$, then $y \in A$ also. By connectedness, $a, z \in A$, hence $h(a) = h(z) = \max_G h > 0$, contradicting our assumption. Thus $h \leq 0$, and an application of this result to $-h$ also yields $h \geq 0$. $\qquad\qquad\square$

This proves that given boundary conditions $h(a) = x$ and $h(z) = y$, if there is a function harmonic on $G \setminus \{a, z\}$ with these boundary conditions, it is unique. To prove that a harmonic function with given boundary values exists, observe that the conditions (21) in the definition of harmonic functions form a system of linear equations with the same number of equations as unknowns, namely (number of nodes in G) $- 2$; for such a system, uniqueness of solutions implies existence.

A more informative way to prove existence is via the probabilistic interpretation of harmonic functions and voltages. Consider the Markov chain on the nodes of G with transition probabilities

$$p_{xy} = \mathbf{P}(X_{n+1} = y \mid X_n = x) = \frac{c_{xy}}{\pi_x}.$$

This process is called the **weighted random walk** on G with edge weights $\{c_e\}$, or the Markov chain associated to the network $\langle G, \{c_e\} \rangle$. This Markov chain is reversible with respect to the measure π:

$$\pi_x p_{xy} = c_{xy} = \pi_y p_{yx}.$$

A special case is the **simple random walk** on G, which has transition probabilities

$$p_{xy} = \frac{1}{\deg(x)} \qquad \text{for } y \sim x$$

and corresponds to the weighted walk with conductances $c_{xy} = 1$ for $y \sim x$.

To get a voltage with boundary values 0 and 1 at z and a respectively, set

$$V_x^* = \mathbf{P}_x(\{X_n\} \text{ hits } a \text{ before } z),$$

where $\mathbf{P}_x$ is the probability for the walk started at node x. For arbitrary boundary values V_a and V_z, define

$$V_x = V_z + V_x^*(V_a - V_z).$$

Until now, we have focused on *undirected graphs*. Now we need to consider also **directed graphs**. An edge in a directed graph is an *ordered* pair of nodes (x, y), which we denote by $\vec{e} = \vec{xy}$.

A **flow** θ from a to z, previously discussed when the underlying graph is a tree, is a function on oriented edges which is antisymmetric, $\theta(\vec{xy}) = -\theta(\vec{yx})$, and which obeys **Kirchhoff's node law** $\sum_{w \sim v} \theta(\vec{vw}) = 0$ at all $v \notin \{a, z\}$. This is just the requirement "flow in equals flow out" for any node $\neq a, z$. Despite notational differences, it is easily seen that these definitions generalize the ones given earlier for trees.

Observe that it is only flows that are defined on oriented edges. Conductance and resistance are defined for unoriented edges; we may of course define them on oriented edges by $c_{\vec{xy}} = c_{\vec{yx}} = c_{xy}$ and $r_{\vec{xy}} = r_{\vec{yx}} = r_{xy}$.

Given a voltage V on the network, the **current flow** associated with V is defined on oriented edges by

$$I(\vec{e}) = \frac{V_y - V_x}{r_e}, \qquad \text{where } \vec{e} = \vec{xy}.$$

Notice that I is antisymmetric and satisfies the node law at every $x \notin \{a, z\}$:

$$\sum_{y \sim x} I(\vec{xy}) = \sum_{y \sim x} c_{xy}(V_y - V_x) = 0.$$

Thus the node law for the current is equivalent to the harmonicity of the voltage.

The current flow also satisfies the **cycle law**: if the edges $\vec{e}_1, \ldots, \vec{e}_m$ form a cycle, i.e., $\vec{e}_i = \overrightarrow{x_{i-1}x_i}$ and $x_n = x_0$, then

$$\sum_{i=1}^{m} r_{e_i} I(\vec{e}_i) = 0.$$

Finally, by definition, a current flow also satisfies **Ohm's law**: if $\vec{e} = \vec{xy}$,

$$r_e I(\vec{e}) = V_y - V_x.$$

The particular values of a voltage function V are less important than the voltage differences, so fix a voltage function V on the network normalized to have $V_z = 0$.

By definition, if θ is an arbitrary flow on oriented edges satisfying Ohm's law $r_{xy}\theta(\vec{xy}) = V_y - V_x$ (with respect to the voltage V), then θ equals the current flow I associated with V.

Define the **strength** of an arbitrary flow θ as

$$\|\theta\| = \sum_{x \sim a} \theta(\vec{ax}).$$

Proposition 8.2 (Node law/cycle law/strength) *If θ is a flow from a to z satisfying the cycle law*

$$\sum_{i=1}^{m} r_{e_i} \theta(\vec{e}_i) = 0$$

for any cycle $\vec{e}_1 \ldots, \vec{e}_m$, and if $\|\theta\| = \|I\|$, then $\theta = I$.

Proof. The function $J = \theta - I$ satisfies the node-law at all nodes and the cycle law. Define

$$h(x) = \sum_{i=1}^{m} J(\vec{e}_i) r_{e_i},$$

where $\vec{e}_i, \ldots, \vec{e}_m$ is an arbitrary path from a to x. By the cycle law, J is well defined. By the node law, it is harmonic everywhere, except possibly at a and z. Now $\|\theta\| = \|I\|$ implies that J is also harmonic at a and z. By the maximum principle, h must be constant. This implies that $J = 0$. $\qquad\qquad\square$

Given a network, the ratio $(V_a - V_z)/\|I\|$, where I is the current flow corresponding to the voltage V, is independent of the voltage V applied to the network. Define the **effective resistance** between vertices a and z as

$$\mathcal{R}(a \leftrightarrow z) := \frac{V_a - V_z}{\|I\|}.$$

We think of effective resistance as follows: replace the whole network by a single edge joining a to z and require that the two networks be equivalent, in the sense that the amount of current flowing from a to z in the new network is the same as in the original network if we apply the same voltage to both.

Next, we discuss the probabilistic interpretation of effective resistance. Denote

$$\mathbf{P}(a \to z) := \mathbf{P}_a(\text{hit } z \text{ before returning to } a).$$

For any vertex x

$$\mathbf{P}_x(\text{hit } z \text{ before } a) = \frac{V_a - V_x}{V_a - V_z}.$$

If $p_{xy} = c_{xy}\pi_x^{-1}$ are the transition probabilities of the Markov chain, then

$$
\begin{aligned}
\mathbf{P}(a \to z) &= \sum_x p_{ax}\mathbf{P}_x(\text{hit } z \text{ before } a) \\
&= \sum_{x \sim a} \frac{c_{ax}}{\pi_a} \frac{V_a - V_x}{V_a - V_z} \\
&= \frac{1}{\pi_a(V_a - V_z)} \sum_{x \sim a} I(a\vec{x}) \\
&= \frac{1}{\pi_a} \frac{\|I\|}{V_a - V_z} \\
&= \frac{1}{\pi_a \mathcal{R}(a \leftrightarrow z)}.
\end{aligned}
$$

Call $[\mathcal{R}(a \leftrightarrow z)]^{-1}$ the **effective conductance**, written as $\mathcal{C}(a \leftrightarrow z)$. Then

$$\mathbf{P}(a \to z) = \frac{1}{\pi_a}\mathcal{C}(a \leftrightarrow z). \tag{22}$$

The **Green function** for the random walk stopped at z, is defined by

$$G(a, x) = \mathbf{E}_a[\# \text{ visits to } x \text{ before hitting } z].$$

(The subscript in $\mathbf{E}_a$ indicates the initial state.) Then $G(a, a) = \pi_a \mathcal{R}(a \leftrightarrow z)$, since the number of visits to a before visiting z has a geometric distribution with parameter $\mathbf{P}(a \to z)$. It is often possible to replace a network by a simplified one without changing quantities of interest, for example the effective resistance between a pair of nodes. The following laws are very useful.

Parallel Law. Conductances in parallel add: Suppose edges e_1 and e_2, with conductances c_1 and c_2 respectively, share vertices v_1 and v_2 as endpoints. Then both edges can be replaced with a single edge of conductance $c_1 + c_2$ without affecting the rest of the network. All voltages and currents in $G \setminus \{e_1, e_2\}$ are unchanged and the current $I(\vec{e})$ equals $I(\vec{e}_1) + I(\vec{e}_2)$. For a proof, check Ohm's and Kirchhoff's laws with $I(\vec{e}) := I(\vec{e}_1) + I(\vec{e}_2)$.

Series Law. Resistances in series add: If $v \in G \setminus \{a, z\}$ is a node of degree 2 with neighbors v_1 and v_2, the edges (v_1, v) and (v, v_2) can be replaced by a single edge (v_1, v_2) of resistance $r_{v_1 v} + r_{v v_2}$. All potentials and currents in $G \setminus \{v\}$ remain the same and the current that flows from v_1 to v_2 equals $I(\overrightarrow{v_1 v}) = I(\overrightarrow{v v_2})$. For a proof, check again Ohm's and Kirchhoff's laws, with $I(\overrightarrow{v_1 v_2}) := I(\overrightarrow{v_1 v}) = I(\overrightarrow{v v_2})$.

Glue. Another convenient operation is to identify vertices having the same voltage, while keeping all existing edges. Because current never flows between vertices with the same voltage, potentials and currents are unchanged.

Example 8.3 Consider a spherically symmetric tree Γ, a tree in which all vertices of Γ_n have the same number of children for all $n \geq 0$. Suppose that all edges at the same distance from the root have the same resistance, that is, $r_e = r_i$ if $|e| = i$, $i \geq 1$. Glue all the vertices in each level; This will not affect effective resistances, so we infer that

$$\mathcal{R}(\rho \leftrightarrow \Gamma_N) = \sum_{i=1}^{N} \frac{r_i}{|\Gamma_i|}$$

and

$$\mathbf{P}(\rho \to \Gamma_N) = \frac{r_1/|\Gamma_1|}{\sum_{i=1}^{N} r_i/|\Gamma_i|} .$$

Therefore the corresponding random walk on Γ is transient iff $\sum_{i=1}^{\infty} r_i/|\Gamma_i| < \infty$. $\triangle$

Theorem 8.4 (Thomson's Principle) *For any finite connected graph,*

$$\mathcal{R}(a \leftrightarrow z) = \inf \{ \mathcal{E}(\theta) : \theta \text{ a unit flow from } a \text{ to } z \},$$

where $\mathcal{E}(\theta) := \sum_e [\theta(e)]^2 r_e$. The unique minimizer in the inf above is the unit current flow.

Note: The sum in $\mathcal{E}(\theta)$ is over unoriented edges, so each edge $\{x, y\}$ is only considered once in the definition of energy. Although θ is defined on oriented edges, it is antisymmetric and hence $\theta(e)^2$ is unambiguous.

Proof. By compactness, there exists flows minimizing $\mathcal{E}(\theta)$ subject to $\|\theta\| = 1$. By Proposition 8.2, to prove that the unit current flow is the unique minimizer, it is enough to verify that any unit flow θ of minimal energy satisfies the cycle law.

Let the edges $\vec{e}_1, \ldots \vec{e}_n$ form a cycle. Set $\gamma(\vec{e}_i) = 1$ for all $1 \le i \le n$ and set γ equal to zero on all other edges. Note that γ satisfies the node law, so it is a flow, but $\sum \gamma(\vec{e}_i) = n \ne 0$. For any $\epsilon \in \mathbf{R}$, we have that

$$0 \le \mathcal{E}(\theta + \epsilon\gamma) - \mathcal{E}(\theta) = \frac{1}{2}\sum_{i=1}^{n}[(\theta(\vec{e}_i) + \epsilon)^2 - \theta(\vec{e}_i)^2]r_{e_i} = \epsilon\sum_{i=1}^{n}r_{e_i}\theta(\vec{e}_i) + O(\epsilon^2).$$

By taking $\epsilon \to 0$ from above and from below, we see that $\sum_{i=1}^{n}r_{e_i}\theta(\vec{e}_i) = 0$, thus verifying that θ satisfies the cycle law.

To complete the proof, we show that the unit current flow I has $\mathcal{E}(I) = \mathcal{R}(a \leftrightarrow z)$:

$$\sum_{e}r_e I(e)^2 = \frac{1}{2}\sum_{x}\sum_{y}r_{xy}\left(\frac{V_y - V_x}{r_{xy}}\right)^2$$

$$= \frac{1}{2}\sum_{x}\sum_{y}c_{xy}(V_y - V_x)^2$$

$$= \frac{1}{2}\sum_{x}\sum_{y}(V_y - V_x)I(\vec{xy}).$$

Since I is antisymmetric,

$$\frac{1}{2}\sum_{x}\sum_{y}(V_y - V_x)I(\vec{xy}) = -\sum_{x}V_x\sum_{y}I(\vec{xy}). \tag{23}$$

Applying the node law and recalling that $\|I\| = 1$, we conclude that the right-hand side of (23) is equal to

$$\frac{V_z - V_a}{\|I\|} = \mathcal{R}(a \leftrightarrow z).$$

$\square$

Let a, z be vertices in a network, and suppose that we add to the network an edge which is not incident to a. How does this affect the escape probability from a to z? Probabilistically the answer is not obvious. In the language of electrical networks, this question is answered by:

Theorem 8.5 (Rayleigh's Monotonicity Law) *If $\{r_e\}$ and $\{r'_e\}$ are sets of resistances on the edges of the same graph G, and if $r_e \le r_{e'}$ for all e, then*

$$\mathcal{R}(a \leftrightarrow z; r) \le \mathcal{R}(a \leftrightarrow z; r').$$

Proof. Note that $\inf_{\theta}\sum_{e}r_e\theta(e)^2 \le \inf_{\theta}\sum_{e}r'_e\theta(e)^2$ and apply Thomson's Principle (Theorem 8.4).

$\square$

Corollary 8.6 *Adding an edge weakly decreases the effective resistance* $\mathcal{R}(a \leftrightarrow z)$. *If the added edge is not incident to* a, *the addition weakly increases the escape probability* $\mathbf{P}(a \to z) = [\pi_a \mathcal{R}(a \leftrightarrow z)]^{-1}$.

Proof. Before we add an edge to a network we can think of it as existing already with $c = 0$ or $r = \infty$. By adding the edge we reduce its resistance to a finite number. □

Thus, combining the relationship (22) and Corollary 8.6 shows that the addition of an edge not incident to a (which we regard as changing a conductance from 0 to 1) cannot decrease the escape probability $\mathbf{P}(a \to z)$.

Exercise 8.7 *Show that* $\mathcal{R}(a \leftrightarrow z)$ *is a concave function of* $\{r_e\}$.

Corollary 8.8 *The operation of gluing vertices cannot increase effective resistance.*

Proof. When we glue vertices together, we take an infimum over a larger class of flows. □

Moreover, if we glue together vertices with different potentials, then effective resistance will strictly decrease.

9 Infinite Networks

For an infinite graph G containing vertex a, let $\{G_n\}$ be a collection of finite connected subgraphs containing a and satisfying $\cup_n G_n = G$. If all the vertices in $G \setminus G_n$ are replaced by a single vertex z_n, then

$$\mathcal{R}(a \leftrightarrow \infty) := \lim_{n \to \infty} \mathcal{R}(a \leftrightarrow z_n \text{ in } G_n \cup \{z_n\}).$$

Now

$$\mathbf{P}(a \to \infty) = \frac{\mathcal{C}(a \leftrightarrow \infty)}{\pi_a}.$$

A flow on G from a to infinity is an antisymmetric edge function obeying the node law at all vertices except a. Thomson's Principle remains valid for infinite networks:

$$\mathcal{R}(a \leftrightarrow \infty) = \inf \{\mathcal{E}(\theta) : \theta \text{ a unit flow from } a \text{ to } \infty\}. \tag{24}$$

Let us summarize the facts in the following proposition.

Proposition 9.1 *Let* $\langle G, \{c_e\} \rangle$ *be a network. The following are equivalent.*

1. *The weighted random walk on the network is transient.*

2. *There is some node* a *with* $\mathcal{C}(a \leftrightarrow \infty) > 0$ *(equivalently,* $\mathcal{R}(a \leftrightarrow \infty) < \infty$*).*

3. *There is a flow* θ *from some node* a *to infinity with* $\|\theta\| > 0$ *and* $\mathcal{E}(\theta) < \infty$.

In particular, any subgraph of a recurrent graph must be recurrent.

Recall that an edge-cutset Π separating a from z is a set of edges so that any path from a to z must include some edge in Π.

Corollary 9.2 (Nash-Williams (1959)) *If $\{\Pi_n\}$ are disjoint edge-cutsets which separate a from z, then*

$$\mathcal{R}(a \leftrightarrow z) \geq \sum_n \left(\sum_{e \in \Pi_n} c_e \right)^{-1}. \tag{25}$$

In an infinite network $\langle G, \{c_e\} \rangle$, the analogous statement with z replaced by ∞ is also valid; in particular, if there exist disjoint edge-cutsets $\{\Pi_n\}$ that separate a from ∞ and satisfy

$$\sum_n \left(\sum_{e \in \Pi_n} c_e \right)^{-1} = \infty,$$

then the weighted random walk on $\langle G, \{c_e\} \rangle$ is recurrent.

Proof. Let θ be a unit flow from a to z. For any n

$$\sum_{e \in \Pi_n} c_e \cdot \sum_{e \in \Pi_n} r_e \theta(e)^2 \geq \left(\sum_{e \in \Pi_n} \sqrt{c_e} \sqrt{r_e} |\theta(e)| \right)^2 = \left(\sum_{e \in \Pi_n} |\theta(e)| \right)^2 \geq \|\theta\|^2 = 1,$$

because Π_n is a cutset and $\|\theta\| = 1$. Therefore

$$\sum_e r_e \theta(e)^2 \geq \sum_n \sum_{e \in \Pi_n} r_e \theta(e)^2 \geq \sum_n \left(\sum_{e \in \Pi_n} c_e \right)^{-1}.$$

$\square$

Example 9.3 ($\mathbf{Z}^2$ is recurrent) Take $r_e = 1$ on $G = \mathbf{Z}^2$ and consider the cutsets consisting of edges joining vertices in $\partial \square_n$ to vertices in $\partial \square_{n+1}$, where $\square_n = [-n, n]^2$. Then by Nash-Williams (25),

$$\mathcal{R}(a \leftrightarrow \infty) \geq \sum_n \frac{1}{4(2n+1)} = \infty.$$

Thus simple random walk on $\mathbf{Z}^2$ is recurrent. Moreover, we obtain a lower bound for the resistance from the center of a square $\square_n = [-n, n]^2$ to its boundary:

$$\mathcal{R}(0 \leftrightarrow \partial \square_n) \geq c \log n.$$

In the next chapter, we will obtain an upper bound of the same type. $\triangle$

The Nash-Williams inequality (25) is useful, but in general is not sharp. For example, for the 3-1 tree in Example 2.6, the effective resistance from the root to ∞ is infinite because the random walk is recurrent, yet the right-hand side of (25) is at most 1 for any sequence of disjoint cutsets (prove this, or see Lyons and Peres 1999).

Example 9.4 (Z^3 is transient) To each directed edge $\vec{e}$ in the lattice Z^3, attach an orthogonal unit square $\square_e$ intersecting $\vec{e}$ at its midpoint m_e. Define $\theta(\vec{e})$ to be the area of the radial projection of $\square_e$ onto the sphere $\partial B(0, \frac{1}{4})$, taken with a positive sign if $\vec{e}$ points in the same direction as the radial vector from 0 to m_e, and with a negative sign otherwise. By considering a unit cube centered at each lattice point and projecting it to $\partial B(0, \frac{1}{4})$, we can easily verify that θ satisfies the node law at all vertices except the origin. Hence θ is a flow from 0 to ∞ in Z^3. It is easy to bound its energy:

$$\mathcal{E}(\theta) \le \sum_n C_1 n^2 \left(\frac{C_2}{n^2}\right)^2 < \infty.$$

By Proposition 9.1, Z^3 is transient. This works for any Z^d, $d \ge 3$. An analytic description of the same flow was given by T. Lyons (1983).

$\triangle$

Exercise 9.5 *Fix $k > 1$. Define the k-**fuzz** of an undirected graph $G = (V, E)$ as the graph $G_k = (V, E_k)$ where for any two distinct vertices $v, w \in V$, the edge $\{v, w\}$ is in E_k iff there is a path of at most k edges in E connecting v to w. Show that for G with bounded degrees, G is transient iff G_k is transient.*

A solution can be found in Doyle and Snell (1984, §8.4).

10   The Method of Random Paths

A **self-avoiding path** from a to z is a sequence of vertices $v_0, \ldots, v_n$ such that $v_0 = a$ and $v_n = z$, adjacent vertices v_{i-1} and v_i are connected by an edge, and $v_i \ne v_j$ for $i \ne j$. If φ and ψ are two self-avoiding paths from a to z, define

$$|\varphi \cap \psi| = \text{number of edges in the intersection of } \varphi \text{ and } \psi.$$

If $\vec{e}$ is the oriented edge pointing from vertex v to w, let $\overleftarrow{e}$ be the reversed edge pointing from w to v. If μ is a measure on the set of self-avoiding paths from a to z, define

$$\mu(e) = \mu(\varphi : \varphi \ni e) = \mu(\varphi : \varphi \ni \vec{e} \text{ or } \varphi \ni \overleftarrow{e}).$$

The Nash-Williams inequality yields lower bounds for effective resistance. For upper bounds the following result is useful. Assume that $r_e = 1$ for all e; the result can be extended easily to arbitrary resistances.

Theorem 10.1 (Method of random paths)

$$\mathcal{R}(a \leftrightarrow z) = \inf_\mu \sum_e [\mu(e)]^2 = \inf_\mu \mathbf{E}_{\mu \times \mu}[\,|\varphi \cap \psi|\,],$$

where the infimum is over all probability measures μ on the set of self-avoiding paths from a to z, and φ and ψ are independent paths with distribution μ. Similarly, if there is a measure μ on infinite self-avoiding paths in a graph G with $\mathbf{E}_{\mu \times \mu}[\,|\varphi \cap \psi|\,] < \infty$, then simple random walk on G is transient.

Remark. The useful direction here is $\mathcal{R}(a \leftrightarrow z) \leq \sum \mu(e)^2$ for all μ.

Proof. The second equality is trivial: write $|\varphi \cap \psi|$ as $\sum_e \mathbf{1}_{\{\varphi \ni e, \psi \ni e\}}$.

Given a probability measure μ on the set of self-avoiding paths from a to z, define

$$\begin{aligned} \theta(\vec{e}) &:= \mu(\varphi : \varphi \ni \vec{e}) - \mu(\varphi : \varphi \ni \overleftarrow{e}) \\ &= \mathbf{E}_\mu \left[\mathbf{1}\{\varphi \ni \vec{e}\} - \mathbf{1}\{\varphi \ni \overleftarrow{e}\} \right]. \end{aligned}$$

By definition, θ is antisymmetric. To see that θ obeys the node law, observe that

$$\sum_{w:w \sim v} \theta(v\vec{w}) = \mathbf{E}_\mu \left[\sum_{w:w \sim v} \mathbf{1}\{\varphi \ni v\vec{w}\} - \mathbf{1}\{\varphi \ni \vec{tw}\} \right].$$

Assume $v \notin \{a, z\}$. If, for a sample path φ, a term in the sum is nonzero, then φ must use either an edge directed *to* v or an edge directed *from* v. But because φ is a self-avoiding walk which terminates at z, it must also use exactly one other edge incident to v, in the first case directed *away* from v and in the second case directed *to* v. Hence the net contribution of φ to the sum is zero. We conclude that θ is a flow.

Clearly, θ is a unit flow, *i.e.*.

$$|\theta| = \sum_{x \sim a} \theta(a\vec{x}) = 1.$$

so we can apply Thomson's principle:

$$\mathcal{R}(a \leftrightarrow z) \leq \sum_e [\theta(e)]^2 \leq \sum_e [\mu(e)]^2.$$

The other inequality $\mathcal{R}(a \leftrightarrow z) \geq \inf_\mu \sum \mu(e)^2$ will not be used in these notes, so we only sketch a proof. Let I denote a unit current flow. Then

$$\mathcal{R}(a \leftrightarrow z) = \sum_e I(e)^2.$$

Notice that a unit current flow is acyclic. Define a Markov chain by making transitions according to the flow I normalized. This chain then defines a measure on paths and $\mu(\vec{e}) = I(\vec{e})$, because I is acyclic. For details, see Lyons and Peres (1999). □

Example 10.2 In $\mathbf{Z}^2$, consider the boundary $\partial \square_n = \{x \in \mathbf{Z}^2 : \|x\|_1 = n\}$ of the square $\square_n = [-n, n]^2$. Using Nash-Williams we have seen that

$$\mathcal{R}(0 \leftrightarrow \square_n) \geq c \log n.$$

Now define a measure μ on self-avoiding paths in $\square_n$ as follows: Pick a ray $\vec{\ell}$ emanating from the origin in a random uniformly distributed direction, and let μ be the distribution of the lattice path that best approximates ℓ. By considering edges e according to their distance from the origin, we also get

$$\sum_e [\mu(e)]^2 \leq \sum_{k=1}^n c_1 k \left(\frac{c_2}{k}\right)^2 \leq C \log n.$$

So in $\mathbf{Z}^2$ we have

$$c \log n \leq \mathcal{R}(0 \leftrightarrow \partial \square_n) \leq C \log n .$$

$\triangle$

Example 10.3 In $\mathbf{Z}^3$, define μ analogously, but this time on the whole infinite lattice. Now

$$\mathcal{R}(0 \leftrightarrow \infty) \leq \sum_k c_1 k^2 \left(\frac{c_2}{k^2} \right)^2 < \infty .$$

$\triangle$

Example 10.4 (Wedges in $\mathbf{Z}^3$) Given a non-negative and non-decreasing function f, consider the wedge

$$W_f = \{(x, y, z) : 0 \leq y \leq x, \ 0 \leq z \leq f(x)\} .$$

By Nash-Williams, the resistance from the origin to ∞ in W_f satisfies

$$\mathcal{R}(0 \leftrightarrow \infty) \geq C \sum_k \frac{1}{k f(k)} .$$

In particular, if this sum diverges, then W_f is recurrent. The converse also holds: $\triangle$

Theorem 10.5 (T. Lyons 1983) *If $\sum_k [k f(k)]^{-1} < \infty$, then the wedge W_f is transient.*

Proof Idea. Choose a random point (U_1, U_2) according to the uniform distribution on $[0, 1]^2$ and find the lattice path closest to $\{(k, U_1 k, U_2 f(k))\}_{k=0}^{\infty}$. The completion of this proof is left as an exercise. $\square$

11 Transience of Percolation Clusters

The graph $\mathbf{Z}^3$ supports a flow of finite energy, described in Example 9.4, and hence simple random walk in three dimensions is transient. Equivalently, if each edge of $\mathbf{Z}^3$ is assigned unit conductance, then the effective conductance from any vertex to infinity is positive. If a finite number of edges are removed, then the random walk on the infinite component of the modified graph is also transient, because the effective conductance remains nonzero.

A much deeper result, first proved by Grimmett, Kesten, and Zhang (1993), is that if $d \geq 3$ and $p > p_c(\mathbf{Z}^d)$, then simple random walk on $\mathcal{C}_\infty(\mathbf{Z}^d, p)$ is transient, where $\mathcal{C}_\infty(\mathbf{Z}^d, p)$ is the unique infinite cluster of Bernoulli(p) percolation on $\mathbf{Z}^d$. Benjamini, Pemantle and Peres (1998) (hereafter referred to as BPP (1998)) gave an alternative proof of this result and extended it to high-density oriented percolation. Their argument uses certain "unpredictable" random paths that have *exponential intersection tails* to construct random flows of finite energy on $\mathcal{C}_\infty(\mathbf{Z}^d, p)$.

Let $G = (V_G, E_G)$ be an infinite graph with all vertices of finite degree and let $v_0 \in V_G$. Denote by $\Upsilon = \Upsilon(G, v_0)$ the collection of infinite oriented paths in G which emanate from v_0. Let $\Upsilon_1 = \Upsilon_1(G, v_0) \subset \Upsilon$ be the set of **paths with unit speed**, *i.e.*, those paths for which the nth vertex is at distance n from v_0.

Let $0 < \zeta < 1$. A Borel probability measure μ on $\Upsilon(G, v_0)$ has **exponential intersection tails** with parameter ζ (in short, EIT(ζ)) if there exists C such that

$$\mu \times \mu \{(\varphi, \psi) : |\varphi \cap \psi| \geq n\} \leq C\zeta^n \tag{26}$$

for all n, where $|\varphi \cap \psi|$ is the number of edges in the intersection of φ and ψ. If such a measure μ exists for some basepoint v_0 and some $\zeta < 1$, then we say that G *admits random paths with* EIT(ζ). By the previous chapter, such a graph G must be transient.

Theorem 11.1 (Cox-Durrett 1983, BPP 1998) *For every $d \geq 3$, there exists $\zeta < 1$ such that the lattice $\mathbf{Z}^d$ admits random paths with EIT(ζ).*

Proof: For $d \geq 4$, we will show (following Cox and Durrett 1983, who attribute the idea to Kesten) that the "uniform distribution" on $\Upsilon_1(\mathbf{Z}^d, 0)$ has the required EIT property; for $d = 3$ such a simple choice cannot work, and we will delay the proof to Chapter 17. Let $d \geq 4$, and define μ to be the distribution of the random walk with i.i.d. increments uniformly distributed on the d standard basis vectors $(1, 0, \ldots, 0), \ldots, (0, \ldots, 0, 1)$. Let $\{X_n\}$ and $\{Y_m\}$ be two independent random walks with distribution μ. It suffices to show that the number of vertex intersections of these two walks has an exponential tail. Since $\|X_n\|_1 = n$ for all n, we can have $X_n = Y_m$ only if $n = m$. The process $\{X_n - Y_n\}$ is a mean 0 random walk in the $d-1$ dimensional sublattice of $\mathbf{Z}^d$ consisting of vectors orthogonal to $(1, 1, \ldots, 1)$, and its increments generate this sublattice. Since $d - 1 \geq 3$, the random walk $\{X_n - Y_n\}$ is transient, and (26) holds with

$$\zeta := \mathbf{P}[\exists n \geq 1 \ \ X_n - Y_n = 0], \quad \text{and} \quad C = 1.$$

$\square$

Proposition 11.2 (BPP 1998) *Suppose that the directed graph G admits random paths with EIT(ζ), and consider Bernoulli(p) percolation on G. If $p > \zeta$ then with probability 1 there is a vertex v in G such that the open cluster $C(v)$ is transient.*

Proof. The hypothesis means that there is some vertex v_0 and a probability measure μ on $\Upsilon = \Upsilon(G, v_0)$ satisfying (26). We will assume here that μ is supported on Υ_1; the general case is treated in BPP (1998).

For $N \geq 1$ and any infinite path $\varphi \in \Upsilon_1(G, v_0)$, denote by φ_N the finite path consisting of the first N edges of φ. Consider the random variable

$$Z_N = \int_{\Upsilon_1} p^{-N} \mathbf{1}_{\{\varphi_N \text{ is open}\}} d\mu(\varphi). \tag{27}$$

Except for the normalization factor p^{-N}, this is the μ-measure of the paths that stay in the open cluster of v_0 for N steps.

Since each edge is open with probability p (independently of other edges), $\mathbf{E}(Z_N) = 1$, but we can say more. Let $\mathcal{B}_N$ be the σ-field generated by the status (open or closed) of all edges on paths φ_N with $\varphi \in \Upsilon_1$. It is easy to check that for each $\varphi \in \Upsilon_1$, the sequence $\{p^{-N} \mathbf{1}_{\{\varphi_N \text{ is open}\}}\}$ is a martingale adapted to the filtration $\{\mathcal{B}_N\}_{N \geq 1}$. Consequently, $\{Z_N\}_{N \geq 1}$ is also a non-negative martingale. By the Martingale Convergence Theorem, $\{Z_N\}$ converges a.s. to a random variable Z_∞. In fact, we now show that $\{Z_N\}$ is bounded in L^2, and hence converges in L^2. Since each edge is open with probability p (independently of other edges), $\mathbf{E}(Z_N) = 1$. The second moment of Z_N satisfies

$$
\begin{aligned}
\mathbf{E}(Z_N^2) &= \mathbf{E} \int_{\Upsilon_1} \int_{\Upsilon_1} p^{-2N} \mathbf{1}_{\{\varphi_N \text{ and } \psi_N \text{ are open}\}} \, d\mu(\varphi) \, d\mu(\psi) \\
&\leq \int_{\Upsilon_1} \int_{\Upsilon_1} p^{-|\varphi \cap \psi|} \, d\mu(\varphi) \, d\mu(\psi) \\
&= \sum_{k=1}^\infty p^{-k} \mu \times \mu\{(\varphi, \psi) : |\varphi \cap \psi| = k\} \, .
\end{aligned}
\tag{28}
$$

By (26), the sum on the right-hand side of (28) is bounded by $\sum_{k=1}^\infty C \left(\frac{\zeta}{p}\right)^k$, which does not depend on N and is finite since $\zeta < p$.

On the event $\{Z_\infty > 0\}$, the cluster $\mathcal{C}(v_0)$ is infinite, and by Cauchy-Schwarz,

$$
\mathbf{P}(|\mathcal{C}(v_0)| = \infty) \geq \mathbf{P}(Z_\infty > 0) \geq \frac{(\mathbf{E}Z_\infty)^2}{\mathbf{E}Z_\infty^2} \, .
$$

Since $\mathbf{E}Z_N^2$ is bounded, by Fatou's Lemma the right-hand side is positive. Thus with positive probability $\mathcal{C}(v_0)$ is infinite, and it remains to prove that $\mathcal{C}(v_0)$ is a.s. transient on this event.

We will construct a flow of finite energy on $\mathcal{C}(v_0)$. For each $N \geq 1$, and every edge $\vec{e}$ directed away from v_0, define

$$
f_N(\vec{e}) = \int_\Upsilon p^{-N} \mathbf{1}_{\{\varphi_N \text{ is open}\}} \mathbf{1}_{\{\vec{e} \in \varphi_N\}} \, d\mu(\varphi) \, .
\tag{29}
$$

If $\vec{e}$ is directed towards v_0, let $f(\vec{e}) = -f(\overleftarrow{e})$, where $\overleftarrow{e}$ is the reversal of $\vec{e}$. Let $B(v_0, N)$ denote the set of all vertices within distance N of v_0. Then f_N is a flow on $\mathcal{C}(v_0) \cap B(v_0, N+1)$ from v_0 to the complement of $B(v_0, N)$, i.e., for any vertex $v \in B(v_0, N)$ except v_0, the incoming flow to v equals the outgoing flow from v. The strength of f_N (the total outflow from v_0) is exactly Z_N.

Next, we estimate the expected energy of f_N by summing over edges directed away from v_0:

$$
\mathbf{E} \sum_{\vec{e}} f_N(\vec{e})^2 = \mathbf{E} \int_{\Upsilon_1} \int_{\Upsilon_1} p^{-2N} \mathbf{1}_{\{\varphi_N, \psi_N \text{ are open}\}} \sum_{\vec{e}} \mathbf{1}_{\{\vec{e} \in \varphi_N\}} \mathbf{1}_{\{\vec{e} \in \psi_N\}} \, d\mu(\varphi) \, d\mu(\psi)
$$

$$\leq \int_{\Upsilon_1}\int_{\Upsilon_1} |\varphi \cap \psi| \, p^{-|\varphi \cap \psi|} \, d\mu(\varphi) \, d\mu(\psi)$$

$$= \sum_{k=1}^{\infty} k p^{-k} \mu \times \mu\{(\varphi, \psi) : |\varphi \cap \psi| = k\}. \tag{30}$$

Again using (26) and $p > \zeta$, from (30) we conclude that

$$\mathbf{E} \sum_{\vec{e}} f_N(\vec{e})^2 \leq \sum_{k=1}^{\infty} k \left(\frac{\zeta}{p}\right)^k = C < \infty, \tag{31}$$

where C does not depend on N.

For each directed edge $\vec{e}$ of G, the sequence $\{f_N(\vec{e})\}$ is a $\{\mathcal{B}_N\}$-martingale which converges a.s. and in L^2 to a nonnegative random variable $f(\vec{e})$. The edge function f is a flow with strength Z_∞ on $\mathcal{C}(v_0)$, and has finite expected energy by (31) and Fatou's Lemma.

Thus

$$\mathbf{P}[\mathcal{C}(v_0) \text{ is transient}] \geq \mathbf{P}[Z_\infty > 0] > 0,$$

so the tail event $\{\exists v : \mathcal{C}(v) \text{ is transient}\}$ must have probability 1 by Kolmogorov's zero-one law. $\qquad\square$

Theorem 11.3 (Grimmett, Kesten and Zhang 1993) *Consider Bernoulli(p) percolation on $\mathbf{Z}^d$, where $d \geq 3$. For all $p > p_c$, the unique infinite cluster is a.s. transient.*

Proof. It follows from Theorem 11.1 and Proposition 11.2 that the infinite cluster is transient if p is close enough to 1.

Recall that a set of graphs **B** is called **increasing** if for any graph G that contains a subgraph in **B**, necessarily G must also be in **B**.

Consider now percolation with *any* parameter $p > p_c$ in $\mathbf{Z}^d$. Following Pisztora (1996), call an open cluster $\mathcal{C}$ contained in some cube Q a *crossing cluster* for Q if for all d directions there is an open path contained in $\mathcal{C}$ joining the left face of Q to the right face. For each v in the lattice $N\mathbf{Z}^d$, denote by $\square_N(v)$ the cube of side-length $5N/4$ in $\mathbf{Z}^d$, centered at v. Let $A_p(N)$ be the set of $v \in N\mathbf{Z}^d$ with the following property: *The cube $\square_N(v)$ contains a crossing cluster $\mathcal{C}$ such that any open cluster in $\square_N(v)$ of diameter greater than $N/10$ is connected to $\mathcal{C}$ by an open path in $\square_N(v)$.*

Proposition 2.1 in Antal and Pisztora (1996), which relies on the work of Grimmett and Marstrand (1990), implies that $A_p(N)$ stochastically dominates site percolation with parameter $p_*(N)$ on the stretched lattice $N\mathbf{Z}^d$, where $p_*(N) \to 1$ as $N \to \infty$. By Liggett, Schonmann and Stacey (1996), it follows that $A_p(N)$ stochastically dominates bond percolation with parameter $p^*(N)$ on $N\mathbf{Z}^d$, where $p^*(N) \to 1$ as $N \to \infty$. This domination means that for any increasing Borel set of graphs **B**, the probability that the subgraph of open sites under independent bond percolation with parameter $p^*(N)$ lies in **B**, is at most $\mathbf{P}[A_p(N) \in \mathbf{B}]$. If N is sufficiently large, then the infinite cluster determined by bond percolation with parameter $p^*(N)$ on the lattice $N\mathbf{Z}^d$, is a.s. transient. The set of subgraphs of $N\mathbf{Z}^d$ that contain a transient subgraph is increasing, so $A_p(N)$ contains a transient subgraph $\hat{A}_p(N)$ with probability 1. Observe that $\hat{A}_p(N)$ is isomorphic to a subgraph of the "$3N^d$-fuzz" of the infinite cluster C_p in the original lattice, so by Rayleigh's monotonicity principle, we conclude that C_p is also transient a.s. (See Ex. 9.5, or §8.4 in Doyle and Snell (1984) for the definition and properties of the k-fuzz of a graph.) Alternatively, it can be verified that $\hat{A}_p(N)$ is "roughly isometric" to a subgraph of C_p, and therefore C_p is transient a.s. (see Soardi 1994). □

Remark. Hiemer (1998) proved a renormalization theorem for oriented percolation, that allowed him to extend the result of [6] on transience of oriented percolation clusters in $\mathbf{Z}^d$ for $d \geq 3$, from the case of high p to the whole supercritical phase for oriented percolation.

Recall that a collection of edges Π is a **cutset separating** v_0 **from** ∞ if any infinite self-avoiding path emanating from v_0 must intersect Π. Nash-Williams proved that if $\{\Pi_n\}_{n=1}^\infty$ is a sequence of disjoint cutsets separating v_0 from infinity in a connected transient graph, then $\sum_n |\Pi_n|^{-1} < \infty$.

The following extension of Theorem 11.3 provides finer information about the permissible growth rates of cutsets on supercritical infinite percolation clusters.

Exercise 11.4 *Show that for $d \geq 2$,*

$$\inf\{q : \exists \text{ a flow } f \neq 0 \text{ from } 0 \text{ to } \infty \text{ on } \mathbf{Z}^d \text{ with } \sum |f(e)|^q < \infty\} = \frac{d}{d-1}.$$

Theorem 11.5 (Levin and Peres 1998) *Let $C_\infty(\mathbf{Z}^d, p)$ be the infinite cluster of Bernoulli(p) percolation on $\mathbf{Z}^d$. Then for $d \geq 3$ and $p > p_c(\mathbf{Z}^d)$, a.s.,*

$$\inf\{q : \exists \text{ a flow } f \neq 0 \text{ from } 0 \text{ to } \infty \text{ on } C_\infty(\mathbf{Z}^d, p) \text{ with } \sum_e |f(e)|^q < \infty\} = \frac{d}{d-1}.$$

Corollary 11.6 *Let $d \geq 3$ and $p > p_c(\mathbf{Z}^d)$. With probability one, if $\{\Pi_n\}$ is a sequence of disjoint cutsets in the infinite cluster $C_\infty(\mathbf{Z}^d, p)$ that separate a fixed vertex v_0 from ∞, then $\sum_n |\Pi_n|^{-\beta} < \infty$ for all $\beta > \frac{1}{d-1}$.*

Proof. Pick $\beta > \frac{1}{d-1}$, and let f be a unit flow on $C_\infty(\mathbf{Z}^d, p)$ with $\sum |f(e)|^{1+\beta} < \infty$, which exists by Theorem 11.5. Observe first that

$$\mathcal{E}_{1+\beta}(f) = \sum_{e \in E_G} |f(e)|^{1+\beta} \geq \sum_n \sum_{e \in \Pi_n} |f(e)|^{1+\beta},$$

since the $\{\Pi_n\}$ are disjoint. By Jensen's inequality, for all $n \geq 1$,

$$\frac{1}{|\Pi_n|} \sum_{e \in \Pi_n} |f(e)|^{1+\beta} \geq \left(\frac{1}{|\Pi_n|} \sum_{e \in \Pi_n} |f(e)|\right)^{1+\beta} = |\Pi_n|^{-1-\beta}.$$

Multiplying by $|\Pi_n|$ and summing over n establishes the Corollary. □

Remark. Theorem 11.5 was recently sharpened by Hoffman and Mossel.

12 Subperiodic Trees

For a tree Γ, let Γ^v denote the subtree of Γ rooted at vertex v that contains all descendants of v. Γ is N-**subperiodic** if for any vertex $v \in \Gamma$ there exists a 1-1 adjacency preserving map $f \colon \Gamma^v \to \Gamma^{f(v)}$ with $|f(v)| \leq N$.

Example 12.1 *Examples of subperiodic trees.*

- b-ary trees for any integer $b \geq 2$.

- The Fibonacci tree Γ_{fib} described in Exercise 5.6.

- The tree of all self-avoiding walks in $\mathbf{Z}^d$.

- Directed covers of finite connected directed graphs: to every directed path of length n in the graph corresponds $v \in \Gamma$ with $|v| = n$; extensions of the path correspond to descendants of v.

- Universal covers of undirected graphs: to every non-backtracking path of length n in the graph corresponds $v \in \Gamma$ with $|v| = n$; extensions correspond to descendants, as above.

$\triangle$

Suppose that $b \geq 2$ is an integer. For a closed nonempty set $\Lambda \subseteq [0, 1]$, define a tree $\Gamma(\Lambda, b)$ as follows. Consider the system of b-adic subintervals of $[0, 1]$; those which have a non-empty intersection with Λ form the vertices of the tree. Two vertices are connected by an edge if one of the corresponding intervals is contained in the other and their orders differ by one (i.e., the ratio of lengths is b). The root of this tree is $[0, 1]$. Clearly, $\Gamma([0, 1], b)$ is the usual b-ary tree. If $b\Lambda(\text{mod } 1) \subset \Lambda$, i.e., Λ is invariant under the transformation $x \mapsto bx(\text{mod } 1)$, then $\Gamma(\Lambda, b)$ is 0-subperiodic.

Theorem 12.2 (Furstenberg 1967) *For Γ which is subperiodic, $\mathrm{gr}(\Gamma)$ exists and $\mathrm{gr}(\Gamma) = \mathrm{br}(\Gamma)$. Furthermore,*

$$\inf_{\Pi} S(\mathrm{br}(\Gamma), \Pi) > 0 \,,$$

where $S(\lambda, \Pi) = \sum_{v \in \Pi} \lambda^{-|v|}$ for a cutset Π.

Corollary 12.3 (Furstenberg's formulation) *Let $\Lambda \subseteq [0, 1]$ be a compact set. If $b\Lambda(\text{mod } 1) \subseteq \Lambda$, then*

$$\dim_H(\Lambda) = \dim_M(\Lambda) = \beta$$

for some β, and moreover, $\mathcal{H}^\beta(\Lambda) > 0$, where $\mathcal{H}^\beta$ denotes β-dimensional Hausdorff measure.

Proof of Theorem 12.2. We will give the proof for Γ 0-subperiodic. The N-subperiodic case can be reduced to the 0-subperiodic case; this reduction is left as an exercise. Assume first that Γ has no leaves.

Suppose that for some finite cutset Π,

$$S(\lambda, \Pi) < 1. \tag{32}$$

Denote $d = \max_{v \in \Pi} |v|$. By 0-subperiodicity, for any $v \in \Pi$, there exists a cutset $\Pi(v)$ of Γ^v such that

$$\sum_{w \in \Pi(v)} \lambda^{-(|w|-|v|)} < 1.$$

In other words,

$$\sum_{w \in \Pi(v)} \lambda^{-|w|} < \lambda^{-|v|}.$$

Replace v in Π by the vertices in $\Pi(v)$ to obtain a new cutset $\tilde{\Pi}$ in Γ with $S(\lambda, \tilde{\Pi}) < 1$. Given n, repeat this kind of replacement for every vertex v in the current cutset with $|v| \leq n$ to get a cutset Π^* such that all vertices $u \in \Pi^*$ satisfy $n \leq |u| \leq n + d$. Then

$$|\Gamma_n| \lambda^{-n-d} \leq S(\lambda, \Pi^*) < 1.$$

This inequality depends on the assumption of no leaves. Thus $|\Gamma_n| < \lambda^{n+d}$ for all n, whence $\overline{\mathrm{gr}}(\Gamma) \leq \lambda$. Since (32) holds for any $\lambda > \mathrm{br}(\Gamma)$, we infer that $\overline{\mathrm{gr}}(\Gamma) < \lambda$. Therefore

$$\overline{\mathrm{gr}}(\Gamma) \leq \mathrm{br}(\Gamma) \leq \underline{\mathrm{gr}}(\Gamma).$$

Finally, consider $\lambda_1 = \mathrm{br}(\Gamma)$. If $S(\lambda_1, \Pi) < 1$ for some finite cutset Π, then we could find $\lambda < \lambda_1$ such that $S(\lambda, \Pi) < 1$, and the preceding argument would yield that $\overline{\mathrm{gr}}(\Gamma) \leq \lambda < \lambda_1$, a contradiction. Thus for all cutsets Π,

$$S(\mathrm{br}(\Gamma), \Pi) \geq 1.$$

If Γ has leaves, create a modified tree Γ' by attaching to each leaf an infinite path. Γ' is periodic as well, and so the theorem can be applied to it, yielding $\mathrm{br}(\Gamma') = \mathrm{gr}(\Gamma')$. But since $\mathrm{br}(\Gamma) = \mathrm{br}(\Gamma')$ and $\overline{\mathrm{gr}}(\Gamma) \leq \overline{\mathrm{gr}}(\Gamma')$, we have

$$\mathrm{br}(\Gamma) \leq \underline{\mathrm{gr}}(\Gamma) \leq \overline{\mathrm{gr}}(\Gamma) \leq \overline{\mathrm{gr}}(\Gamma') = \mathrm{br}(\Gamma') = \mathrm{br}(\Gamma),$$

and hence $\mathrm{gr}(\Gamma) = \mathrm{br}(\Gamma)$. □

Exercise 12.4 *Construct a subperiodic tree with superlinear polynomial growth (more precisely, construct a subperiodic tree T such that $|T_n| \| to\infty$ as $n \to \infty$, but $|T_n| = O(n^d)$ for some $d < \infty$.*

(Hint: build a subtree of the binary tree where all finite paths are labeled by words in the Morse sequence $0110100110010110\ldots$. This sequence is obtained by iterating the substitution $0 \mapsto 01$, $1 \mapsto 10$. Alternatively, use a lexicographic spanning tree in $\mathbf{Z}^d$, as described in the next chapter.)

Exercise 12.5 *Does every subperiodic tree with exponential growth have a subtree without leaves that has bounded pipes?*

(Hint: Consider the subtree T of the binary tree T_2, containing all self-avoiding paths from the root in T_2 with the property that for every $n > 100$, any n^2 consecutive edges on the path contain a run of n consecutive left turns.)

13   The Random Walks RW_λ

For a graph G, fix an origin o, and define $|e|$ as the length of a shortest path from o to an end-vertex of e. We will define a family of processes RW_λ as weighted random walks on G. Specifically, each edge e is assigned conductance $\lambda^{-|e|}$. We will mostly consider the case where Γ is a tree and o is the root ρ, although we will also consider these processes defined on Cayley graphs of groups. By fine tuning λ, we obtain random walks that explore the graph better than the simple random walk. The following result is stronger than Theorem 2.9 mentioned in Chapter 2.

Theorem 13.1 (R. Lyons 1990) RW_λ *is transient on a tree Γ if $\lambda < \mathrm{br}(\Gamma)$, and recurrent if $\lambda > \mathrm{br}(\Gamma)$.*

Proof. If $\lambda > \mathrm{br}(\Gamma)$, then for any ϵ there exists a cutset Π such that $\sum_{v\in\Pi} \lambda^{-|v|} < \epsilon$. By Nash-Williams (for just one cutset)

$$\mathcal{R}(\rho \leftrightarrow \infty) \geq \frac{1}{\sum\limits_{v\in\Pi} \lambda^{-|v|}} > \frac{1}{\epsilon}.$$

Letting $\epsilon \downarrow 0$ shows that $\mathcal{R}(\rho \leftrightarrow \infty)$ is infinite, and hence the walk is recurrent.

If $\lambda < \mathrm{br}(\Gamma)$ choose $\lambda < \lambda_* < \mathrm{br}(\Gamma)$ so that there exists a unit flow θ from ρ to ∞ with $\theta(e) \leq C\lambda_*^{-|e|}$. Then

$$\mathcal{E}(\theta) = \sum_e r_e[\theta(e)]^2 \leq \sum_n \lambda^n \sum_{|e|=n} \theta(e) C\lambda_*^{-|e|} = C\sum_n \left(\frac{\lambda}{\lambda_*}\right)^n \sum_{|e|=n} \theta(e) < \infty,$$

since θ is a unit flow. $\qquad\qquad\square$

Let G be a countable group with a finite set of generators $S = \langle g_1, \ldots g_m \rangle$. With every generator we include its inverse, so $S = S^{-1}$. The **Cayley graph** of G has as vertices the elements of the group, and contains an (unoriented) edge between u and v if $u = g_i v$ for some $g_i \in S$. Each element $g \in G$ can be represented as a word in the generators, $g = g_{i(1)} \cdots g_{i(m)}$; let $|g|$ be the minimal length of words which represent g, and let $G_n = \{g \in G : |g| = n\}$. The growth $\mathrm{gr}(G) := \lim_n |G_n|^{1/n}$ exists for such groups, and the group is of exponential growth if $\mathrm{gr}(G) > 1$.

Corollary 13.2 (R. Lyons 1995) RW_λ *on the Cayley graph of a group G of exponential growth is transient for $\lambda < \mathrm{gr}(G)$ and recurrent for $\lambda > \mathrm{gr}(G)$.*

Proof. The second statement follows from the Nash-Williams inequality. For the first, we will show that random walk on a subgraph is transient; by Rayleigh's Monotonicity Principle, this is enough. We will use the lexicographic spanning tree Γ in G. Assign g its lexicographically minimal representation $g = g_{i(1)} \cdots g_{i(m)}$ where $m = |g|$ and if $g = g_{j(1)} \cdots g_{j(m)}$ is another representation of g, then at the smallest k such that $i(k) \neq j(k)$ we have $i(k) < j(k)$. The edge gh is in Γ if $\big| |g| - |h| \big| = 1$ and either g is an initial segment of h or h is an initial segment of g. Let the identity be the root. Since there is a unique path from the root to any element in Γ, and Γ contains all elements of G, it is indeed a spanning tree. One can check that it is 0–subperiodic.

Observe that $|\Gamma_n| = |G_n|$, so $\mathrm{gr}(\Gamma) = \mathrm{gr}(G)$. Since Γ is subperiodic, Theorem 12.2 implies that $\mathrm{br}(\Gamma) = \mathrm{gr}(G)$. By Theorem 13.1, for $\lambda < \mathrm{gr}(G)$ the biased walk RW_λ is transient on Γ, hence also on G. $\qquad\square$

Open Problem 1 *For $1 < \lambda < \mathrm{gr}(G)$, is it true that*

$$\mathrm{speed}(\mathrm{RW}_\lambda) := \lim_{n\to\infty} \frac{|X_n|}{n} > 0, \qquad a.s. \ ?$$

Here $|v|$ denotes the distance of v from the identity.

We remark that there exist groups of exponential growth where the speed of simple random walk is 0 a.s. An example is the simple random walk on the lamplighter group; see Lyons, Pemantle and Peres (1996).

14 Capacity

In Chapter 6 we considered capacity on the boundary of a tree. We now generalize the definition to any metric space X equipped with the Borel σ-field $\mathcal{B}$. A **kernel** F is a measurable function $F : X \times X \to [0,\infty]$. For a measure μ on $(X, \mathcal{B})$, the **energy** of μ in the kernel F is defined as

$$\mathcal{E}_F(\mu) = \int_X \int_X F(x,y) d\mu(x) d\mu(y).$$

We will mostly consider F of the form $F(x,y) = f(|x - y|)$ for f non-negative and non-increasing; we write $\mathcal{E}_f$ for $\mathcal{E}_F$ in this case. Define the **capacity** of a set Λ in the kernel F as

$$\mathrm{Cap}_F(\Lambda) = \left[\inf_{\mu:\mu(\Lambda)=1} \mathcal{E}_F(\mu) \right]^{-1}.$$

The first occurrence of capacity in probability theory was the following result.

Theorem 14.1 (Kakutani 1944a, 1944b) *If $\Lambda \subset \mathbf{R}^d$ is compact with $0 \notin \Lambda$ and B is a Brownian motion, then*

$$\mathbf{P}_0(B \text{ hits } \Lambda) > 0 \ \text{ if and only if } \ \mathrm{Cap}_G(\Lambda) > 0,$$

where G is the Green kernel

$$G(x,y) \;=\; \begin{cases} |x-y|^{2-d} & d \geq 3, \\ \log^+\left(|x-y|^{-1}\right) & d = 2. \end{cases}$$

R. Lyons discovered connections between capacity and percolation on trees, already discussed in Chapter 6. Let $\{p_e\}$ be a set of probabilities indexed by the edges of a tree Γ. Let $\mathrm{path}(v)$ denote the unique path from the root to v, and let F be the kernel

$$F(x,y) \;=\; \prod_{e \in \mathrm{path}(x \wedge y)} p_e^{-1}. \tag{33}$$

If $p_e \equiv p$, then $F(x,y) = p^{-|x \wedge y|}$. More generally, if $\mathbf{P}$ is the probability measure corresponding to independent $\{p_e\}$ percolation, then $F(x,y) = [\mathbf{P}(\rho \leftrightarrow x \wedge y)]^{-1}$. A. H. Fan proved that on an infinite tree of bounded degree, $\mathbf{P}(\rho \leftrightarrow \partial\Gamma) > 0$ iff $\mathrm{Cap}_F(\partial\Gamma) > 0$. This was sharpened by R. Lyons to a quantitative estimate.

Theorem 14.2 (R. Lyons 1992) *Let $\mathbf{P}$ be the probability measure corresponding to independent $\{p_e\}$ percolation on a tree Γ and F the kernel defined in (33). Then*

$$\mathrm{Cap}_F(\partial\Gamma) \;\leq\; \mathbf{P}(\rho \leftrightarrow \partial\Gamma) \;\leq\; 2\mathrm{Cap}_F(\partial\Gamma). \tag{34}$$

Consider Brownian motion in dimension $d \geq 3$. One obstacle to obtaining quantitative estimates for Brownian hitting probabilities with capacity in Green's kernel is translation invariance of that kernel: If B is a Brownian motion started at the origin, then $\mathbf{P}(B \text{ hits } \Lambda + x)$ becomes small as $x \to \infty$. If we had a scale invariant kernel instead, we would have more hope, as $\mathbf{P}(B \text{ hits } c\Lambda) = \mathbf{P}(B \text{ hits } \Lambda)$ for any $c > 0$. Hence we use capacity in the **Martin kernel**

$$K(x,y) \;=\; \frac{G(x,y)}{G(0,y)} \;=\; \left(\frac{|y|}{|x-y|}\right)^{d-2} \tag{35}$$

for $d \geq 3$.

Theorem 14.3 (Benjamini, Pemantle, and Peres 1995) *Let B be a Brownian motion in $\mathbf{R}^d$ for $d \geq 3$, started at the origin. Let K be the Martin kernel defined in (35). Then for any closed set Λ in $\mathbf{R}^d$,*

$$\frac{1}{2}\mathrm{Cap}_K(\Lambda) \;\leq\; \mathbf{P}_0(B \text{ hits } \Lambda) \;\leq\; \mathrm{Cap}_K(\Lambda).$$

Remark. An analogous statement holds for planar Brownian motion, provided it is killed at an appropriate finite stopping time (e.g., an independent exponential time, or the first exit from a bounded domain) and the corresponding Green function $G(x,y)$ is used to define the Martin Kernel.

Theorem 14.4 (BPP 1995) *Let $\{X_n\}$ be a transient Markov chain on a countable state space S with initial state $\rho \in S$, and set*

$$G(x,y) \;=\; \mathbf{E}_x\left[\sum_{n=0}^{\infty} \mathbf{1}_{\{y\}}(X_n)\right] \;\; and \;\; K(x,y) \;=\; \frac{G(x,y)}{G(\rho,y)}\,.$$

Then for any initial state ρ and any subset Λ of S,

$$\frac{1}{2}\mathrm{Cap}_K(\Lambda) \leq \mathbf{P}_\rho(\{X_n\}\ hits\ \Lambda) \leq \mathrm{Cap}_K(\Lambda)\,.$$

Exercise 14.5 *Verify the analogous result for the stable-$\frac{1}{2}$ subordinator and the kernel*

$$G(s,t) := \begin{cases} (t-s)^{-1/2} & 0 < s \leq t, \\ 0 & s > t > 0. \end{cases}$$

Problem: Find the class of Markov processes for which the above estimate (for suitable kernel G and resulting K) holds.

Proof of Theorem 14.4. To prove the right-hand inequality, we may assume that the hitting probability is positive. Let $\tau = \inf\{n : X_n \in \Lambda\}$ and let ν be the measure $\nu(A) = \mathbf{P}_\rho(\tau < \infty$ and $X_\tau \in A)$. In general, ν is a sub-probability measure, as τ may be infinite. By the Markov property, for $y \in \Lambda$,

$$\int_\Lambda G(x,y)d\nu(x) = \sum_{x\in\Lambda} \mathbf{P}_\rho(X_\tau = x)G(x,y) = G(\rho,y)\,,$$

whence $\int_\Lambda K(x,y)d\nu(x) = 1$. Therefore $\mathcal{E}_K(\nu) = \nu(\Lambda)$, $\mathcal{E}_K(\nu/\nu(\Lambda)) = [\nu(\Lambda)]^{-1}$; consequently, since $\nu/\nu(\Lambda)$ is a probability measure,

$$\mathrm{Cap}_K(\Lambda) \;\geq\; \nu(\Lambda) \;=\; \mathbf{P}_\rho(\{X_n\}\ hits\ \Lambda)\,.$$

This yields one inequality. Note that the Markov property was used here.

For the reverse inequality, we use the second moment method. Given a probability measure μ on Λ, set

$$Z = \int_\Lambda \sum_{n=0}^{\infty} \mathbf{1}_{\{y\}}(X_n)\frac{d\mu(y)}{G(\rho,y)}\,.$$

$\mathbf{E}_\rho[Z] = 1$, and the second moment satisfies

$$\mathbf{E}_\rho[Z^2] \;=\; \mathbf{E}_\rho\int_\Lambda\int_\Lambda \sum_{m=0}^{\infty}\sum_{n=0}^{\infty} \mathbf{1}_{\{x\}}(X_m)\mathbf{1}_{\{y\}}(X_n)\frac{d\mu(x)d\mu(y)}{G(\rho,x)G(\rho,y)}$$

$$\leq\; 2\mathbf{E}_\rho\int_\Lambda\int_\Lambda \sum_{m\leq n} \mathbf{1}_{\{x\}}(X_m)\mathbf{1}_{\{y\}}(X_n)\frac{d\mu(x)d\mu(y)}{G(\rho,x)G(\rho,y)}\,.$$

Observe that

$$\sum_{m=0}^{\infty}\mathbf{E}_\rho\sum_{n=m}^{\infty} \mathbf{1}_{\{x\}}(X_m)\mathbf{1}_{\{y\}}(X_n) = \sum_{m=0}^{\infty}\mathbf{P}_\rho(X_m = x)G(x,y) = G(\rho,x)G(x,y)\,.$$

Hence

$$\mathbf{E}_\rho[Z^2] \le 2 \int_\Lambda \int_\Lambda \frac{G(x,y)}{G(\rho,y)} d\mu(x) d\mu(y) = 2\mathcal{E}_K(\mu),$$

and therefore

$$\mathbf{P}_\rho(\{X_n\} \text{ hits } \Lambda) \ge \mathbf{P}_\rho(Z > 0) \ge \frac{(\mathbf{E}_\rho[Z])^2}{\mathbf{E}_\rho[Z^2]} \ge \frac{1}{2\mathcal{E}_K(\mu)}.$$

We conclude that $\mathbf{P}_\rho(\{X_n\} \text{ hits } \Lambda) \ge \frac{1}{2}\mathrm{Cap}_K(\Lambda)$. □

The upper bound on $\mathbf{P}(\rho \leftrightarrow \partial\Gamma)$ obtained by the first moment method (8) is not sharp enough to prove Theorem 14.2. For example, take the binary tree with Bernoulli(p) percolation for $p = \frac{1}{2}$; if $\Gamma_n = \{v : |v| \le n\}$, then the first-moment method yields an upper bound of 1 for any n, while $\mathrm{Cap}_F(\partial\Gamma_n) = 2(n+2)^{-1}$. However, we can use Theorem 14.4 to give a short proof of Theorem 14.2.

Proof of Theorem 14.2. The first inequality was already proven in Proposition 7.1.

It remains to prove the right-hand inequality in (34). Assume first that Γ is finite. There is a Markov chain $\{V_k\}$ hiding here: Embed Γ in the lower half-plane, with the root at the origin. The random set of $r \ge 0$ leaves that survive the percolation may be enumerated from left to right as $V_1, V_2, \ldots, V_r$. The key observation is that the random sequence $\rho, V_1, V_2, \ldots, V_r, \Delta, \Delta, \ldots$ is a Markov chain on the state space $\partial\Gamma \cup \{\rho, \Delta\}$, where ρ is the root and Δ is a formal absorbing cemetery.

Indeed, given that $V_k = x$, all the edges on the unique path from ρ to x are retained, so that survival of leaves to the right of x is determined by the edges strictly to the right of the path from ρ to x, and is thus conditionally independent of $V_1, \ldots, V_{k-1}$. This verifies the Markov property, so Theorem 14.4 may be applied.

The transition probabilities for the Markov chain above are complicated, but it is easy to write down the Green kernel. Clearly, $G(\rho, y)$ equals the probability that y survives percolation, so

$$G(\rho, y) = \prod_{e \in \mathrm{path}(y)} p_e.$$

If x is to the left of y, then $G(x, y)$ is equal to the probability that the range of the Markov chain contains y given that it contains x, which is just the probability of y surviving given that x survives. Therefore,

$$G(x, y) = \prod_{e \in \mathrm{path}(y) \setminus \mathrm{path}(x)} p_e,$$

and hence

$$K(x, y) = \frac{G(x, y)}{G(\rho, y)} = \prod_{e \in \mathrm{path}(x \wedge y)} p_e^{-1}.$$

Now $G(x, y) = 0$ for x on the right of y; thus (keeping the diagonal in mind)

$$F(x, y) \le K(x, y) + K(y, x)$$

for all $x, y \in \partial\Gamma$, and therefore

$$\mathcal{E}_F(\partial\Gamma) \leq 2\mathcal{E}_K(\partial\Gamma).$$

Now apply Theorem 14.4 to $\Lambda = \partial\Gamma$:

$$\mathrm{Cap}_F(\partial\Gamma) \geq \frac{1}{2}\mathrm{Cap}_K(\partial\Gamma) \geq \frac{1}{2}\mathbf{P}(\{V_k\} \text{ hits } \partial\Gamma) = \frac{1}{2}\mathbf{P}(\rho \leftrightarrow \partial\Gamma).$$

This establishes the upper bound for finite Γ.

The inequality for general Γ follows from the finite case by taking limits. $\qquad\square$

Remark. The inequality (34) was recently sharpened by Marchal [68].

The notation $\mathcal{E}$ has appeared twice, once as a functional on *flows* and once as a functional on *measures*. As discussed following Lemma 2.10, measures on the boundary of a tree correspond to flows on the tree; we shall see that the energy of a measure on $\partial\Gamma$ is (up to an additive constant) the same as the energy of the corresponding flow on Γ: Given a measure μ on $\partial\Gamma$, let θ be the corresponding flow: $\theta(uv) = \mu(\xi : v \in \xi)$, where u is the parent of v. Observe that

$$\mathcal{E}(\theta) = \sum_e r_e\theta(e)^2 = \sum_e r_e \int_{\partial\Gamma}\int_{\partial\Gamma} \mathbf{1}_{\{\xi \ni e\}}\mathbf{1}_{\{\eta \ni e\}}d\mu(\xi)d\mu(\eta).$$

Moving the sum inside the integral, the above equals

$$\int_{\partial\Gamma}\int_{\partial\Gamma} \sum_e \mathbf{1}_{\{e \in \xi \cap \eta\}}r_e d\mu(\xi)d\mu(\eta) = \int_{\partial\Gamma}\int_{\partial\Gamma} \sum_{e \leq \xi \wedge \eta} r_e d\mu(\xi)d\mu(\eta).$$

By the series law for resistances, we are left with

$$\mathcal{E}(\theta) = \int_{\partial\Gamma}\int_{\partial\Gamma} \mathcal{R}(\rho \leftrightarrow \xi \wedge \eta)d\mu(\xi)d\mu(\eta). \tag{36}$$

Now if

$$1/\mathcal{C}(\rho \leftrightarrow v) + 1 = 1/\mathbf{P}(\rho \leftrightarrow v), \tag{37}$$

then substituting in (36) yields

$$\mathcal{E}_K(\mu) = 1 + \mathcal{E}(\theta), \tag{38}$$

where $K(\xi, \eta) = 1/\mathbf{P}(\rho \leftrightarrow \xi \wedge \eta)$. By taking infimum on both sides of (38) and applying Thomson's Principle, we can rewrite Theorem 14.2: If the correspondence (37) holds for resistances $\{r_e\}$ and an independent $\{p_e\}$ percolation $\mathbf{P}$, then

$$\frac{1}{1 + \mathcal{R}(\rho \leftrightarrow \infty)} \leq \mathbf{P}(\rho \leftrightarrow \infty) \leq \frac{2}{1 + \mathcal{R}(\rho \leftrightarrow \infty)}. \tag{39}$$

It is easily checked that in the case of Bernoulli(p) percolation, the correspondence (37) is preserved by taking $c_e = (1 - p)^{-1}p^{|v|}$, where e is the edge connecting v to its parent. In this case the weighted random walk on the resulting network is $RW_{1/p}$. Thus, (39) implies that percolation occurs at p if and only if $RW_{1/p}$ is transient.

Consider a Cantor set Λ in the unit interval and the corresponding tree $\Gamma(\Lambda, b)$. We shall see that simple random walk on this tree is transient iff Λ, considered as a subset of $\mathbf{R}^2$, is non-polar for Brownian motion. In particular, transience of $\Gamma(\Lambda, b)$ is independent of b. The following theorem can be found in Benjamini and Peres (1992) in a special case, and in Pemantle and Peres (1995b) in general.

Theorem 14.6 *Let Γ be a subtree of the b^d-adic tree and let $f : (0, \infty) \to (0, \infty)$ be a non-increasing function with $f(0+) = \infty$. Let Ψ be the canonical map from the boundary of the b^d-adic tree to $[0, 1]^d$; Ψ^{-1} is base-b representation of points in $[0, 1]^d$. Let $\mathrm{dist}(v, w) = b^{-|v \wedge w|}$ for $v, w \in \partial\Gamma$ and let $\mathrm{dist}(x, y)$ be Euclidean distance for $x, y \in [0, 1]^d$. Then*

$$\mathrm{Cap}_f(\partial\Gamma) \asymp \mathrm{Cap}_f(\Psi(\partial\Gamma)),$$

where Cap_f stands for capacity in the kernel $F(x, y) = f(\mathrm{dist}(x, y))$. This means there exist constants c and C, depending on b and d only, such that

$$c \, \mathrm{Cap}_f(\Psi(\partial\Gamma)) \leq \mathrm{Cap}_f(\partial\Gamma) \leq C \, \mathrm{Cap}_f(\Psi(\partial\Gamma)).$$

Exercise 14.7 *Consider Bernoulli(p) percolation on an infinite tree Γ. Prove that*

$$\mathbf{P}_p(\text{component of } \rho \text{ is transient}) > 0 \quad \text{iff} \quad \mathbf{P}_{\{p_e\}}(\rho \leftrightarrow \partial\Gamma) > 0,$$

where $p_e = \frac{k}{k+1}p$ when $|e| = k$.

Hint: An infinite tree T is transient iff $\mathrm{Cap}_{|x \wedge y|}(\partial T) > 0$. The kernel $|x \wedge y|$ is obtained by applying $f(r) = -\log_b r$ to the distance between x and y.

Proof of Theorem 14.6 For $v \in \Gamma$, let $\mu(v) = \mu(\xi : \xi \ni v)$. We will prove that $\mathcal{E}_f(\mu) \asymp \mathcal{E}_f(\mu\Psi^{-1})$, i.e.,

$$c(b, d) \leq \frac{\mathcal{E}_f(\mu)}{\mathcal{E}_f(\mu\Psi^{-1})} \leq C(b, d) \tag{40}$$

for some constants $0 < c(b, d) \leq C(b, d) < \infty$, depending on b and d only. This will yield $\mathrm{Cap}_f(\partial\Gamma) \asymp \mathrm{Cap}_f(\Psi(\partial\Gamma))$, proving the theorem.

Let

$$h(k) = \begin{cases} f(b^{-k}) - f(b^{1-k}), & k \geq 1 \\ f(1), & k = 0. \end{cases}$$

In the following, write $u \leq w$ if w is a descendant of u. Then

$$\mathcal{E}_f(\mu) = \int_{\partial\Gamma} \int_{\partial\Gamma} \sum_{k=0}^{|x \wedge y|} h(k) \, d\mu(x)d\mu(y) = \sum_{k=0}^{\infty} h(k) \int\int_{|x \wedge y| \geq k} d\mu(x)d\mu(y).$$

Breaking up the region of integration and observing that $x \wedge y \geq v$ iff $x \geq v$ and $y \geq v$, the above is equal to

$$\sum_{k=0}^{\infty} h(k) \sum_{|v|=k} \iint_{x \wedge y \geq v} d\mu(x) d\mu(y) = \sum_{k=0}^{\infty} h(k) \sum_{|v|=k} [\mu(v)]^2 = \sum_{k=0}^{\infty} h(k) S_k,$$

where $S_k = S_k(\mu) = \sum_{|v|=k} [\mu(v)]^2$. Note that

$$\sum_{|v|=k+1} [\mu(v)]^2 \leq \sum_{|v|=k} [\mu(v)]^2 \leq b^d \sum_{|v|=k+1} [\mu(v)]^2,$$

i.e., $S_{k+1} \leq S_k \leq b^d S_{k+1}$.

We claim that in $[0,1]^d$,

$$\mathcal{E}_f(\mu \Psi^{-1}) \leq \int_{\Psi(\partial \Gamma)} \int_{\Psi(\partial \Gamma)} \sum_{k=0}^{\infty} h(k) 1_{\{k : b^{1-k} \geq |x-y|\}} d\mu \Psi^{-1}(x) d\mu \Psi^{-1}(y).$$

This holds because for the largest k yielding a non-zero term in the sum above, $b^{-k} < |x - y|$ and thus the sum is bounded below by $f(|x - y|)$.

For vertices v, w at the same level of Γ, set $\chi(v,w) = 1$ iff $\Psi(v)$ and $\Psi(w)$ are the same or adjacent subcubes of $[0,1]^d$, and $\chi(v,w) = 0$ otherwise. Then

$$\mu \times \mu\{(\xi, \eta) : |\Psi(\xi) - \Psi(\eta)| \leq b^{1-k}\} \leq \sum_{|v|=k-1} \sum_{|w|=k-1} \mu(v) \mu(w) \chi(v, w). \tag{41}$$

Now use the standard inequality $2\mu(v)\mu(w) \leq [\mu(v)]^2 + [\mu(w)]^2$ and the fact that the number of cubes adjacent to a given cube is bounded above by 3^d, to deduce that

$$\mu \times \mu\{(\xi, \eta) : |\Psi(\xi) - \Psi(\eta)| \leq b^{1-k}\} \leq 3^d S_{k-1} \leq 3^d b^d S_k.$$

It follows that

$$\mathcal{E}_f(\mu \Psi^{-1}) \leq (3b)^d \sum_k h(k) S_k = (3b)^d \mathcal{E}_f(\mu).$$

For the reverse inequality, choose l so that $b^l \geq \sqrt{d}$. Then $|v \wedge w| = k + l$ implies that $|\Psi(v) - \Psi(w)| \leq b^{-k}$, and consequently

$$\mathcal{E}_f(\mu \Psi^{-1}) \geq \sum_{k=0}^{\infty} f(b^{-k}) \mu \times \mu\{|v \wedge w| = k + l\}$$

$$= \sum_{k=0}^{\infty} f(b^{-k}) [S_{k+l}(\mu) - S_{k+l+1}(\mu)].$$

Using summation-by-parts shows that the right-hand side above is equal to

$$\sum_{k=0}^{\infty} h(k) S_{k+l}(\mu) \geq b^{-dl} \sum_{k=0}^{\infty} h(k) S_k(\mu) = b^{-dl} \mathcal{E}_f(\mu).$$

$\square$

15 Intersection-Equivalence

This Chapter follows Peres (1996). Throughout this chapter we work in $[0, 1]^d$ and all processes considered are started according to the uniform measure on $[0, 1]^d$, unless otherwise indicated.

Lemma 15.1 *If B is a Brownian path (killed at an exponential time for $d = 2$), then*

$$\mathbf{P}(B \cap \Lambda \neq \emptyset) \asymp \mathrm{Cap}_g(\Lambda)$$

for any Borel set Λ, where

$$g(r) = \begin{cases} \log^+(r^{-1}) & \text{if } d = 2 \\ r^{2-d} & \text{if } d > 2 \end{cases} . \tag{42}$$

Proof. (for $d \geq 3$). Denote by K the Martin kernel, see (35). By Theorem 14.3,

$$\mathbf{P}(B \text{ hits } \Lambda) = \int_{[0,1]^d} \mathbf{P}_0(B \text{ hits } \Lambda - x)dx \geq \frac{1}{2} \int_{[0,1]^d} \mathrm{Cap}_K(\Lambda - x)dx .$$

Because $\mathcal{E}_K(\mu) \leq C_d \mathcal{E}_g(\mu)$ for any measure μ on $[0, 1]^d$, the right-hand side above is bounded below by

$$\frac{1}{2C_d} \int_{[0,1]^d} \mathrm{Cap}_g(\Lambda - x)dx = \frac{1}{2C_d}\mathrm{Cap}_g(\Lambda) .$$

The upper-bound is a consequence of the probabilistic potential theory developed by Hunt and Doob. There exists a finite measure ν such that

$$\mathbf{P}_x(B \text{ hits } \Lambda) = \int_\Lambda g(|x - y|)d\nu(y) \qquad \text{and} \qquad \nu(\Lambda) = \mathrm{Cap}_g(\Lambda) .$$

(see, e.g., Chung (1973).) Then

$$\mathbf{P}(B \text{ hits } \Lambda) = \int_{[0,1]^d} \mathbf{P}_x(B \text{ hits } \Lambda)dx = \int_\Lambda \int_{[0,1]^d} g(|x - y|)dx d\nu(y) \leq C_d \nu(\Lambda) ,$$

where C_d is a constant depending only on d. Note that this proof extends to any initial distribution π for $B(0)$ with a bounded density; more generally a bounded Greenian potential suffices. $\qquad\square$

Shizuo Kakutani, generalizing a question of Paul Lévy, asked which compact sets Λ satisfy $\mathbf{P}(\Lambda \cap B_1 \cap B_2 \neq \emptyset) > 0$, where B_1, B_2 are independent Brownian paths in $\mathbf{R}^d$ ($d = 2$ or 3)?

Evans (1987) and Tongring (1988) gave a partial answer:

$$\text{If } \mathrm{Cap}_{g^2}(\Lambda) > 0, \text{ then } \mathbf{P}(\Lambda \cap B_1 \cap B_2 \neq \emptyset) > 0 . \tag{43}$$

They also found a necessary condition involving the Hausdorff measure of Λ. Later Fitzsimmons and Salisbury (1989) gave the full answer: $\text{Cap}_{g^2}(\Lambda) > 0$ is necessary as well as sufficient in (43). Furthermore, in dimension 2, their very general results yield the equivalence

$$\text{Cap}_{g^k}(\Lambda) > 0 \quad \Leftrightarrow \quad P(\Lambda \cap B_1 \cap \ldots \cap B_k \neq \emptyset) > 0. \tag{44}$$

This led Chris Bishop to make the following insightful conjecture:

Conjecture 2 (Bishop) *Let B denote a Brownian path. Then for any nonincreasing gauge f and any closed set Λ, the event that $\text{Cap}_f(\Lambda \cap B) > 0$ has positive probability iff $\text{Cap}_{f_g}(\Lambda) > 0$.*

We will present a proof of this below. Applying Kakutani's Theorem 14.1 to $\Lambda' = \Lambda \cap B_1$ and B_2 shows that

$$\mathbf{P}(\Lambda \cap B_1 \cap B_2 \neq \emptyset) > 0 \quad \Leftrightarrow \quad \text{Cap}_g(\Lambda \cap B_1) > 0 \text{ with positive probability.} \tag{45}$$

Bishop's Conjecture (with $f = g$) along with (45) imply that

$$\text{Cap}_{g^2}(\Lambda) > 0 \quad \Leftrightarrow \quad \mathbf{P}(\Lambda \cap B_1 \cap B_2 \neq \emptyset) > 0.$$

Hence Bishop's Conjecture and Kakutani's Theorem together give (44).

Theorem 15.2 *Let f be a non-negative and non-increasing function. Consider independent $\{p_e\}$ percolation on the 2^d-ary tree, with $p_e = p_k$ whenever $|e| = k$ and with $p_1 \ldots p_k = 1/f(2^{-k})$. Let $Q_d(f) \subseteq [0,1]^d$ be the set corresponding to $\partial\Gamma$ in $[0,1]^d$, where Γ is the component of the root in this percolation. (This component may be finite, whence $Q_d(f) = \emptyset$.) Then, for any closed set $\Lambda \subseteq [0,1]^d$,*

$$\text{Cap}_f(\Lambda) \asymp \mathbf{P}(\Lambda \cap Q_d(f) \neq \emptyset). \tag{46}$$

For $f = g$ in particular, $Q_d(f)$ is intersection-equivalent to Brownian motion, i.e.,

$$\mathbf{P}(\Lambda \cap Q_d(g) \neq \emptyset) \asymp \mathbf{P}(\Lambda \cap B \neq \emptyset). \tag{47}$$

Proof. By Theorem 14.2,

$$\mathbf{P}(\Lambda \cap Q_d(f) \neq \emptyset) = \mathbf{P}_{\{p_e\}}(\rho \leftrightarrow \partial\Gamma(\Lambda, 2)) \asymp \text{Cap}_f(\partial\Gamma(\Lambda, 2)), \tag{48}$$

where the constants in $\asymp$ are universal, namely 1 and 2. Theorem 14.6 with $b = 2$ yields

$$\text{Cap}_f(\partial\Gamma(\Lambda, 2)) \asymp \text{Cap}_f(\Lambda), \tag{49}$$

where the constants in $\asymp$ depend on d. Combining (48) and (49) establishes (46).

Finally, use (46) and Lemma 15.1 to prove (47). $\qquad\square$

Corollary 15.3 *Let f and h be non-negative and non-increasing functions. If a random closed set A in $[0,1]^d$ satisfies*

$$\mathbf{P}(A \cap \Lambda \neq \emptyset) \asymp \mathrm{Cap}_h(\Lambda) \tag{50}$$

for all closed $\Lambda \subseteq [0,1]^d$, then

$$\mathbf{P}(\mathrm{Cap}_f(A \cap \Lambda) > 0) > 0 \ \text{ if and only if } \ \mathrm{Cap}_{fh}(\Lambda) > 0 \tag{51}$$

for all closed $\Lambda \subseteq [0,1]^d$. In particular, Bishop's conjecture is true.

Proof. Enlarge the probability space where A is defined to include independent limit sets of fractal percolations $Q_d(f)$ and $\tilde{Q}_d(h)$. By Theorem 15.2

$$\mathbf{P}(A \cap \Lambda \cap Q_d(f) \neq \emptyset \,|\, A) > 0 \ \text{ if and only if } \ \mathrm{Cap}_f(A \cap \Lambda) > 0,$$

it follows that

$$\mathbf{P}(\,\mathrm{Cap}_f(A \cap \Lambda) > 0\,) > 0 \ \text{ if and only if } \ \mathbf{P}(A \cap \Lambda \cap Q_d(f) \neq \emptyset) > 0. \tag{52}$$

Conditioning on $Q_d(f)$ and then using (50) with $\Lambda \cap Q_d(f)$ in place of Λ gives

$$\mathbf{P}(A \cap \Lambda \cap Q_d(f) \neq \emptyset) > 0 \ \text{ if and only if } \ \mathbf{P}(\,\mathrm{Cap}_h(\Lambda \cap Q_d(f)) > 0\,) > 0. \tag{53}$$

Conditioning on $Q_d(f)$ and applying Theorem 15.2 yields

$$\mathbf{P}(\,\mathrm{Cap}_h(\Lambda \cap Q_d(f)) > 0\,) > 0 \ \text{ if and only if } \ \mathbf{P}(\Lambda \cap Q_d(f) \cap \tilde{Q}_d(h) \neq \emptyset) > 0. \tag{54}$$

Since $Q_d(f) \cap \tilde{Q}_d(h)$ has the same distribution as $Q_d(fh)$, Theorem 15.2 implies that

$$\mathbf{P}(\Lambda \cap Q_d(f) \cap \tilde{Q}_d(h) \neq \emptyset) > 0 \ \text{ if and only if } \ \mathrm{Cap}_{fh}(\Lambda) > 0. \tag{55}$$

Combining (52),(53),(54), and (55) proves (51). □

Corollary 15.4 *Suppose $\{A_i\}$ are independent random closed sets in $[0,1]^d$ satisfying*

$$\mathbf{P}(A_i \cap \Lambda \neq \emptyset) \asymp Cap_{g_i}(\Lambda)$$

for all closed $\Lambda \subseteq [0,1]^d$ and some g_i non-negative and non-increasing. Then

$$\mathbf{P}(A_1 \cap \ldots \cap A_k \cap \Lambda \neq \emptyset) > 0 \ \Leftrightarrow \ \mathrm{Cap}_{g_1 \ldots g_k}(\Lambda) > 0.$$

Example 15.5 *A.s., two independent Brownian paths in $\mathbf{R}^4$ do not intersect.*

This is a well-known result of Dvoretsky, Erdős and Kakutani (1950); we will show how it follows from intersection-equivalence. Let B_1 and B_2 be two independent Brownian paths in $\mathbf{R}^4$, started uniformly in the cube $[0,1]^4$ and intersected with that cube. Each is intersection-equivalent to $Q_4(g)$, and thus

$$\mathbf{P}([0,1]^4 \cap B_1 \cap B_2 \neq \emptyset) \asymp \mathbf{P}(Q_4(g) \cap \tilde{Q}_4(g) \neq \emptyset), \tag{56}$$

where $\tilde{Q}_4(g)$ is an independent copy of $Q_4(r^{-2})$. Because $Q_4(g) \cap \tilde{Q}_4(g)$ has the same distribution as $Q_4(g^2)$,

$$\mathbf{P}([0,1]^4 \cap B_1 \cap B_2 \neq \emptyset) \asymp \mathbf{P}(Q_4(g^2) \neq \emptyset). \tag{57}$$

Since the edge probabilities in the percolation corresponding to $g^2(r) = r^{-4}$ are all $p_k = 1/16$, the tree corresponding to $Q_4(g^2)$ is a critical branching process and thus dies out almost surely:

$$\mathbf{P}(Q_4(g^2) \neq \emptyset) = 0. \tag{58}$$

Putting together (57) and (58) shows that the two paths never intersect. $\quad\triangle$

Corollary 15.6 (Lawler (1982, 1985), Aizenman (1985)) *Let B_1 and B_2 be independent Brownian paths intersected with $[0,1]^d$, considered as sets in $[0,1]^d$. Then*

$$\mathbf{P}(\mathrm{dist}(B_1, B_2) < \epsilon) \asymp \begin{cases} 1 & d \leq 3 \\ \frac{1}{-\log \epsilon} & d = 4 \\ \epsilon^{d-4} & d > 4 \end{cases}.$$

Proof. We will prove the cases $d \geq 4$; the other cases are handled similarly. Let g be the Greenian potential (42), and write $Q_d(p)$ instead of $Q_d(g)$, where $p = 2^{2-d}$. For a closed set C and $\epsilon > 0$, let C^ϵ be the set of points within distance ϵ from a point in C. Conditioning on B_2^ϵ and applying Theorem 15.2 gives

$$\mathbf{P}(\mathrm{dist}(B_1, B_2) < \epsilon) = \mathbf{P}(B_1 \cap B_2^\epsilon \neq \emptyset) \asymp \mathbf{P}(Q_d(p) \cap B_2^\epsilon \neq \emptyset). \tag{59}$$

Now conditioning on $[Q_d(p)]^\epsilon$ and again applying Theorem 15.2 yields

$$\mathbf{P}(Q_d(p) \cap B_2^\epsilon \neq \emptyset) = \mathbf{P}([Q_d(p)]^\epsilon \cap B_2 \neq \emptyset) \asymp \mathbf{P}([Q_d(p)]^\epsilon \cap \tilde{Q}_d(p) \neq \emptyset), \tag{60}$$

where $\tilde{Q}_d(p)$ is an independent copy of $Q_d(p)$.

Combining (59) and (60) shows that

$$\mathbf{P}(\mathrm{dist}(B_1, B_2) < \epsilon) \asymp \mathbf{P}([Q_d(p)]^\epsilon \cap \tilde{Q}_d(p) \neq \emptyset). \tag{61}$$

Next let $\epsilon/2 < 2^{-k} \leq \epsilon$ and choose ℓ so that $2^\ell \geq \sqrt{d}$. Then $\mathbf{P}([Q_d(p)]^\epsilon \cap \tilde{Q}_d(p) \neq \emptyset)$ is at most the probability that $Q_d(p)$ and $\tilde{Q}_d(p)$ both intersect the interior of the same binary cube of side-length $2^{-(k+\ell)}$, and this is bounded below by

$$c^2 \cdot \mathbf{P}(\text{the construction leading to } Q_d(p^2) \text{ survives for } k + \ell \text{ generations}), \tag{62}$$

where $c = 1 - q > 0$ is the probability of survival of the (supercritical) branching process associated to the construction of $Q_d(p)$.

The probability in (62) may be estimated via standard branching process arguments, but we use percolation instead. Consider Bernoulli(p^2) percolation on the 2^d-ary tree T and write the probability as $\mathbf{P}_{p^2}(\rho \leftrightarrow T_{k+\ell})$. Since the minimal energy measure on ∂T_k is the uniform measure μ, Theorem 14.2 yields that

$$\frac{1}{\mathbf{P}_p(\rho \leftrightarrow T_k)} \asymp \frac{1}{\mathrm{Cap}_F(T_k)} = \mathcal{E}_F(\mu),$$

where $F(v, w) = p^{-|v \wedge w|}$. We have

$$\mathcal{E}_F(\mu) = 1 + \sum_{v,w \in T_k} \sum_{j=1}^{k} (p^{-j} - p^{1-j})\mu(v)\mu(w) = 1 + \sum_{j=1}^{k} \sum_{|v \wedge w| \geq j} (p^{-j} - p^{1-j})\mu(v)\mu(w).$$

Since $|v \wedge w| \geq j$ if and only if $|v| \geq j$ and $|w| \geq j$,

$$\mathcal{E}_F(\mu) = 1 + \sum_{j=1}^{k} (p^{-j} - p^{1-j}) \left(\sum_{|v| \geq j} \mu(v)\right)^2 = 1 + \sum_{j=1}^{k} (p^{-j} - p^{1-j}) 2^{-dj},$$

where the last equality holds because μ is the uniform measure. We conclude that $\mathcal{E}_F(\mu) \asymp \sum_{j=1}^{k} (p2^d)^{-j}$ and

$$\frac{1}{\mathbf{P}_p(\rho \leftrightarrow T_k)} \asymp \begin{cases} k & \text{if } p = 2^{-d} \\ (2^d p)^{-k} & \text{if } p < 2^{-d}. \end{cases}$$

Recall that $p = 2^{2-d}$ and hence the probability in (62) is equal to

$$\mathbf{P}_{p^2}(\rho \leftrightarrow T_{k+\ell}) \asymp \begin{cases} (k+\ell)^{-1} \asymp |\log \epsilon|^{-1} & \text{if } d = 4, \text{ because } p^2 = 2^{-d} \text{ for } d = 4, \\ 2^{(4-d)(k+\ell)} \asymp \epsilon^{d-4} & \text{if } d > 4. \end{cases}$$

For the reverse inequality, recall (61):

$$\mathbf{P}(\mathrm{dist}(B_1, B_2) < \epsilon) \asymp \mathbf{P}([Q_d(p)]^\epsilon \cap \tilde{Q}_d(p) \neq \emptyset).$$

Let $Q_d^{k-1}(p)$ denote the union of all binary cubes of side-length 2^{1-k} in the $(k-1)$th step of the construction of $Q_d(p)$, and recall that $\epsilon/2 < 2^{-k} \leq \epsilon$. Then $[Q_d(p)]^\epsilon$ is contained in the union of 3^d translates $Q_d^{k-1}(p) + x$ of $Q_d^{k-1}(p)$ and therefore the probability $\mathbf{P}([Q_d(p)]^\epsilon \cap \tilde{Q}_d(p) \neq \emptyset)$ is bounded above by

$$3^d \mathbf{P}(\text{the construction leading to } Q_d(p^2) \text{ survives to the } (k-1)\text{th generation}).$$

The proof is now concluded by using the previous calculation for this probability. $\square$

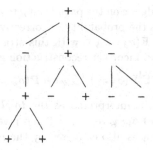

Figure 4: Tree with $+,-$ spins at the vertices.

16 Reconstruction for the Ising Model on a Tree

This chapter follows Evans, Kenyon, Peres and Schulman (1998).

Consider the following broadcast process. At the root ρ of a tree T, a random ± 1 valued "spin" σ_ρ is chosen uniformly. This spin is then propagated, with error, throughout the tree as follows: For a fixed $\epsilon \in (0, 1/2]$, each vertex receives the spin at its parent with probability $1 - \epsilon$, and the opposite spin with probability ϵ. These events at the vertices are statistically independent. This model has been studied in information theory, mathematical genetics and statistical physics; some of the history is described below.

Suppose we are given the spins that arrived at some fixed set of vertices W of the tree. Using the optimal reconstruction strategy (maximum likelihood), the probability of correctly reconstructing the original spin at the root is clearly at least $1/2$; denote this probability by $\frac{1+\Delta}{2}$. We will establish a lower bound for $\Delta = \Delta(T, W, \epsilon)$ in terms of the the effective electrical conductance from the root ρ to W (Theorem 16.2), and an upper bound for Δ which is the maximum flow from ρ to W for certain edge capacities (Theorem 16.3.) When T is an infinite tree, these bounds allow us to determine (in Theorem 16.1) the critical parameter ϵ_c so that, denoting the nth level of T by T_n, we have

$$\lim_{n\to\infty} \Delta(T, T_n, \epsilon) \begin{cases} > 0 & \text{if } \epsilon < \epsilon_c \\ = 0 & \text{if } \epsilon > \epsilon_c. \end{cases} \tag{63}$$

As we explain below, vanishing of the above limit is equivalent to extremality of the "free boundary" limiting Gibbs state for the ferromagnetic Ising model. For the special case of regular trees, the problem of determining ϵ_c was open for two decades, and was finally solved in 1995 by Bleher, Ruiz and Zagrebnov [12].

The random spins $\{\sigma_v\}$ that label the vertices of T as described above, can be constructed from independent variables $\{\eta_e\}$ labeling the edges of T, as follows. For each edge e, let $\mathbf{P}[\eta_e = -1] = \epsilon = 1 - \mathbf{P}[\eta_e = 1]$. Let σ_ρ be a uniformly chosen spin, and for any other vertex v let

$$\sigma_v := \sigma_\rho \prod_e \eta_e, \tag{64}$$

where the product is over all edges e on the path from ρ to v. Given $\sigma_W = \{\sigma_v : v \in W\}$, the strategy which maximizes the probability of correctly reconstructing σ_ρ, is to decide according to the sign of $\mathbf{E}(\sigma_\rho \mid \sigma_W)$; with this strategy, the difference between the probabilities of correct and incorrect reconstruction is

$$\Delta(T, W, \epsilon) = \mathbf{E}\big|\mathbf{P}(\sigma_\rho = 1 \mid \sigma_W) - \mathbf{P}(\sigma_\rho = -1 \mid \sigma_W)\big|. \tag{65}$$

Alternatively, $\Delta(T, W, \epsilon)$ can be interpreted as the *total variation distance* between the conditional distributions of σ_W given $\sigma_\rho = 1$ and given $\sigma_\rho = -1$; see below. The dependence between σ_ρ and σ_W is also captured by the *mutual information*

$$I(\sigma_\rho; \sigma_W) := \sum_{x,y} \mathbf{P}[\sigma_\rho = x, \sigma_W = y] \log \frac{\mathbf{P}[\sigma_\rho = x, \sigma_W = y]}{\mathbf{P}[\sigma_\rho = x]\mathbf{P}[\sigma_W = y]}.$$

Theorem 16.1 *Let T be an infinite tree with root ρ, and suppose its vertices are assigned random spins $\{\sigma_v\}$, using the flip probability $\epsilon < 1/2$ as in (64). Consider the problem of reconstructing σ_ρ from the spins at the n'th level T_n of T.*

(i) *If $1 - 2\epsilon > \mathrm{br}(T)^{-1/2}$ then $\inf_{n \geq 1} \Delta(T, T_n, \epsilon) > 0$ and $\inf_{n \geq 1} I(\sigma_\rho; \sigma_{T_n}) > 0$.*

(ii) *If $1 - 2\epsilon < \mathrm{br}(T)^{-1/2}$ then $\inf_{n \geq 1} \Delta(T, T_n, \epsilon) = 0$ and $\inf_{n \geq 1} I(\sigma_\rho; \sigma_{T_n}) = 0$.*

The tail field of the random variables $\{\sigma_v\}_{v \in T}$ contains events with probability strictly between 0 and 1 in case (i), but not in case (ii).

Thus in the notation of (63), $\epsilon_c = (1 - \mathrm{br}(T)^{-1/2})/2$. As mentioned above, this was already known when T is a $b+1$-regular tree (for which $\mathrm{br}(T) = b$). Theorem 16.1 is considerably more general. Simple examples show that at criticality, when $1 - 2\epsilon = \mathrm{br}(T)^{-1/2}$, asymptotic solvability of the reconstruction problem is not determined by the branching number; in this case there is a sharp capacity criterion, proved in [75], that we will not develop here. To see the relevance of the quantity $1 - 2\epsilon$ appearing in Theorem 16.1, note the following equivalent construction of the random variables $\{\sigma_v\}$: Perform independent bond percolation on T with parameter $\gamma = 1 - 2\epsilon$ (the probability of open bonds), and independently assign to each of the resulting percolation clusters a uniform random spin (the same spin is assigned to all vertices in each cluster). This is a special case of the Fortuin-Kasteleyn random cluster representation of the Ising model (see, *e.g.*, [32]); on a tree, it is elementary to verify the equivalence of this representation with the construction (64).

The following two theorems contain estimates of reconstruction probability and mutual information, that imply Theorem 16.1.

Theorem 16.2 *Let T be a tree with root ρ, and let W be a finite set of vertices in T. Given $\epsilon \in (0, 1/2]$, denote $\gamma := 1 - 2\epsilon$, and consider the electrical network obtained by assigning to each edge e of T the resistance $(1 - \gamma^2)\gamma^{-2|e|}$. Then*

$$\left.\begin{array}{c} \Delta(T, W, \epsilon) \\ I(\sigma_\rho; \sigma_W) \end{array}\right\} \geq \frac{1}{1 + \mathcal{R}(\rho \leftrightarrow W)}, \tag{66}$$

where $\mathcal{R}$ denotes effective resistance.

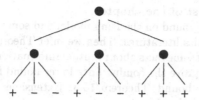

Figure 5: Majority vote can disagree with maximum likelihood.

The proof of this theorem is based on reconstruction by weighted majority vote, i.e., reconstruction according to the sign of an unbiased linear estimator of the root spin. We relate the variance of such an estimator to the energy of a corresponding unit flow from ρ to W. We find it quite surprising that on any infinite tree, reconstruction using such linear estimators has the same threshold as maximum-likelihood reconstruction.

Next, we present an upper bound on Δ and $I(\sigma_\rho; \sigma_W)$. Say that a set of vertices W_1 **separates** ρ **from** W if any path from ρ to W intersects W_1. For a vertex v of T, denote by $|v|$ the number of edges on the path from v to ρ.

Theorem 16.3 *Let W be a finite set of vertices in the tree T. For any set of vertices W_1 that separates the root ρ from W, we have*

$$\Delta(T, W, \epsilon)^2 \leq 2\Big(1 - \prod_{v \in W_1} \sqrt{1 - \gamma^{2|v|}}\Big) \leq 2 \sum_{v \in W_1} \gamma^{2|v|} \tag{67}$$

and

$$I(\sigma_\rho; \sigma_W) \leq \sum_{v \in W_1} I(\sigma_\rho; \sigma_v) \leq \sum_{v \in W_1} \gamma^{2|v|}. \tag{68}$$

In view of the mincut-maxflow theorem, (68) is an upper bound on mutual information in terms of the maximum flow in a capacitated network. Theorem 16.3 is proved by comparing the given tree T to a "stringy tree" $\widehat{T}$ which has an isomorphic set of paths from the root to the vertices of W_1, but these paths are pairwise edge-disjoint. We show that $\Delta(T, W, \epsilon) \leq \Delta(\widehat{T}, W_1, \epsilon)$ by constructing a noisy channel that maps the spins on W_1 in $\widehat{T}$ to the spins on W in T.

Symmetric trees: Recall that a tree T is **spherically symmetric** if for every $n \geq 1$, all vertices in T_n have the same degree. For such a tree, the effective resistance from the root to level n is easily computed, and we infer from Theorems 16.1-16.3 that

$$\Big(2 + 2(1 - \gamma^2) \sum_{k=1}^{n} \frac{\gamma^{-2k}}{|T_k|}\Big)^{-1} \leq I(\sigma_\rho; \sigma_{T_n}) \leq \inf_{k \leq n} |T_k| \gamma^{2k} \tag{69}$$

and $(1 - 2\epsilon_c)^{-2} = \liminf_n |T_n|^{1/n}$.

The example in Figure 5 shows that even on a regular tree, majority vote can disagree with maximum likelihood when the spin configuration σ_{T_n} is given.

Given the boundary data in Figure 5, the root spin σ_ρ is more likely to be -1 than $+1$ provided that ϵ is sufficiently small, since $\sigma_\rho = +1$ requires 4 spin flips, while $\sigma_\rho = -1$ requires only 3 spin flips.

Organization of the rest of the chapter.
Next, we present background on the Ising model and some references to the statistical mechanics and genetics literatures. Then we infer Theorem 16.1 from Theorems 16.2-16.3. After collecting some facts about mutual information and distances between probability measures, we prove the conductance lower bound for reconstruction, Theorem 16.2, and the upper bound, Theorem 16.3. Extensions and unsolved problems are discussed at the end of the chapter.

Background

Let G be a finite graph with vertex set V. In the *ferromagnetic Ising model* with no external field on G, the *interaction strength* $J > 0$ and the *temperature* $t > 0$ determine a **Gibbs distribution** $\mathcal{G} = \mathcal{G}_{J,t}$ on $\{\pm 1\}^V$ which is defined by

$$\mathcal{G}(\sigma) = Z(t)^{-1} \exp(\sum_{u \sim v} J\sigma_u \sigma_v / t), \tag{70}$$

where the normalizing factor $Z(t)$ is called the *partition function*. If the graph G is a tree, then this is equivalent to the Markovian propagation description in the beginning of the chapter, for an appropriate choice of the error parameter ϵ. Indeed, if $u \sim v$ are adjacent vertices in a finite tree with $\sigma_u = \sigma_v$, then flipping all the spins on one side of the edge connecting u and v will multiply the probability in (70) by $e^{-2J/t}$. Thus if we define ϵ by

$$\frac{\epsilon}{1 - \epsilon} = e^{-2J/t}, \tag{71}$$

then the distributions defined by (64) and (70) coincide. For an infinite graph G, a weak limit point of the Gibbs distributions (70) on finite subgraphs $\{G_n\}$ exhausting G, (possibly with boundary conditions imposed on $\sigma_{\partial G_n}$), is called a (limiting) **Gibbs state** on G. See Georgii [30] for more complete definitions, using the notion of specification.

For any infinite graph with bounded degrees, the limiting Gibbs state is unique at sufficiently high temperatures, i.e., the limit from finite subgraphs exists and does not depend on boundary conditions. When $G = T$ is a tree, this means that

$$\lim_{n \to \infty} \mathbf{E}[\sigma_\rho \mid \sigma_{T_n} \equiv 1] = 0 \tag{72}$$

at high temperatures. Some graphs admit a phase transition: below a certain critical temperature, multiple Gibbs states appear and the limit in (72) is strictly positive. The critical temperature t_c^+ for this transition on a regular tree T was determined in 1974 by Preston [79]; his result was generalized in 1989 by Lyons [59] who showed that $\tanh(J/t_c^+) = \mathrm{br}(T)^{-1}$; in the equivalent Markovian description, the critical parameter ϵ_c^+ for an all $+$ boundary to affect σ_ρ in the limit, satisfies $1 - 2\epsilon_c^+ = \mathrm{br}(T)^{-1}$.

In general, a Gibbs state is extremal (or "pure") iff it has a trivial tail, see Georgii ([30], Theorem 7.7). The tree-indexed Markov chain (64) on an infinite tree T is the limit of the Gibbs distributions (70) on finite subtrees, with no boundary conditions imposed; hence it is called the **free boundary Gibbs state** on T. In 1975 Spitzer

([82], Theorem 4) claimed that on a $b+1$-regular tree $T^{(b)}$, the free boundary Gibbs states are extremal at any temperature. A counterexample, due to T. Kamae, was published in 1977 (see Higuchi [42]). Kamae showed that the sum of spins on $T_n^{(b)}$, normalized by its L^2 norm, converges to a non-constant tail-measurable function, provided that $1-2\epsilon > b^{-1/2}$. In 1978, this result was put in a broader context by Moore and Snell [69], who showed it followed from the 1966 results of Kesten and Stigum [51] on multi-type branching processes. Moore and Snell noted that it was open whether the free boundary Gibbs state on $T^{(b)}$ is extremal when $b^{-1} < 1 - 2\epsilon \leq b^{-1/2}$. Chayes, Chayes, Sethna and Thouless [14] successfully analyzed a closely related spin-glass model on T_b; by a gauge transformation, this is equivalent to the Ising model with i.i.d. uniform $\{\pm 1\}$ boundary conditions. Although these boundary conditions are quite different from a free boundary, they turn out to have the same critical temperature. Bleher, Ruiz and Zagrebnov [12] adapted the recursive methods of Chayes et al [14] to the extremality problem, and showed that the free boundary Gibbs state on $T^{(b)}$ is extremal whenever $1 - 2\epsilon \leq b^{-1/2}$. Shortly thereafter, a more streamlined argument was found by Ioffe [44]. Theorem 16.1 was first established in [24]. After learning of that result, Ioffe [45] found an elegant alternative proof for the upper bound.

Genetic reconstruction and parsimony

Tree-indexed Markov chains as in the introduction have been studied in the Mathematical Biology literature by Cavender [13], by Steel and Charleston [84], and others. In that literature the two "spins" are often called "colors", and correspond to traits of individuals, species, or DNA sequences. The "broadcasting errors" (color changes along edges) represent mutations, and one attempts to infer traits of ancestors from those of an observable population.

Proof of Theorem 16.1

(i) From $\gamma = 1 - 2\epsilon > \mathrm{br}(T)^{-1/2}$ it follows that

$$\mathcal{R}(\rho \leftrightarrow \infty) := \sup_n \mathcal{R}(\rho \leftrightarrow T_n) < \infty$$

when each edge e is assigned conductance $\gamma^{2|e|}$; see (39) and Theorem 2.8. Therefore by (66),

$$\inf_{n \geq 1} \Delta(T, T_n, \epsilon) \geq \inf_{n \geq 1} \frac{1}{1 + \mathcal{R}(\rho \leftrightarrow T_n)} \geq \frac{1}{1 + \mathcal{R}(\rho \leftrightarrow \infty)} > 0$$

and similarly $\inf_{n \geq 1} I(\sigma_\rho; \sigma_{T_n}) > 0$, as asserted. In particular, σ_ρ is not independent of the tail field of $\{\sigma_v\}$, so this tail field is not trivial.

(ii) If $\gamma = 1 - 2\epsilon < \mathrm{br}(T)^{-1/2}$ then $\inf_\Pi \sum_{v \in \Pi} \gamma^{2|v|} = 0$, so Theorem 16.3 implies that $\inf_{n \geq 1} \Delta(T, T_n, \epsilon) = 0$ and $\inf_{n \geq 1} I(\sigma_\rho; \sigma_{T_n}) = 0$.

Next, fix a finite set of vertices W_0. For each $w \in W_0$ and $n > |w|$, denote by $T_n(w)$ the set of vertices in T_n which connect to ρ via w. Then Lemma 16.4(iii) implies that for sufficiently large n,

$$I(\sigma_{W_0}; \sigma_{T_n}) \leq \sum_{w \in W_0} I(\sigma_{W_0}, \sigma_{T_n(w)}) = \sum_{w \in W_0} I(\sigma_w, \sigma_{T_n(w)}), \qquad (73)$$

since the conditional distribution of $\sigma_{T_n(w)}$ given σ_{W_0} is the same as its conditional distribution given σ_w.

For any finite W_0, the right-hand side of (73) tends to 0 as $n \to \infty$; It follows that the tail of $\{\sigma_v\}$ is trivial.

□

Mutual Information: Definition and Properties

Let X, Y be random variables defined on the same probability space which take finitely many values. The **entropy** of X is defined by

$$H(X) := -\sum_x \mathbf{P}[X = x] \log \mathbf{P}[X = x]$$

and the **mutual information** $I(X; Y)$ between X and Y is defined to be

$$I(X; Y) := H(X) + H(Y) - H(X, Y) = \sum_{x,y} \mathbf{P}[X = x, Y = y] \log \frac{\mathbf{P}[X = x, Y = y]}{\mathbf{P}[X = x]\mathbf{P}[Y = y]}.$$

We collect a few basic properties of mutual information in the following lemma. See, e.g., Cover and Thomas [15] §2.

Lemma 16.4 (i) $I(X; Y) \geq 0$, with equality iff X and Y are independent;

(ii) *Data processing inequality: If $X \mapsto Y \mapsto Z$ form a Markov chain (i.e., X and Z are conditionally independent given Y), then $I(X; Y) \geq I(X; Z)$.*

(iii) *Subadditivity: If $Y_1, \ldots, Y_n$ are conditionally independent given X, then $I(X; (Y_1, \ldots, Y_n)) \leq \sum_{j=1}^n I(X; Y_j)$.*

The assumption of conditional independence in part (iii) cannot be omitted, as is shown by standard examples of 3 dependent random variables which are pairwise independent (e.g., Boolean variables satisfying $X = Y_1 + Y_2$ mod 2). Nevertheless, inequality (68) in Theorem 16.3 extends (iii) to a setting where this conditional independence need not hold.

Distances between probability measures

Let ν_+ and ν_- be two probability measures on the same space Ω. (In our application Ω is finite, but it is convenient to use notation that applies more generally.) Set $\nu := \frac{\nu_+ + \nu_-}{2}$ and denote $f = \frac{d\nu_+}{d\nu}$, $g = \frac{d\nu_-}{d\nu}$, so that $f + g \equiv 2$ identically. Suppose that ξ is uniform in $\{\pm 1\}$, and X has distribution ν_ξ. Inferring ξ from X is a basic problem of Bayesian hypothesis testing. (In our application, ξ will be the root spin σ_ρ, and X will be some function of the spin configuration σ_W on a finite vertex set W.)

There are several important notions of distance between ν_+ and ν_-, that can be related to this inference problem:

- **Total variation distance** $D_V(\nu_+, \nu_-) := \frac{1}{2}\int |f - g|\, d\nu$ can be interpreted as the difference between the probabilities of correct and erroneous inference. Indeed, among all functions $\hat\xi$ of the observations, the probability of error $\mathbf{P}[\hat\xi \neq \xi]$ is minimized by taking $\hat\xi = 1$ if $f(X) \geq g(X)$, and $\hat\xi = -1$ otherwise. We then have

$$\Delta := \mathbf{P}[\hat\xi = \xi] - \mathbf{P}[\hat\xi \neq \xi] = \frac{1}{2}\Big(\int \hat\xi f\, d\nu - \int \hat\xi g\, d\nu\Big) = \frac{1}{2}\int |f - g|\, d\nu. \quad (74)$$

- χ^2 **distance** $D_\chi(\nu_+, \nu_-) := \frac{1}{2}\{\int (f - g)^2\, d\nu\}^{1/2}$ represents the L^2 norm of the conditional expectation $\mathbf{E}(\xi\,|\,X) = \frac{1}{2}(f(X) - g(X))$.

- **Mutual information** between ξ and X,

$$D_I(\nu_+, \nu_-) := I(\xi; X) = \frac{1}{2}\int (f \log f + g \log g)\, d\nu \quad (75)$$

is a symmetrized version of the Kullback-Leibler divergence (see Vajda [86]).

- **The Hellinger distance**

$$D_H(\nu_+, \nu_-) := \int (\sqrt{f} - \sqrt{g})^2\, d\nu = 2\Big(1 - \int \sqrt{fg}\, d\nu\Big). \quad (76)$$

derives its importance from the simple behavior of the Hellinger integrals

$$\mathrm{Int}_H(\nu_+, \nu_-) := \int \sqrt{fg}\, d\nu$$

for product measures:

$$\mathrm{Int}_H(\nu_+ \times \mu_+,\ \nu_- \times \mu_-) = \mathrm{Int}_H(\nu_+, \nu_-)\mathrm{Int}_H(\mu_+, \mu_-). \quad (77)$$

These distances appear in different sources under different names and with different normalizations. We collect here some well known inequalities between them, that will be useful below. For more on this topic, see, e.g., Le Cam [56] or Vajda [86].

Lemma 16.5 *With the notation above,*

(i) $D_\chi^2 \leq D_V \leq D_\chi \leq \sqrt{D_H}$

(ii) $D_\chi^2 \leq D_I \leq 2D_\chi^2$

(iii) *If ν_+ and ν_- are measures on $\mathbb{R}$, then*

$$\Big\{\int x\, d(\nu_+ - \nu_-)\Big\}^2 = \Big\{\int x[f(x) - g(x)]\, d\nu\Big\}^2 \leq 4\int x^2\, d\nu \cdot D_\chi^2.$$

Proof.

(i) The left-hand inequality follows from $|f(x) - g(x)| \leq 2$, and the middle inequality from Cauchy-Schwarz. The right-hand inequality follows from the identity $f - g = (\sqrt{f} - \sqrt{g}) \cdot (\sqrt{f} + \sqrt{g})$ and the concavity relation $\frac{\sqrt{f} + \sqrt{g}}{2} \leq \sqrt{\frac{f+g}{2}} = 1$.

(ii) Setting $\psi = (f - g)/2$, the assertion follows from the pointwise inequalities

$$\frac{\psi^2}{2} \leq \frac{1+\psi}{2} \log(1+\psi) + \frac{1-\psi}{2} \log(1-\psi) \leq \psi^2. \tag{78}$$

Here the left-hand inequality is verified for $\psi \in [0,1)$ by comparing second derivatives, and the right-hand inequality follows from $\log(1+y) \leq y$.

(iii) This is just the Cauchy-Schwarz inequality.

$\square$

Finally, we interpret the data processing inequality in terms of distances. Suppose that we are given transition probabilities on the state space, *i.e.*, a stochastic matrix M (the entries of M are nonnegative and the row sums are all 1). Write $M^*\mu(y) := \sum_x M(x,y)\mu(x)$. Then Lemma 16.4 (ii) implies that

$$D_I(M^*\nu_+, M^*\nu_-) \leq D_I(\nu_+, \nu_-).$$

An analogous inequality holds for total variation:

$$\begin{aligned} D_V(M^*\nu_+, M^*\nu_-) &= \frac{1}{2}\sum_y |M^*\nu_+(y) - M^*\nu_-(y)| \\ &\leq \frac{1}{2}\sum_y \sum_x M(x,y)|\nu_+(x) - \nu_-(x)| \\ &= \frac{1}{2}\sum_x |\nu_+(x) - \nu_-(x)| = D_V(\nu_+, \nu_-). \end{aligned} \tag{79}$$

Conductance lower bounds: Proof of Theorem 16.2

Recall that each edge e was assigned the resistance

$$R(e) := (1 - \gamma^2)\gamma^{-2|e|}. \tag{80}$$

Say that a set of vertices W is an **antichain** if no vertex in W is a descendant of another.

Lemma 16.6 *Let W be a finite antichain in T. For any unit flow μ from ρ to W, the weighted sum*

$$S_\mu := \sum_{v \in W} \frac{\mu(v)\sigma_v}{\gamma^{|v|}} \tag{81}$$

satisfies $\mathbf{E}[S_\mu \,|\, \sigma_\rho] = \sigma_\rho$ *and*

$$\mathbf{E}[S_\mu^2] = \mathbf{E}[S_\mu^2 \,|\, \sigma_\rho] = 1 + \sum_e R(e)\mu(e)^2 \,. \tag{82}$$

Consequently,

$$\min_\mu \mathbf{E}[S_\mu^2] = 1 + \mathcal{R}(\rho \leftrightarrow W)\,, \tag{83}$$

and the minimum is attained precisely when μ *is the unit current flow from* ρ *to* W.

Proof. From the product representation (64), we infer that

$$\mathbf{E}[\sigma_v \,|\, \sigma_\rho] = \sigma_\rho \gamma^{|v|}$$

for any vertex v. The formula for $\mathbf{E}[S_\mu \,|\, \sigma_\rho]$ follows by linearity. For any two vertices v, w in T, denote by $\texttt{path}(v, w)$ the path from v to w. Also, write $\texttt{path}(v)$ for $\texttt{path}(\rho, v)$. Clearly,

$$\mathbf{E}[\sigma_v \sigma_w] = \gamma^{|\texttt{path}(v,w)|} = \gamma^{|v|+|w|-2|v \wedge w|}\,, \tag{84}$$

where $v \wedge w$, the **meeting point** of v and w, is the vertex farthest from the root ρ on $\texttt{path}(v) \cap \texttt{path}(w)$. The percolation representation can also be invoked to justify (84).

It is now easy to determine the second moment of S_μ:

$$\mathbf{E}[S_\mu^2] = \sum_{v,w \in W} \frac{\mu(v)\mu(w)}{\gamma^{|v|}\gamma^{|w|}} \mathbf{E}[\sigma_v \sigma_w] = \sum_{v,w \in W} \frac{\mu(v)\mu(w)}{\gamma^{2|v \wedge w|}}\,, \tag{85}$$

Next, insert the identity

$$\gamma^{-2|u|} = 1 + \sum_{e \in \texttt{path}(u)} R(e)$$

with $u = v \wedge w$, into (85). Changing the order of summation, and using the fact that W is an antichain, we obtain

$$\mathbf{E}[S_\mu^2] = 1 + \sum_e R(e) \sum_{v,w \in W} \mathbf{1}_{\{e \in \texttt{path}(v \wedge w)\}} \mu(v)\mu(w)\,. \tag{86}$$

Since $\texttt{path}(v \wedge w) = \texttt{path}(v) \cap \texttt{path}(w)$ and

$$\sum_{v,w \in W} \mathbf{1}_{\{e \in \texttt{path}(v \wedge w)\}} \mu(v)\mu(w) = \Big(\sum_{v \in W} \mathbf{1}_{\{e \in \texttt{path}(v)\}} \mu(v) \Big)\Big(\sum_{w \in W} \mathbf{1}_{\{e \in \texttt{path}(w)\}} \mu(w) \Big) = \mu(e)^2\,,$$

(86) is equivalent to (82). Finally, (83) follows from Thomson's principle. □

Proof of Theorem 16.2: We may assume that W is an antichain. (Otherwise, remove from W all vertices which have an ancestor in W.) Let μ be the unit current flow from ρ to W for the resistances $R(e)$ as in the preceding lemma, and let S_μ be the weighted sum (81). In order to apply Lemma 16.5, denote by ν_+ the conditional

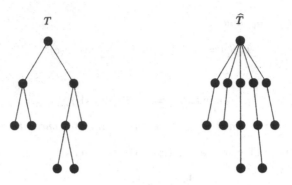

Figure 6: A tree T and the corresponding stringy tree $\hat{T}$.

distribution of S_μ given that $\sigma_\rho = 1$; define ν_- analogously by conditioning that $\sigma_\rho = -1$, so that $\nu = (\nu_+ + \nu_-)/2$ is the unconditioned distribution of S_μ. We then have by Lemma 16.5(iii) that

$$D_\chi^2(\nu_+, \nu_-) \geq \frac{\left\{\int x d(\nu_+ - \nu_-)\right\}^2}{4 \int x^2 \, d\nu} = \frac{(\mathbf{E}[S_\mu \mid \sigma_\rho = 1] - \mathbf{E}[S_\mu \mid \sigma_\rho = -1])^2}{4\mathbf{E}[S_\mu^2]}.$$

Applying Lemma 16.6, we deduce that

$$D_\chi^2(\nu_+, \nu_-) \geq \frac{1}{1 + \mathcal{R}(\rho \leftrightarrow W)}. \tag{87}$$

By Lemma 16.5, the difference $\Delta = \Delta(T, W, \epsilon)$ between the probabilities of correct and incorrect reconstruction, satisfies $\Delta = D_V(\nu_+, \nu_-) \geq D_\chi^2(\nu_+, \nu_-)$, and the mutual information between σ_ρ and σ_W also satisfies $I(\sigma_\rho; \sigma_W) = D_I(\nu_+, \nu_-) \geq D_\chi^2(\nu_+, \nu_-)$. In conjunction with (87), this completes the proof. □

Mincut upper bound: Proof of Theorem 16.3

Definition. A **noisy tree** is a tree with flip probabilities labeling the edges. The *stringy tree* $\hat{T}$ associated with a finite noisy tree T is the tree which has the same set of root-leaf paths as T but in which these paths act as independent channels. More precisely, for every root-leaf path in T, there exists an identical (in terms of length and flip probabilities on the edges) root-leaf path in $\hat{T}$, and in addition, all the root-leaf paths in $\hat{T}$ are edge-disjoint.

Theorem 16.7 *Given a finite noisy tree T with leaves W, let $\hat{T}$, with leaves $\widehat{W}$ and root $\hat{\rho}$, be the stringy tree associated with T. There is a channel which, for $\xi \in \{\pm 1\}$, transforms the conditional distribution $\sigma_{\widehat{W}} \mid (\sigma_{\hat{\rho}} = \xi)$ into the conditional distribution $\sigma_W \mid (\sigma_\rho = \xi)$. Equivalently, we say that $\hat{T}$ dominates T.*

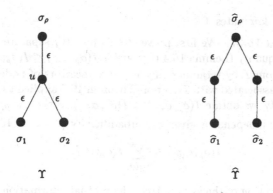

Figure 7: Υ is dominated by $\hat{\Upsilon}$.

Remark A *channel* is formally defined as a stochastic matrix describing the conditional distribution $\mathbf{P}(Y \mid X)$ of the output variable Y given the input X, see [15]. Often a channel is realized by a relation of the form $Y = f(X, Z)$, where f is a deterministic function and Z is a random variable (representing the "noise") which is independent of X.

Proof: We only establish a key special case of the theorem: namely, that the tree Υ shown in Figure 7, is dominated by the corresponding stringy tree $\hat{\Upsilon}$. The general case is derived from it by first allowing the flip probabilities to vary from edge to edge, and then applying an inductive argument; see [25] for details.

Given $0 \le \alpha \le 1$, to be specified below, we define the channel as follows:

$$\sigma_1^* = \hat{\sigma}_1$$
$$\sigma_2^* = \begin{cases} \hat{\sigma}_2 & \text{with probability } \alpha \\ \hat{\sigma}_1 & \text{with probability } 1 - \alpha \end{cases}$$

To prove that $(\hat{\sigma}_\rho, \sigma_1^*, \sigma_2^*)$ has the same distribution as $(\sigma_\rho, \sigma_1, \sigma_2)$, it suffices to show that the means of corresponding products are equal. (This is a special case of the fact that the characters on any finite Abelian group G form a basis for the vector space of complex functions on G.) By symmetry

$$\mathbf{E}(\sigma_\rho) = \mathbf{E}(\sigma_1) = \mathbf{E}(\sigma_2) = \mathbf{E}(\sigma_\rho \sigma_1 \sigma_2) = \mathbf{E}(\hat{\sigma}_\rho) = \mathbf{E}(\sigma_1^*) = \mathbf{E}(\sigma_2^*) = \mathbf{E}(\hat{\sigma}_\rho \sigma_1^* \sigma_2^*) = 0$$

and thus we only need to check pair correlations. Clearly, $\mathbf{E}(\hat{\sigma}_\rho \sigma_1^*) = \mathbf{E}(\sigma_\rho \sigma_1)$ and $\mathbf{E}(\hat{\sigma}_\rho \hat{\sigma}_1) = \gamma^2$, whence $\mathbf{E}(\hat{\sigma}_\rho \sigma_2^*) = \gamma^2 = \mathbf{E}(\sigma_\rho \sigma_2)$ for any choice of α. Finally, since $\mathbf{E}(\sigma_1^* \hat{\sigma}_2) = \gamma^4 < \gamma^2 = \mathbf{E}(\sigma_1 \sigma_2)$ and

$$\mathbf{E}(\sigma_1^* \hat{\sigma}_1) = 1 > \gamma^2 ,$$

we can choose $\alpha \in [0, 1]$ so that $\mathbf{E}(\sigma_1^* \sigma_2^*) = \mathbf{E}(\sigma_1 \sigma_2)$; explicitly,

$$\alpha = (1 - \gamma^2)/(1 - \gamma^4) . \tag{88}$$

This proves that $\widehat{\Upsilon}$ dominates Υ. □

Proof of Theorem 16.3: We first prove (68). Since W_1 separates ρ from W, the data processing inequality (Lemma 16.4 (ii)) yields $I(\sigma_\rho; \sigma_W) \le I(\sigma_\rho; \sigma_{W_1})$. Let T_1 be the tree obtained from T by retaining only W_1 and ancestors of nodes in W_1. Let $\widehat{T_1}$ be the stringy tree associated with T_1. From Theorem 16.7 applied to T_1 and the data processing inequality, we obtain $I(\sigma_\rho; \sigma_{W_1}) \le I(\sigma_{\hat\rho}; \sigma_{\widehat{W}_1})$. Since the spins on leaves of $\widehat{T_1}$ are conditionally independent given $\sigma_{\hat\rho}$, subadditivity (Lemma 16.4 (iii)) gives

$$I(\sigma_{\hat\rho}; \sigma_{\widehat{W}_1}) \le \sum_{\hat{v} \in \widehat{W}_1} I(\sigma_{\hat\rho}; \sigma_{\hat{v}}).$$

But due to the definition of the stringy tree, the mutual information between $\sigma_{\hat\rho}$ and $\sigma_{\hat{v}}$ is identical to the mutual information between σ_ρ and σ_v in T_1, hence the left inequality in (68).

Since $\mathbf{E}(\sigma_\rho \sigma_v) = \gamma^{|v|}$ for each v, the right-hand inequality in (68) follows from the right-hand inequality in (78).

We now turn to the total variation inequality (67). Recall that $\Delta(T, W, \epsilon)$, the difference between the probabilities of correct and incorrect reconstruction, equals $D_V(\nu_+^W, \nu_-^W)$, the total variation distance between the two distributions of the spins on W given $\sigma_\rho = \pm 1$.

By (79), Theorem 16.7, and Lemma 16.5,

$$D_V(\nu_+^W, \nu_-^W) \le D_V(\nu_+^{W_1}, \nu_-^{W_1}) \le D_V(\nu_+^{\widehat{W}_1}, \nu_-^{\widehat{W}_1}) \le \sqrt{D_H(\nu_+^{\widehat{W}_1}, \nu_-^{\widehat{W}_1})}.$$

Now, $D_H(\nu_+^{\widehat{W}_1}, \nu_-^{\widehat{W}_1})$ on the stringy tree $\widehat{T_1}$ is easily calculated using the multiplicative property of Hellinger integrals: $\nu_+^{\widehat{W}_1}$ is just the product over $w \in \widehat{W}_1$ of ν_+^w, the distribution of σ_w given $\sigma_\rho = 1$, and similarly $\nu_-^{\widehat{W}_1} = \prod_w \nu_-^w$. Since $\mathrm{Int}_H(\nu_+^w, \nu_-^w) = \sqrt{1 - \gamma^{2|w|}}$, the left-hand inequality in (67) follows; the right-hand inequality there is a consequence of the standard inequality $\prod(1 - x_j) \ge 1 - \sum x_j$. □

Remarks and unsolved problems

1. **Reconstruction at criticality.** It is shown in [12, 44] that on infinite regular trees, $\lim_n \Delta(T, T_n, \epsilon_c) = 0$. On general trees, Theorem 16.2 implies that finite effective resistance from the root to infinity (when each edge at level ℓ is assigned the resistance $(1-2\epsilon)^{-2\ell}$) is sufficient for $\lim_n \Delta(T, T_n, \epsilon) > 0$. In [75], a recursive method is used to show this condition is also necessary.

2. **Multi-colored trees and the Potts model.** The most natural generalization of the two-state tree-indexed Markov chain model studied in this chapter involves multicolored trees, where the coloring propagates according to any finite state tree-indexed Markov chain. For instance, if this Markov chain is defined by a $q \times q$ stochastic matrix where all entries off the main diagonal equal ϵ, then

the q-state Potts model arises. The proof of Theorem 16.2 extends to general Markov chains, and shows that the tail of the tree-indexed chain is nontrivial if $\mathrm{br}(T) > \lambda_2^{-2}$, where λ_2 is the second eigenvalue of the transition matrix (*e.g.* for the q-state Potts model, $\lambda_2 = 1 - q\epsilon$). However, unpublished calculations of E. Mossel indicate that this lower bound is not sharp in general. Furthermore, we do not know a reasonable upper bound on mutual information between root and boundary variables. In particular, it seems that the critical parameter for tail triviality in the Potts model on a regular tree is not known.

3. **An information inequality.** Theorem 16.3 implies that the spins in the ferromagnetic Ising model on a tree satisfy

$$I(\sigma_v; \sigma_W) \leq \sum_{w \in W} I(\sigma_v; \sigma_w),$$

for any vertex v and any finite set of vertices W. Does this inequality hold on other graphs as well?

More generally, are there natural assumptions (*e.g.*, positive association) on random variables $X, Y_1, \ldots Y_n$ that imply the inequality $I(X; (Y_1, \ldots, Y_n)) \leq \sum_{j=1}^n I(X, Y_j)$?

17 Unpredictable Paths in Z and EIT in $\mathbf{Z}^3$

The goal of this chapter is to complete the proof of Theorem 11.1, by exhibiting a probability measure on directed paths in $\mathbf{Z}^3$ that has exponential intersection tails. We construct the required measure in three dimensions from certain nearest-neighbor stochastic processes on $\mathbf{Z}$ that are "less predictable than simple random walk".

For a sequence of random variables $S = \{S_n\}_{n \geq 0}$ taking values in a countable set V, we define its **predictability profile** $\{\mathrm{PRE}_S(k)\}_{k \geq 1}$ by

$$\mathrm{PRE}_S(k) = \sup \mathbf{P}[S_{n+k} = x \mid S_0, \ldots, S_n], \tag{89}$$

where the supremum is over all $x \in V$, all $n \geq 0$, and all histories $S_0, \ldots, S_n$.

Thus $\mathrm{PRE}_S(k)$ is the maximal chance of guessing S correctly k steps into the future, given the past of S. Clearly, the predictability profile of simple random walk on $\mathbf{Z}$ is asymptotic to $Ck^{-1/2}$ for some $C > 0$.

Theorem 17.1 (Benjamini, Pemantle, and Peres 1998) *For any $\alpha < 1$ there exists an integer-valued stochastic process $\{S_n\}_{n \geq 0}$ such that $|S_n - S_{n-1}| = 1$ a.s. for all $n \geq 1$ and*

$$\mathrm{PRE}_S(k) \leq C_\alpha k^{-\alpha} \quad \text{for some } C_\alpha < \infty, \text{ for all } k \geq 1. \tag{90}$$

After Theorem 17.1 was proven in BPP (1998), Häggström and Mossel (1998) constructed processes with lower predictability profile. They showed that if f is non-decreasing and $\sum_k (f(k)k)^{-1} < \infty$, then there is a nearest-neighbor process S on $\mathbf{Z}$

with $\text{PRE}_S(k) \leq C f(k) k^{-1}$. (For example, $f(k) = \log^{1+\epsilon}(k)$ satisfies this summability condition.)

Hoffman (1998) proved that this result is sharp: if a nondecreasing function f satisfies $\sum_k (f(k)k)^{-1} = \infty$, then there is no nearest–neighbor process on $\mathbf{Z}$ with predictability profile bounded by $O(f(k)k^{-1})$.

We prove Theorem 17.1 using the Ising model on a tree. We follow Häggström and Mossel (1998), who improved the original argument from BPP (1998). The following lemma is the engine behind the proof. Let T be the b-adic tree of depth N, and fix $0 < \epsilon < 1/4$. We will assign to the vertices of T ± 1 labels $\{\sigma(v)\}_{v \in T}$ according to an Ising model (see Chapter 16). For the root ρ, set $\sigma(\rho) = 1$, and for a vertex w with parent v, let

$$\sigma(w) = \begin{cases} \sigma(v) & \text{with probability } 1 - \epsilon \\ -\sigma(v) & \text{with probability } \epsilon \end{cases} .$$

Lemma 17.2 *Denote by $Y_N := \sum_{v \in T_N} \sigma(v)$ the sum of the spins at level N. There exists $C_b < \infty$ such that for all $N \geq 1$ and all $x \in \mathbf{Z}$,*

$$P[Y_N = x] \leq \frac{C_b}{\epsilon[b(1 - 2\epsilon)]^N} .$$

Proof. By decomposing the sum Y_{M+1} into b parts corresponding to the subtrees of depth M rooted at the first level, we get

$$Y_{M+1} = \sum_{j=1}^{b} \sigma(v_j) Y_M^{(j)} ,$$

where $\{\sigma(v_j)\}_{j=1}^{b}$ are b i.i.d. spins with

$$\sigma(v_j) = \begin{cases} +1 & \text{with probability } 1 - \epsilon \\ -1 & \text{with probability } \epsilon \end{cases} ,$$

and $\{Y_M^{(j)}\}_{j=1}^{b}$ are i.i.d. variables with the distribution of Y_M, independent of these spins. Consequently, the characteristic functions

$$\hat{Y}_M(\lambda) = \mathbf{E}(e^{i\lambda Y_M})$$

satisfy the recursion

$$\begin{aligned} \hat{Y}_{M+1}(\lambda) &= ((1 - \epsilon)\hat{Y}_M(\lambda) + \epsilon\hat{Y}_M(-\lambda))^b \\ &= (\Re\hat{Y}_M(\lambda) + i(1 - 2\epsilon)\Im\hat{Y}_M(\lambda))^b \end{aligned} \tag{91}$$

where $\Re$ denotes real part, and $\Im$ imaginary part. For $\theta_n(\lambda) := \arg \hat{Y}_n(\lambda)$, define

$$J_n := \left\{ 0 \leq \lambda \leq \frac{\pi}{2} : \theta_k(\lambda) < \frac{\pi}{2b}, \; k = 0, \cdots, n-1 \right\}$$

and

$$I_n := J_n \setminus J_{n+1} .$$

We will evaluate the integral of $\hat{Y}_N(\lambda)$ over $(0, \pi/2]$ by using the decomposition

$$[0, \frac{\pi}{2}] = \left(\bigcup_{k=0}^{N-1} I_k \right) \bigcup J_N .$$

Rewrite (91) as

$$\hat{Y}_{M+1}(\lambda) = |\hat{Y}_M(\lambda)|^b [\cos \theta_M(\lambda) + i(1 - 2\epsilon) \sin \theta_M(\lambda)]^b, \tag{92}$$

and infer, for $0 \le \theta_M(\lambda) \le \frac{\pi}{2b}$, that

$$\theta_{M+1}(\lambda) = b \arctan \left((1 - 2\epsilon) \tan \theta_M(\lambda) \right).$$

Since arctan is concave in $[0, \infty)$ and $\arctan 0 = 0$,

$$\arctan \left((1 - 2\epsilon)\alpha \right) \ge (1 - 2\epsilon) \arctan(\alpha)$$

for any $\alpha \ge 0$. Therefore

$$\text{If } 0 \le \theta_M(\lambda) \le \frac{\pi}{2b}, \text{ then } \frac{\pi}{2} \ge b\theta_M(\lambda) \ge \theta_{M+1}(\lambda) \ge b(1 - 2\epsilon)\theta_M(\lambda). \tag{93}$$

If $\lambda \in I_n$, then applying (93) for $M = n - 1$ shows that

$$\frac{\pi}{2} \ge \theta_n(\lambda) \ge \frac{\pi}{2b} . \tag{94}$$

Using (92) with $M = n$ together with (94), we find that for $\lambda \in I_n$,

$$|\hat{Y}_{n+1}(\lambda)| \le \left(\cos^2(\frac{\pi}{2b}) + (1 - 2\epsilon)^2 \sin^2(\frac{\pi}{2b}) \right)^{\frac{b}{2}} \le \left(1 - 2\epsilon \sin^2(\frac{\pi}{2b}) \right)^{\frac{b}{2}} \le e^{-\varrho\epsilon b}, \tag{95}$$

where $\varrho := \sin^2(\frac{\pi}{2b})$. Inductive use of (92) for $\lambda \in I_n$ and $N > n$ gives

$$|\hat{Y}_N(\lambda)| \le e^{-\varrho\epsilon b^{N-n}}. \tag{96}$$

Since $\theta_0(\lambda) \equiv \lambda$, (93) implies that $\theta_k(\lambda) \ge b^k (1 - 2\epsilon)^k |\lambda|$ for $\lambda \in J_n$ and $k \le n$. Therefore

$$|I_n| \le |J_n| \le \frac{\pi}{2b^n(1 - 2\epsilon)^n}, \tag{97}$$

By (96),

$$\frac{1}{2\pi} \int_{-\pi}^{\pi} |\hat{Y}_N(\lambda)| \, d\lambda = \frac{2}{\pi} \int_0^{\frac{\pi}{2}} |\hat{Y}_N(\lambda)| \, d\lambda \le \frac{2}{\pi} \left(\sum_{k=0}^{N-1} |I_k| e^{-\varrho\epsilon b^{N-k}} + |J_N| \right).$$

Inserting (97) yields

$$\frac{1}{2\pi} \int_{-\pi}^{\pi} |\hat{Y}_N(\lambda)| \, d\lambda \le \frac{1}{(1 - 2\epsilon)^N} \left(\sum_{k=0}^{N-1} b^{-k} e^{-\varrho\epsilon b^{N-k}} + b^{-N} \right). \tag{98}$$

In order to evaluate the sum in the right hand side of (98), we define

$$r = \max\{k \ : \ \varrho \epsilon b^{N-k} > 1\}.$$

Separating the contributions of $k \geq r$ and $k < r$, we obtain that

$$\sum_{k=r}^{N-1} b^{-k} e^{-\varrho \epsilon b^{N-k}} + b^{-N} \leq \sum_{k=r}^{N} b^{-k} \leq b^{-r} \sum_{k=0}^{\infty} b^{-k} \tag{99}$$

and

$$\sum_{k=0}^{r-1} b^{-k} e^{-\varrho \epsilon b^{N-k}} \leq \sum_{k=0}^{r-1} b^{-k} e^{-b^{r-k}} \leq b^{-r} \sum_{k=0}^{\infty} b^{k} e^{-b^{k}}. \tag{100}$$

Furthermore, since $\varrho \epsilon b^{N-r-1} \leq 1$, we have that

$$b^{-r} \leq \frac{1}{\varrho \epsilon b^{N-1}}. \tag{101}$$

Combining (98), (99), (100), and (101) we see that

$$\frac{1}{2\pi} \int_{-\pi}^{\pi} |\widehat{Y}_N(\lambda)| \, d\lambda \leq \frac{C_b}{b^N \epsilon (1 - 2\epsilon)^N},$$

where

$$C_b = \frac{b(\sum_{k=0}^{\infty} b^{-k} + \sum_{k=0}^{\infty} b^k e^{-b^k})}{\varrho}$$

and ϱ was defined after (95). Using the inversion formula we conclude that

$$\mathbf{P}[Y_N = x] = \frac{1}{2\pi} \int_{-\pi}^{\pi} \widehat{Y}_N(\lambda) e^{-i\lambda x} \, d\lambda \leq \frac{1}{2\pi} \int_{-\pi}^{\pi} |\widehat{Y}_N(\lambda)| \, d\lambda \leq \frac{C_b}{b^N \epsilon (1 - 2\epsilon)^N}.$$

$\square$

Proof of Theorem 17.1. For all $N > 0$, we will define a process S up to time $M = 2^N$ with the required properties. A process defined for all times will then exist by consistency of the finite dimensional distributions.

Fix a small $\epsilon > 0$. We assign spins $\{\sigma_v\}$ to the vertices of the binary tree T of depth N, according to the Ising model (described before Lemma 17.2) with error rate ϵ, but we take σ_ρ to be random uniform in $\{\pm 1\}$, rather than fixing it. Enumerate the vertices at depth N from left to right as $v_0, v_1, \ldots, v_M$, and set

$$S_n = \sum_{k=1}^{n} \sigma(v_k).$$

We claim that $\{S_n\}$ has the desired predictability profile. To see this, fix $0 \leq n < M$ and $0 < k \leq M - n$. Observe that $S_{n+k} = S_n + \sum_{j=n+1}^{n+k} \sigma(v_j)$. If we now take the unique h satisfying $2^{h+1} \leq k < 2^{h+2}$, there will exist a vertex w at level $N - h$ for which all of the descendants at depth N are in the set $\{v_{n+1}, \ldots, v_{n+k}\}$. It follows (by conditioning on the spins of all v_i which are not descendants of w and on the spin of w) that

$$\sup_{x \in \mathbf{Z}} \mathbf{P}[S_{n+k} = x | S_0, \ldots, S_n] \leq \sup_{x \in \mathbf{Z}} \mathbf{P}[Y_h = x]. \tag{102}$$

Applying Lemma 17.2 and (102) we get

$$\mathrm{PRE}_S(k) \leq \frac{C_b}{2^h \epsilon (1 - 2\epsilon)^h}, \tag{103}$$

and the proof is complete. □

The process S serves as a building block for $\mathbf{Z}^d$-valued processes whose predictability profiles are controlled.

Corollary 17.3 *For each $\frac{1}{2} < \alpha < 1$, there is a $\mathbf{Z}^d$-valued process $\Phi = \Phi^{\alpha,d}$ such that the random edge sequence $\{\Phi_{n-1}\Phi_n\}_{n\geq1}$ is in Υ_1, and*

$$\forall k \geq 1 \quad \mathrm{PRE}_\Phi(k) \leq C(\alpha, d) k^{-(d-1)\alpha}. \tag{104}$$

Proof. Let $W_k^r = (S_k^{(r)} + k)/2$ for $r = 1, \ldots, d-1$, where $S^{(r)}$ are independent copies of the process described in Theorem 17.1. For $r = 1, \ldots, d-1$, define clocks

$$t_r(n) := \lfloor \frac{n+d-1-r}{d-1} \rfloor,$$

and let $D(n) := n - \sum_{r=1}^{d-1} W_{t_r(n)}^r$.

Write $\Phi_n = (W_{t_1(n)}^1, \ldots, W_{t_{d-1}(n)}^{d-1}, D(n))$. It is then easy to see that

$$\mathrm{PRE}_\Phi(k) \leq \left[\mathrm{PRE}_S(\lfloor \frac{k}{d-1} \rfloor) \right]^{d-1} \leq \left(\frac{C_\alpha k}{d-1} \right)^{-\alpha(d-1)} \leq C(\alpha, d) k^{-\alpha(d-1)}.$$

□

The last ingredient we need to prove that $\mathbf{Z}^3$ admits paths with exponential intersection tails is the following.

Lemma 17.4 *Let $\{\Gamma_n\}$ be a sequence of random variables taking values in a countable set V. If the predictability profile (defined in (89)) of Γ satisfies $\sum_{k=1}^\infty \mathrm{PRE}_\Gamma(k) < \infty$, then there exist $C < \infty$ and $0 < \theta < 1$ such that for any sequence $\{v_n\}_{n\geq0}$ in V and all $\ell \geq 1$,*

$$\mathbf{P}[\#\{n \geq 0 : \Gamma_n = v_n\} \geq \ell] \leq C\theta^\ell. \tag{105}$$

Proof. Choose m large enough so that $\sum_{k=1}^\infty \mathrm{PRE}_\Gamma(km) = \beta < 1$, whence for any sequence $\{v_n\}_{n\geq0}$,

$$\mathbf{P}\Big[\exists k \geq 1 : \Gamma_{n+km} = v_{n+km} \Big| \Gamma_0, \ldots, \Gamma_n\Big] \leq \beta \quad \text{for all } n \geq 0. \tag{106}$$

If n is replaced by a stopping time τ and the σ-field generated by $\Gamma_0, \ldots, \Gamma_n$ is replaced by the usual stopping time σ-field, then (106) remains valid. This can be seen by decomposing the probability according to the value of τ, and checking that the bound holds in each case. Hence, it follows by induction on $r \geq 1$ that for all $j \in \{0, 1, \ldots, m-1\}$,

$$P[\#\{k \geq 1 : \Gamma_{j+km} = v_{j+km}\} \geq r] \leq \beta^r. \tag{107}$$

If $\#\{n \geq 0 : \Gamma_n = v_n\} \geq \ell$ then there must be some $j \in \{0, 1, \ldots, m-1\}$ such that

$$\#\{k \geq 1 : \Gamma_{j+km} = v_{j+km}\} \geq \ell/m - 1.$$

Thus the inequality (105), with $\theta = \beta^{1/m}$ and $C = m\beta^{-1}$, follows from (107). $\square$

Proof of Theorem 11.1 for $d = 3$: The process Φ constructed in Corollary 17.3 with $\alpha > 1/2$ and $d = 3$ satisfies $\sum_k \mathrm{PRE}_\Phi(k) < \infty$, and hence by Lemma 17.4, the distribution μ of the edge sequence $\{\Phi_{n-1}\Phi_n\}_{n=1}^\infty$ has exponential intersection tails. $\square$

18 Tree-Indexed Processes

Label the vertices of a tree Γ by a collection of i.i.d. real random variables $\{X_v\}_{v\in\Gamma}$. Given Γ and the collection $\{X_v\}_{v\in\Gamma}$, we define the **tree-indexed random walk** $\{S_v\}_{v\in\Gamma}$ by

$$S_v = \sum_{w \leq v} X_w,$$

where $w \leq v$ means that v is a descendant of w.

The simple case where Γ is a binary tree and $X_v = \pm 1$ with probabilities p and $1 - p$ was considered by Dubins and Freedman (1967).

We want to determine the speed of tree-indexed random walks, or at least recognize when the speed is positive.

There are several ways to define speed for tree-indexed walks and the answers depend on the definition used. Here are three notions of speed.

Definitions of Speed

- **Cloud Speed**

$$s_{\mathrm{cloud}} := \varlimsup_n \frac{1}{n} \max_{|v|=n} S_v;$$

- **Burst Speed**

$$s_{\mathrm{burst}} := \sup_{\xi\in\partial\Gamma} \varlimsup_{\substack{v\in\xi}} \frac{S_v}{|v|};$$

- **Sustainable Speed**

$$s_{\mathrm{sust}} := \sup_{\xi\in\partial\Gamma} \varliminf_{\substack{v\in\xi}} \frac{S_v}{|v|};$$

These speeds are a.s. constant by Kolmogorov's zero-one law. The first two were studied by Benjamini and Peres (1994b), while the third was studied earlier by Lyons and Pemantle (1992).

Assumptions. Throughout this chapter we will assume that each variable

$$X_v \text{ is not a.s. constant, } \mathbf{E}[X_v] = 0 \text{ and } \mathbf{E}[e^{\lambda X_v}] < \infty \text{ for all } \lambda > 0. \qquad (108)$$

These assumptions can be relaxed, but they make the ideas of the proofs more transparent.

In general, $s_{\text{cloud}} \geq s_{\text{burst}} \geq s_{\text{sust}}$. The following examples shows that the inequalities may be strict.

Example 18.1 Consider the 3-1 tree Γ in Example 2.6. It follows from Theorem 18.4 below that on this tree

$$s_{\text{cloud}} > 0 \quad \text{but} \quad s_{\text{burst}} = s_{\text{sust}} = 0.$$

Example 18.2 Let $n_1 < n_2 < \ldots$ be a sequence of positive integers. Construct a tree Γ as follows: The first n_1 levels of Γ are as in the 3-1 tree. To each vertex v in the n_1-th level of Γ, attach a copy of the first $n_2 - n_1$ levels of the 3-1 tree, with v as its root. Continue by attaching a copy of the first $n_{k+1} - n_k$ levels of the 3-1 tree to each vertex at level n_k of Γ. For any choice $\{n_i\}$, the tree Γ has positive packing dimension; in particular, $\dim_M(\partial\Gamma) = \dim_P(\partial\Gamma) = \log 2$. However, if the n_i increase sufficiently fast, then the Hausdorff dimension of $\partial\Gamma$ is 0, as in the 3-1 tree. Thus in this case Theorem 18.4 yields that $s_{\text{cloud}} \geq s_{\text{burst}} > 0$, but $s_{\text{sust}} = 0$.

Notation. Denote by $\{\tilde{S}_n\}_{n\geq 0}$ the ordinary random walk indexed by the non-negative integers with i.i.d. increments distributed like X_v. Let $I(\cdot)$ be the *rate function* for the random walk $\{\tilde{S}_n\}$, defined by

$$I(a) = \lim_{n\to\infty} \frac{1}{n} \log \mathbf{P}(\tilde{S}_n > na) \quad (a > 0).$$

Theorem 18.3 (Hammersley (1974), Kingman (1975), Biggins (1977)) *Let Γ be a GW tree with mean $m > 1$. Suppose that the vertices of Γ are labeled by random variables X_v that satisfy (108). On the event that Γ survives, a.s. all speeds coincide and equal $s^* := \sup\{s : I(s) \leq \log m\}$.*

Proof. The inequality $s_{\text{cloud}} \leq s^*$ is easy: By the definition of s^*, for any $\epsilon > 0$ there is $\delta > 0$ such that $I(s^* + \epsilon) > \log m + \delta$. Therefore,

$$\mathbf{P}(\tilde{S}_n > n(s^* + \epsilon)) \leq e^{-n(\log m + \delta)} = m^{-n}e^{-n\delta}.$$

Consequently,

$$\mathbf{P}(S_v > n(s^* + \epsilon) \text{ for some } v \in \Gamma_n \mid \text{non-extinction }) \leq \frac{m^n}{1-q} m^{-n}e^{-n\delta},$$

where q is the probability of extinction. The proof is concluded by invoking the Borel-Cantelli Lemma.

For the reverse inequality, let $a < s^*$ be given. Using the strict monotonicity of the rate function and the definition of s^*, choose ϵ so that $I(a) + 2\epsilon < \log m$. For each $k \geq 1$ and $M \in [1, \infty]$, we define a new embedded branching process as follows: start from the root of Γ, and take the set of offspring $\Gamma(v, k, M)$ of a vertex v to consist of all its descendants w in Γ that satisfy

- $|w| = |v| + k$ in Γ ;

- $S_w > S_v + ka$.

- $S_u > S_v - M$ for all u on the path from v to w.

(Here $M = \infty$ means the last requirement holds automatically.) Since $\mathbf{E}|\Gamma(v, k, \infty)| = m^k \mathbf{P}[\tilde{S}_k > ka]$, the definition of I yields that for sufficiently large k,

$$\mathbf{E}|\Gamma(v, k, \infty)| \geq m^k e^{-k[I(a)+\epsilon]} > 2 .$$

By choosing M large, we can ensure that the embedded process has mean offspring

$$\mathbf{E}|\Gamma(v, k, M)| \geq \frac{1}{2} m^k e^{-k[I(a)+\epsilon]} > 1 .$$

Thus for large k, M, the embedded process is supercritical. Therefore $s_{\text{sust}} > a$ with positive probability. Since

$$\{\Gamma : \Gamma \text{ finite or } s_{\text{sust}} \leq a \text{ on } \Gamma\}$$

is an inherited property, Proposition 3.2 implies that $\mathbf{P}[s_{\text{sust}} > a \mid \text{survival}] = 1$. Hence, given survival, we have that a.s.,

$$s^* \geq s_{\text{cloud}} \geq s_{\text{burst}} \geq s_{\text{sust}} \geq s^* . \qquad \square$$

We have already encountered two of the following definitions:

- The *upper Minkowski* dimension of $\partial\Gamma$, written $\overline{\dim}_M(\partial\Gamma)$, is $\log \overline{\mathrm{gr}}(\Gamma)$.

- The *Hausdorff* dimension of $\partial\Gamma$, written $\dim_H(\partial\Gamma)$, is $\log \mathrm{br}(\Gamma)$.

- The **Packing** dimension of $\partial\Gamma$, is defined by

$$\dim_P(\partial\Gamma) := \inf \{ \sup_i \dim_M(\partial\Gamma^{(i)}) \},$$

where the infimum extends over all countable collections $\{\Gamma^{(i)}\}$ of subtrees of Γ such that $\partial\Gamma \subseteq \bigcup_i \partial\Gamma^{(i)}$.

Theorem 18.4 *Suppose that* Γ *is an infinite tree without leaves, and the vertices of* Γ *are labeled by random variables* X_v *that satisfy (108). Then*

(i) $s_{\text{cloud}} > 0 \Leftrightarrow \overline{\dim}_M(\partial\Gamma) > 0$.

(ii) $s_{\text{burst}} > 0 \Leftrightarrow \dim_P(\partial\Gamma) > 0$.

(iii) $s_{\text{sust}} > 0 \Leftrightarrow \dim_H(\partial\Gamma) > 0$.

Proof. (i) The implication "$\Rightarrow$" is easy: By Cramér's theorem on large deviations, (108) implies that $I(a) > 0$ for any $a > 0$. Therefore

$$\sum_n \mathbf{P}(S_v > na \text{ for some } v \in \Gamma_n) \le \sum_n |\Gamma_n| \, \mathbf{P}(\tilde{S}_n > an) \le \sum_n |\Gamma_n| \, e^{-nI(a)}.$$

which is finite since $\overline{\dim}_M(\partial\Gamma) = 0$ means that Γ has subexponential growth. Thus by Borel-Cantelli

$$\mathbf{P}(\{S_v > na \text{ for some } v \in \Gamma_n\} \text{ i.o. }) = 0$$

for any $a > 0$.

For the implication "$\Leftarrow$", observe that because we assumed Γ has no leaves, there exists at least one descendant in Γ_{2n} for each $v \in \Gamma_n$. Denote the leftmost such descendant by $w(v)$. The $|\Gamma_n|$ paths from vertices $v \in \Gamma_n$ to the corresponding $w(v)$ are disjoint. Since $\overline{\dim}_M(\partial\Gamma) > 0$, if we choose ϵ sufficiently small, then

$$|\Gamma_n| > e^{n[I(2\epsilon) + 2\epsilon]} \quad \text{for infinitely many } n \tag{109}$$

By Cramér's theorem, $\mathbf{P}(\tilde{S}_n > 2n\epsilon) > e^{-n[I(2\epsilon) + \epsilon]}$ for large n.

Write $\Gamma'_n = \{v \in \Gamma_n : S_v > -n\epsilon\}$. By the Weak Law of Large Numbers,

$$|\Gamma_n|^{-1} \mathbf{E}|\Gamma'_n| = \mathbf{P}(\tilde{S}_n > -n\epsilon) \longrightarrow 1$$

and therefore $\mathbf{P}(|\Gamma'_n| < |\Gamma_n|/2) \longrightarrow 0$. Denote

$$A_n := \{\exists w \in \Gamma_{2n} : S_w > n\epsilon\}.$$

Then

$$\mathbf{P}[A_n^c] \le \mathbf{P}(|\Gamma'_n| < |\Gamma_n|/2) + \mathbf{P}(|\Gamma'_n| \ge |\Gamma_n|/2 \text{ and } S_{w(v)} - S_v \le 2n\epsilon \; \forall v \in \Gamma'_n).$$

The right-hand side is at most

$$\mathbf{P}(|\Gamma'_n| < |\Gamma_n|/2) + \left(1 - e^{-n[I(2\epsilon) + \epsilon]}\right)^{|\Gamma_n|/2},$$

which tends to zero along a subsequence of n values by (109). Taking stock, we infer that $\mathbf{P}(A_n \text{ i.o.}) \ge \lim_n \mathbf{P}(A_n) = 1$, so $s_{\text{cloud}} \ge \epsilon/2$ a.s.

(ii) The implication "⇒" is easy again: if $\dim_P(\partial\Gamma) = 0$, then given $\epsilon > 0$ we can find a cover $\bigcup_i \partial\Gamma^{(i)}$ of $\partial\Gamma$ with $\overline{\dim}_M(\partial\Gamma^{(i)}) \leq \epsilon$ for all i. As in the proof of (i),

$$s_{\text{cloud}}(\Gamma^{(i)}) \leq \epsilon'$$

for some ϵ' and all i. Whence

$$s_{\text{burst}}(\Gamma^{(i)}) \leq s_{\text{cloud}}(\Gamma^{(i)}) \leq \epsilon'$$

for all i and so $s_{\text{burst}}(\Gamma) \leq \epsilon'$. Here ϵ' can be made arbitrarily small because ϵ may be taken arbitrarily small.

For the reverse implication "⇐", let $d = \dim_P(\partial\Gamma) > 0$. Pick $\epsilon > 0$ small and let

$$\Gamma' = \{v \in \Gamma : \dim_P(\Gamma(v)) > d - \epsilon\};$$

here $\Gamma(v) = \{w \in \Gamma : w \leq v \text{ or } w \geq v\}$.

Now $\rho \in \Gamma'$, so $\Gamma' \neq \emptyset$ and $\dim_P(\partial\Gamma') > d - \epsilon$. Actually, it is easy to see from the definition of packing dimension that

$$\overline{\dim}_M(\partial\Gamma'(v)) > d - \epsilon \quad \text{for all } v \in \Gamma'.$$

By (i) and the definition of cloud-speed, with probability one we can find for each $v \in \Gamma'$ a vertex $w = f(v) \in \Gamma'(v)$ with $w > v$ and $S_w > |w|\beta$ for some fixed $\beta > 0$. The sequence $\rho, f(\rho), f(f(\rho)), \ldots$ is a sequence of vertices $\{v_j\}_{j \geq 0}$ along a ray of Γ such that

$$\frac{S_{v_i}}{|v_i|} > \beta, \qquad \text{for all } i \geq 1.$$

(iii) was proved by Lyons and Pemantle (1992) in the following sharp form:

$$I(s_{\text{sust}}) = \log \text{br}(\Gamma) = \dim_H(\partial\Gamma).$$

(For the other speed notions there is no analogous exact formula.)

The inequality $I(s_{\text{sust}}) \leq \log \text{br}(\Gamma)$ is proved using the first-moment method (see the proof of Theorem 5.4). For the other inequality, fix a so that $I(a) < \dim_H(\partial\Gamma)$ and then choose k such that $\mathbf{P}\left(\tilde{S}_k > ka\right) > \text{br}(\Gamma)^k$. Consider a compressed tree $\Gamma[k]$ whose ℓth level is the $k\ell$th level of Γ, with the induced partial order. It is easy to see that $\dim_H(\partial\Gamma[k]) = k\dim_H(\partial\Gamma)$. Define a general percolation on $\Gamma[k]$ in which the edge $\overrightarrow{vw}$ is retained if $S_w - S_v > ka$. This general percolation process is not independent; however, for each fixed k, it is quasi-independent. By proposition 7.1, this percolation survives with positive probability, whence $s_{\text{sust}} \geq a$. It follows that $I(s_{\text{sust}}) \geq \log \text{br}(\Gamma)$. □

Exercise 18.5 *Suppose that Γ is an infinite tree without leaves, and its vertices are labeled by i.i.d. variables $X_v \sim N(0,1)$. Denote $d = \dim_M(\partial\Gamma)$. Prove that*

$$\sqrt{d/2} \leq s_{\text{cloud}} \leq \sqrt{2d}$$

and both bounds can be achieved.

Hint: Use the ideas in the proof of (i) and optimize, or see [9]. These bounds were sharpened by Benassi (1996).

Consider an infinite tree Γ again, label its vertices by i.i.d. real-valued random variables $\{X_v\}_{v\in\Gamma}$, and let $\{S_v\}_{v\in\Gamma}$ be the corresponding tree-indexed random walk. The following question is mostly open.

Open Problem 2 (Bouncing Rays) *Suppose that there a.s. exists a ray $\xi \in \partial\Gamma$ such that $\liminf\limits_{v\in\xi} S_v > -\infty$. Must there a.s. exist a ray $\xi' \in \partial\Gamma$ with $\lim\limits_{v\in\xi'} S_v = +\infty$?*

The only cases for which the answer is known (Pemantle and Peres 1995a) are when

- $X_v = \pm 1$ with probability $1/2$ each, or when

- $X_v \sim N(0,1)$.

In these cases there is an exact capacity criterion on the tree for the property to hold. Even in these special cases the proofs are complicated.

19 Recurrence for Tree-Indexed Markov Chains

This chapter is based on Benjamini and Peres (1994a). For a tree Γ and a vertex v, denote by Γ^v the subtree consisting of v and its descendants. We are given a countable state-space G and a set of transition probabilities $\{p(x,y): x, y \in G\}$. the induced Γ-indexed Markov chain is a collection of G-valued random variables $\{S_v\}_{v\in\Gamma}$, with some initial state $S_\rho := x_0 \in G$ and finite-dimensional distributions specified by the following requirement: if $w \in \Gamma$ and v is the parent of w, then

$$\mathbf{P}\big(S_w = y \mid S_v = x,\, S_u \text{ for } u \notin \Gamma_v\big) = \mathbf{P}(S_w = y \mid S_v = x) = p(x,y).$$

We may think of the state-space G as a graph, with vertices the elements of G and an edge between x and y iff $p(x,y) > 0$. If $p = \{p(x,y): x, y \in G\}$ is irreducible, i.e., for any $x, y \in G$ there exists an n such that $p^n(x,y) > 0$, then the associated graph is connected.

Definitions. A tree-indexed Markov chain is **recurrent** if it returns infinitely often to its starting point with positive probability:

$$\mathbf{P}(S_v = S_\rho \text{ for infinitely many } v \in \Gamma) > 0.$$

A stronger requirement is **ray-recurrence**: $\{S_v\}_{v\in\Gamma}$ is ray-recurrent if

$$\mathbf{P}(\exists\, \xi \in \partial\Gamma : S_v = S_\rho \text{ for infinitely many } v \in \xi) > 0.$$

In general, recurrence does not imply ray-recurrence (even when $G = \mathbf{Z}^3$). Indeed, the 3-1 tree has exponential growth (which yields recurrence for $G = \mathbf{Z}^d$), yet it has a countable boundary (which precludes ray-recurrence on any transient G).

The probabilities in the definitions of recurrence and ray-recurrence may lie strictly between 0 and 1, even when the indexing tree is a binary tree. If G is a group and the transition probabilities are G-invariant, then there are zero–one laws for both notions of recurrence.

Given a state space G, an irreducible stochastic matrix $p = \{p(x, y) : x, y \in G\}$ and a finite subset F of G, write $\rho(p_F)$ for the spectral radius of the substochastic matrix $p_F = \{p(x, y) : x, y \in F\}$. We then define

$$\rho(G, p) = \sup_{F \text{ finite}} \rho(p_F).$$

Then

$$\mathbf{P}(\exists\, \xi \in \partial\Gamma \text{ with bounded trajectory}) > 0 \;\Leftrightarrow\; \mathrm{br}(\Gamma) > \frac{1}{\rho(G, p)}.$$

Simple random walk on Z has spectral radius 1, but we can make a quantitative statement on rays with bounded trajectories: For the Γ-indexed simple random walk on $\mathbf{Z}$,

$$\mathrm{br}(\Gamma) > \frac{1}{\cos\left(\pi/(b+1)\right)}$$

is sufficient for the existence of a ray with trajectory in $\{0, 1, \ldots, b-1\}$ to have positive probability, and

$$\mathrm{br}(\Gamma) \geq \frac{1}{\cos\left(\pi/(b+1)\right)}$$

is necessary.

Finally, we note that recurrence of a Γ-indexed Markov chain on G is related to a comparison of the Minkowski dimension of Γ and the spectral radius of G, while ray-recurrence is related to a comparison of packing dimension and spectral radius. In particular, $\dim_M(\partial\Gamma) < -\log[\rho(G, p)]$ implies non-recurrence and $\dim_P(\partial\Gamma) < -\log[\rho(G, p)]$ implies non-ray-recurrence.

More details on the notions described in this chapter, and some amusing examples, can be found in [8, 9]. Benjamini and Schramm [10] give an application of tree-indexed Markov chains to a problem in discrete geometry.

20   Dynamical Percolation

This chapter is based on Häggström, Peres, and Steif (1997).

Consider Bernoulli(p) percolation on an infinite graph G. Recall that each edge is, independently, open with probability p. As before, $\mathbf{P}_{G,p} = \mathbf{P}_p$ will denote this product measure. Write $\mathcal{C}$ for the event that there exists an infinite open cluster. Recall that by Kolmogorov's 0-1 law, the probability of $\mathcal{C}$ is, for fixed G and p, either 0 or 1. As remarked previously, there exists a critical probability $p_c = p_c(G) \in [0, 1]$ such that

$$\mathbf{P}_p(\mathcal{C}) = \begin{cases} 0 & \text{for } p < p_c \\ 1 & \text{for } p > p_c. \end{cases}$$

At $p = p_c$ we can have either $\mathbf{P}_p(\mathcal{C}) = 0$ or $\mathbf{P}_p(\mathcal{C}) = 1$, depending on G.

In this chapter we consider a dynamical variant of percolation. Given $p \in (0,1)$, we want the set of open edges to evolve so that at any fixed time $t \geq 0$, the distribution of this set is $\mathbf{P}_p$. The most natural way to accomplish this is to let the distribution at time 0 be given by $\mathbf{P}_p$, and to let each edge change its status (open or closed) according to a continuous time, stationary 2-state Markov chain, independently of all other edges. For an edge e of G, write $\eta_t(e) = 1$ if e is open at time t, and $\eta_t(e) = 0$ otherwise. The entire configuration of open and closed edges at time t, denoted η_t, can then be regarded as an element of $X = \{0,1\}^E$ (where E is the edge set of G). The evolution of η_t is a Markov process, and can be viewed as the simplest type of particle system. Each edge flips (changes its value) at rate

$$\lambda(\eta_t, e) = \begin{cases} p & \text{if } \eta_t(e) = 0 \\ 1 - p & \text{if } \eta_t(e) = 1 \end{cases}$$

and the probability that two edges flip simultaneously is 0. Write $\boldsymbol{\Psi}_{G,p}$ (or $\boldsymbol{\Psi}_p$) for the underlying probability measure of this Markov process, and write $\mathcal{C}_t$ for the event that there is an infinite cluster of open edges in η_t. Since $\mathbf{P}_p$ is a stationary measure for this Markov process, Fubini's theorem implies that

$$\begin{cases} \boldsymbol{\Psi}_p(\mathcal{C}_t \text{ occurs for Lebesgue a.e. } t) = 1 & \text{if } \mathbf{P}_p(\mathcal{C}) = 1 \\ \boldsymbol{\Psi}_p(\neg\mathcal{C}_t \text{ occurs for Lebesgue a.e. } t) = 1 & \text{if } \mathbf{P}_p(\mathcal{C}) = 0 \end{cases}$$

where $\neg\mathcal{C}_t$ denotes the complement of $\mathcal{C}_t$. The main question studied here is the following,

Question 20.1 *For which graphs can the quantifier "for a.e. t" in the above statements be replaced by "for every t"?*

For $p \neq p_c$, the answer is all graphs.

Proposition 20.2 *For any graph G we have*

$$\begin{cases} \boldsymbol{\Psi}_p(\mathcal{C}_t \text{ occurs for every } t) = 1 & \text{if } p > p_c(G) \\ \boldsymbol{\Psi}_p(\neg\mathcal{C}_t \text{ occurs for every } t) = 1 & \text{if } p < p_c(G). \end{cases} \tag{110}$$

Notation: For $0 \leq a \leq b < \infty$ and any edge e of a graph G, we abbreviate

$$\inf_{[a,b]} \eta(e) := \inf_{t \in [a,b]} \eta_t(e).$$

and write $C_{[a,b]}^{\inf}$ for the event that there is an infinite cluster of edges with $\inf_{[a,b]} \eta(e) = 1$. Analogously, define $\sup_{[a,b]} \eta$, and let $C_{[a,b]}^{\sup}$ be the event that there is an infinite cluster of edges with $\sup_{[a,b]} \eta(e) = 1$.

Proof. (i) Suppose $p > p_c$. Let $0 < \epsilon < p - p_c$ and observe that for every edge e,

$$\boldsymbol{\Psi}_p\Big\{ \inf_{[0,\epsilon]} \eta(e) = 1 \Big\} = p \exp(-(1-p)\epsilon) > p - \epsilon > p_c.$$

Since the events $\{ \inf_{[0,\epsilon]} \eta(e) = 1 \}$ are mutually independent as e ranges over the edges of G, it follows from the definition of p_c that $\Psi_p\left[C_{[0,\epsilon]}^{\inf} \right] = 1$ and therefore

$$\Psi_p\big(C_t \text{ occurs for all } t \in [0,\epsilon] \big) = 1 \,.$$

Repeating the argument for the intervals $[k\epsilon, (k+1)\epsilon]$ with integer k and using countable additivity, we obtain the supercritical part of the proposition.

(ii) A similar argument proves that for $p < p_c$ there is never an infinite open cluster. We take $\epsilon \in (0, p_c - p)$ and find that

$$\Psi_p\Big\{ \sup_{[0,\epsilon]} \eta(e) = 1 \Big\} = 1 - (1-p)\exp(-p\epsilon) < p + p\epsilon < p_c \,. \tag{111}$$

Therefore $\Psi_p\big(C_{[0,\epsilon]}^{\sup} \big) = 0$, whence there is a.s. no infinite cluster for any $t \in [0,\epsilon]$. Countable additivity concludes the argument. $\qquad\square$

At the critical value $p_c(G)$ the situation is more delicate.

Theorem 20.3 *There exists a graph G_1 with the property that at $p = p_c(G)$ we have $\mathbf{P}_{G,p}(C) = 0$ but $\Psi_{G,p}\big(\cup_{t>0} C_t \big) = 1$. (The latter probability is 0 or 1 for any graph.) There also exists a graph G_2 such that for $p = p_c(G_2)$ we have $\mathbf{P}_{G_2,p}(C) = 1$, yet $\Psi_{G_2,p}(\cap_{t>0} C_t) = 0$.*

The graphs for which percolation problems have been studied most extensively are the lattices $\mathbf{Z}^d$, and trees. On $\mathbf{Z}^2$, the critical value p_c is $1/2$ and $\mathbf{P}_{p_c}(C) = 0$ (see Kesten (1980)); for $d > 2$ the precise value of $p_c(\mathbf{Z}^d)$ is not known. Hara and Slade (1994) showed that $\mathbf{P}_{p_c}(C) = 0$ for $\mathbf{Z}^d$ if $d \geq 19$, and it is certainly believed that this holds for all d.

Theorem 20.4 *Let G be either the integer lattice $\mathbf{Z}^d$ with $d \geq 19$ or a regular tree. Then $\Psi_{G,p_c}(\neg C_t \text{ occurs for every } t) = 1$.*

Remark. It is not known whether $G = \mathbf{Z}^2$ can be included in Theorem 20.4. Let $\theta(p)$ denote the $\mathbf{P}_p$-probability that the origin is in an infinite open cluster. The proof of Theorem 20.4 for $G = Z^d$ with $d \geq 19$ uses more information than just $\theta(p_c) = 0$; it also uses that θ has a finite right derivative at p_c. In $\mathbf{Z}^2$ it is known that $\theta(p_c) = 0$, but Kesten and Zhang proved that the right derivative of θ is infinite at p_c.

Next, we consider dynamical percolation on general trees. In Chapter 14, we proved R. Lyons' criterion for $\mathbf{P}_p(C) > 0$ in terms of effective electrical resistance (see (39)); effective resistance is easy to calculate on trees using the parallel and series laws. Here we obtain such a criterion for dynamical percolation.

For an infinite tree Γ with root ρ, as before we write Γ_n for the set of vertices at distance exactly n from ρ, the nth level of Γ. Recall that a tree is spherically symmetric if all vertices on the same level have equally many children.

Theorem 20.5 *Let* $\{\eta_t\}$ *be a dynamical percolation process with parameter* $0 < p < 1$ *on an infinite tree* Γ. *Assign each edge between levels* $n - 1$ *and* n *of* Γ *the resistance* p^{-n}/n. *If in the resulting resistor network the effective resistance from the root to infinity is finite, then* $\Psi_{\Gamma,p}$-*a.s. there exist times* $t > 0$ *such that* Γ *has an infinite open cluster, while if this resistance is infinite, then a.s. there are no such times. In particular, if* Γ *is spherically symmetric, then*

$$\Psi_{\Gamma,p}(\cup_{t>0}\mathcal{C}_t) = 1 \text{ if and only if } \sum_{n=1}^{\infty} \frac{p^{-n}}{n|\Gamma_n|} < \infty. \tag{112}$$

Recall R. Lyons' criterion for the percolation probability on a general tree Γ to be positive: Suppose that $0 < p < 1$ and assign each edge between levels $n - 1$ and n resistance p^{-n}. Then $\mathbf{P}_{\Gamma,p}(\mathcal{C}) > 0$ iff the resulting effective resistance from the root to infinity is finite. Thus a spherically symmetric tree Γ with $p = p_c(\Gamma) \in (0, 1)$, has $\Psi_{\Gamma,p}(\cup_{t>0}\mathcal{C}_t) = 1$ but $\mathbf{P}_{\Gamma,p}(\mathcal{C}) = 0$ iff the series in (112) converges but $\sum_{n=1}^{\infty} \frac{p^{-n}}{|\Gamma_n|} = \infty$.

In the course of the proof of Theorem 20.5, we obtain bounds for the probability that there exists a time $t \in [0, 1]$ for which there is an open path in η_t from the root to the nth level Γ_n. For example, on the regular tree $\mathbf{T}^k$ with $p = 1/k$, this probability is bounded between constant multiples of $1/\log n$. (The probability under $\mathbf{P}_{1/k}$ that an open path exists from ρ to the nth level of $\mathbf{T}^k$, is bounded between constant multiples of $1/n$; this follows from Kolmogorov's theorem on critical branching processes, see Athreya and Ney (1972).) For a general tree these bounds, given in Theorem 20.9, can be expressed in terms of the effective resistance from the root to Γ_n, and the ratio of the upper and lower bounds is an absolute constant.

For a graph with $\Psi_{G,p}(\cup_{t>0}\mathcal{C}_t) = 1$ but $\mathbf{P}_{G,p}(\mathcal{C}) = 0$, the set of percolating times at criticality has zero Lebesgue measure, so it is natural to ask for its Hausdorff dimension. For spherically symmetric trees there is a complete answer.

Theorem 20.6 *Let* $p \in (0, 1)$ *and let* Γ *be a spherically symmetric tree. If the set of times* $\{t \in [0, \infty) : \mathcal{C}_t \text{ occurs}\}$ *is a.s. nonempty, then* Ψ_p-*a.s. this set has Hausdorff dimension*

$$\sup\left\{ \alpha \in [0, 1] : \sum_{n=1}^{\infty} \frac{p^{-n}n^{\alpha-1}}{|\Gamma_n|} < \infty \right\}.$$

(*Note that this series converges for* $\alpha = 0$ *by* (112).)

Here are some interesting trees with $\Psi_{T,p}(\cup_{t>0}\mathcal{C}_t) = 1$ but $\mathbf{P}_{T,p}(\mathcal{C}) = 0$:

Example 20.7 Let Γ be the spherically symmetric tree where each vertex on level n has 4 children if $n = 1, 2, 4 \ldots$ is a power of 2, and 2 children otherwise. Then it is easily seen that $n2^n \leq |\Gamma_n| \leq 2n2^n$ for all $n > 0$. Combining Theorem 20.6 with the result of R. Lyons quoted after Theorem 20.5, we see that $\Psi_{1/2}$-a.s. the set of times for which percolation occurs on Γ has Hausdorff dimension 1 but Lebesgue measure 0. $\triangle$

Example 20.8 Let $0 < p, \beta < 1$, and suppose that Γ is a spherically symmetric tree with $|\Gamma_n| = p^{-n} n^{\beta+o(1)}$ as $n \to \infty$. Then Theorem 20.6 implies that Ψ_p-a.s. the set of times for which percolation occurs on Γ has Hausdorff dimension β. $\triangle$

Since we will introduce an auxiliary random killing time τ, we denote the underlying probability measure $\mathbf{P}$ rather than Ψ_p. The event that there is an open path from the root to $\partial\Gamma$ in η_t is denoted $\{\rho \overset{t}{\leftrightarrow} \partial\Gamma\}$.

Theorem 20.9 *Consider dynamical percolation $\{\eta_t\}$ with parameter $0 < p < 1$ on a tree Γ which is either finite or infinite with $\mathbf{P}_{\Gamma,p}(C) = 0$. Let τ be a random variable with an exponential distribution of mean 1, which is independent of the process $\{\eta_t\}$. Let*

$$h(n) = \frac{p^{-n}}{n+1} \cdot \frac{1 - p^{n+1}}{1 - p} \quad \text{for } n \geq 0. \tag{113}$$

Then the event $A = \{\exists t \in [0, \tau] : \rho \overset{t}{\leftrightarrow} \partial\Gamma\}$ satisfies for some constant C

$$\frac{1}{2}\mathrm{Cap}_h(\partial\Gamma) \leq \mathbf{P}(A) \leq 2C\mathrm{Cap}_h(\partial\Gamma), \tag{114}$$

Remarks:

(i) It is easy to verify that h is increasing and $h(n) \leq p^{-n}$ for all n. These properties also follow from the interpretation of h given in Lemma 20.10(iii) below. In the sequel, we will sometimes write $h(v)$ instead of $h(|v|)$ when v is a vertex.

(ii) The event A is easier to work with than the perhaps more natural event $B = \{\exists t \in [0, 1] : \rho \overset{t}{\leftrightarrow} \partial\Gamma\}$. Noting that $\mathbf{P}(B) \leq \mathbf{P}(A | \tau > 1) \leq \mathbf{P}(A)/e^{-1}$ and $\mathbf{P}(A) \leq \sum_{k=0}^{\infty} e^{-k}\mathbf{P}(B) = \mathbf{P}(B)/(1 - e^{-1})$, we obtain

$$\frac{1 - e^{-1}}{2}\mathrm{Cap}_h(\partial\Gamma) \leq \mathbf{P}(B) \leq 2eC\mathrm{Cap}_h(\partial\Gamma).$$

We will only prove the lower bound in Theorem 20.9; consult [37] for the other inequality. We will need a lemma concerning the behavior of a pair of paths.

Notation: Denote by $\{v \overset{t}{\leftrightarrow} w\}$ the event that there is an open path in η_t between the vertices v and w. Similarly, when x is a ray of the tree, $\{\rho \overset{t}{\leftrightarrow} x\}$ means that x is open at time t. Thus $\{\rho \overset{t}{\leftrightarrow} \partial\Gamma\} = \bigcup_{x \in \partial\Gamma} \{\rho \overset{t}{\leftrightarrow} x\}$. For $s > 0$ let $T_v(s) := \int_0^s \mathbb{1}_{\{\rho \overset{t}{\leftrightarrow} v\}} dt$ be the amount of time in $[0, s]$ when the path from the root to v is open.

Lemma 20.10 *Let u and w be vertices of Γ. With the notation of Theorem 20.9 in force,*

(i) $\mathbf{E}[T_w(\tau)] = p^{|w|}$

(ii) $\mathbf{E}[T_w(\tau) \mid T_w(\tau) > 0] = \mathbf{E}[T_w(\tau) \mid \rho \overset{0}{\leftrightarrow} w] = h(w)p^{|w|}$

(iii) $\mathbf{P}(T_w(\tau) > 0) = h(w)^{-1}$

(iv) $\mathbf{E}[T_u(\tau)T_w(\tau)] = 2h(u \wedge w)p^{|u|+|w|}$

Proof: Let $q = 1 - p$.

(i) This is immediate from Fubini's Theorem.

(ii) The first equality follows from the lack of memory of the exponential distribution. Verifying the second equality requires a calculation:

$$\mathbf{E}[T_w(\tau) \mid \rho \overset{0}{\leftrightarrow} w] = \int_0^\infty \mathbf{P}(\rho \overset{t}{\leftrightarrow} w \mid \rho \overset{0}{\leftrightarrow} w)\mathbf{P}(\tau > t)\, dt$$

$$= \int_0^\infty (p + qe^{-t})^{|w|} e^{-t}\, dt = \frac{-(p + qe^{-t})^{|w|+1}}{(|w| + 1)q}\Big|_{t=0}^\infty.$$

(iii) The required probability is the ratio of the expectations in (i) and (ii).

(iv) Since the process $\{\eta_t\}$ is reversible,

$$\mathbf{E}[T_u(\tau)T_w(\tau)] = \mathbf{E}\int_0^\tau\int_0^\tau \mathbf{1}_{\{\rho \overset{s}{\leftrightarrow} u\}}\mathbf{1}_{\{\rho \overset{t}{\leftrightarrow} w\}}\, dt\, ds$$

$$= 2\int_0^\infty\int_s^\infty \mathbf{P}(\rho \overset{s}{\leftrightarrow} u)\mathbf{P}\Big(\rho \overset{t}{\leftrightarrow} w \mid \rho \overset{s}{\leftrightarrow} u\Big)e^{-t}\, dt\, ds. \quad (115)$$

Observe that for $t > s$,

$$\mathbf{P}\Big(\rho \overset{t}{\leftrightarrow} w \mid \rho \overset{s}{\leftrightarrow} u\Big) = p^{|w|-|u\wedge w|}\mathbf{P}\Big(\rho \overset{t}{\leftrightarrow} u \wedge w \mid \rho \overset{s}{\leftrightarrow} u \wedge w\Big).$$

Change variables $\tilde{t} = t - s$ in (115) to get that $\mathbf{E}[T_u(\tau)T_w(\tau)]$ equals

$$= 2p^{|w|-|u\wedge w|}\int_0^\infty\int_0^\infty \mathbf{P}(\rho \overset{s}{\leftrightarrow} u)e^{-s-\tilde{t}}\mathbf{P}\Big(\rho \overset{\tilde{t}}{\leftrightarrow} u \wedge w \mid \rho \overset{0}{\leftrightarrow} u \wedge w\Big)d\tilde{t}\, ds$$

$$= 2p^{|w|-|u\wedge w|}\mathbf{E}[T_u(\tau)] \cdot \mathbf{E}[T_{u\wedge w}(\tau) \mid \rho \overset{0}{\leftrightarrow} u \wedge w].$$

Substituting parts (i) and (ii) of the lemma into the last equation proves (iv).

$\square$

Proof of lower bound in Theorem 20.9. We prove the theorem when Γ is a finite tree; the general case then follows by an appropriate limiting procedure. The lower bound on $\mathbf{P}(A)$ is proved via the second moment method. Let μ be a probability measure on $\partial\Gamma$, and consider the random variable

$$Z := \sum_{v \in \partial\Gamma} T_v(\tau)p^{-|v|}\mu(v). \quad (116)$$

Lemma 20.10(i) implies that $\mathbf{E}(Z) = 1$. Part (iv) of the same lemma gives

$$\mathbf{E}[Z^2] = \sum_{v \in \partial\Gamma} \sum_{w \in \partial\Gamma} \mathbf{E}[T_v(\tau)T_w(\tau)]p^{-|v|-|w|}\mu(v)\mu(w) = 2\mathcal{E}_h(\mu) . \tag{117}$$

Using the Cauchy-Schwarz inequality we find that

$$\mathbf{P}(A) \geq \mathbf{P}(Z > 0) \geq \frac{\mathbf{E}[Z]^2}{\mathbf{E}[Z^2]} = \frac{1}{2\mathcal{E}_h(\mu)} .$$

Taking the supremum of the right-hand side over all probability measures μ on $\partial\Gamma$ proves the lower bound on $\mathbf{P}[A]$ in (114). $\square$

We include the statement of one result from Peres and Steif (1998).

Theorem 20.11 *Let Γ be an infinite spherically symmetric tree, $p = p_c(\Gamma) \in (0,1)$ and T^k denote the set of times in $[0,\infty)$ when there are at least k infinite clusters. Suppose that $\mathbf{P}_p(\mathcal{C}) = 0$. Let*

$$\alpha_c := \sup \left\{ \alpha \in [0,1] : \sum_{n=1}^{\infty} \frac{p^{-n}n^{\alpha-1}}{|\Gamma_n|} < \infty \right\} .$$

Then for all k, the Hausdorff dimension of T^k is

$$\max\{0, 1 - k(1 - \alpha_c)\} \quad \Psi_p\text{--a.s.}. \tag{118}$$

21 Stochastic Domination Between Trees

For a tree Γ with total height $N \leq \infty$, label its vertices by i.i.d. random variables $\{X_v\}_{v \in \Gamma}$. If $B \subseteq \mathbf{R}^N$ is a Borel set, we write

$$\mathbf{P}(B; \Gamma) = \mathbf{P}(\exists \xi \in \partial\Gamma : (X_v)_{v \in \xi} \in B) .$$

For two such trees Γ and Γ' of height $N \leq \infty$, labeled by $\{X_v\}_{v \in \Gamma}$ and $\{X'_v\}_{v \in \Gamma'}$ respectively, we say that Γ' **stochastically dominates** Γ if for any Borel set $B \subseteq \mathbf{R}^N$,

$$\mathbf{P}(B; \Gamma) \leq \mathbf{P}(B; \Gamma') .$$

To verify that one tree dominates another, it suffices to consider the case where the X_v are i.i.d. uniform random variables in $[0,1]$, since other random variables can be written as functions of these.

Recall that a tree Γ is spherically symmetric if all vertices in Γ_n have the same number of offspring.

Theorem 21.1 (Pemantle and Peres 1994) *Let Γ' be a spherically symmetric tree and Γ another (arbitrary) tree. Then Γ' stochastically dominates Γ iff $|\Gamma_n| \leq |\Gamma'_n|$ for all $n \geq 1$.*

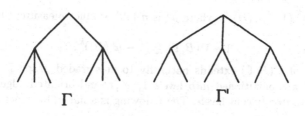

Figure 8: Γ is dominated by Γ'.

Example 21.2 *Two trees of height 2.*

Let Γ be the tree of height 2 in which the root has two offspring and each of these three offspring. Let Γ' be the tree for which the root has three offspring and and each of these two offspring.

Then it is not clear a priori which tree dominates. The result above yields that Γ is dominated by Γ'. △

Stochastic domination between trees is well understood only for trees which are either spherically symmetric or have height two. Already for trees of height three, the domination order is somewhat mysterious, as the following example from Pemantle and Peres (1994) demonstrates.

Example 21.3 *Comparison between a tree T and T with vertices glued.*

Consider the trees T and T' in the next figure, where T' is obtained from T by gluing together the vertices in the first generation.

Intuitively, it seems that T should dominate T', but this is not the case. If

$$B^c = ([0, 1/2] \times [0, 1] \times [0, 2/3]) \cup ([1/2, 1] \times [0, 1/2] \times [0, 1])$$

and the X_v are uniform on $[0, 1]$, then the probability that $(X_{v_1}, X_{v_2}, X_{v_3}) \in B^c$ for all paths (ρ, v_1, v_2, v_3) in T is 1075/7776, while the corresponding probability for T' is only 998/7776. △

A consequence of Theorem 21.1 is that, among all trees of height n with $|\Gamma_n| = k$, the tree $T(n, k)$ consisting of k disjoint paths joined at the root is maximal in the stochastic order. If the common law of the X_v is μ and $B \subseteq \mathbf{R}^n$, then $1 -$

$\mathbf{P}(B; T(n, k)) = (1 - \mu^n(B))^k$, where μ^n is n-fold product measure; thus for any tree Γ of height n,

$$1 - \mathbf{P}(B; \Gamma) \geq (1 - \mu^n(B))^k.$$

The definition of $\mathbf{P}(B; \Gamma)$ extends naturally to any graded graph Γ, a finite graph whose vertices are partitioned into levels $1, \ldots, n$ and oriented edges allowed only between vertices in adjacent levels. The following is a natural conjecture.

Conjecture 3 *For any graded graph Γ of height n, let $K(\Gamma)$ be the number of oriented paths that pass through every level of Γ and let X_v be i.i.d. random variables with common law μ. Then for any $B \subseteq \mathbf{R}^n$,*

$$1 - \mathbf{P}(B; \Gamma) \geq (1 - \mu^n(B))^{K(\Gamma)}.$$

If B is upwardly closed (that is, $\mathbf{x} \in B$ and $\mathbf{y} \geq \mathbf{x}$ coordinate-wise imply $\mathbf{y} \in B$), then the conjecture is an easy consequence of the FKG inequality. The case $n = 2$ corresponds to a bipartite graph; Conjecture 3 for this case is due to Sidorenko (1994), who stated it (and proved it in many special cases) in the following analytic form:

Sidorenko's Conjecture: *Let $f : [0, 1]^2 \to [0, \infty)$ be a nonnegative bounded measurable function and consider the bipartite graph with vertices $X_1, \ldots, X_n$ and $Y_1, \ldots, Y_m$. If E is the edge-set of this graph, then*

$$\int \cdots \int \prod_{X_i \sim Y_j} f(x_i, y_j) dx_1 \ldots dx_n dy_1 \ldots dy_m \geq \left(\iint f(x, y) dx dy \right)^{|E|}. \tag{119}$$

For the bipartite graph consisting of three vertices X, Y, Z and two edges XY and XZ, the conjecture reads

$$\iiint f(x, y) f(x, z) dx dy dz \geq \left(\iint f(x, y) dx dy \right)^2$$

and can be easily proved using the Cauchy-Schwarz inequality.

Exercise 21.4 *Prove Sidorenko's conjecture for the bipartite graph with four vertices and three edges, XY, XZ, and WZ. (Hint: use Hölder's inequality with $p = 3$ and $q = 3/2$.)*

Sidorenko proved his conjecture for bipartite graphs with at most one cycle, and for bipartite graphs where one side has at most four vertices. For general finite bipartite graphs, it is still open whether (119) always holds.

We conclude with yet another **problem:** In the statement of Theorem 16.7 we defined an information-theoretic domination relation between trees. It would be quite interesting to compare that relation with the stochastic domination relation studied in this chapter.

Bibliography

[1] M. Aizenman (1985). The intersection of Brownian paths as a case study of a renormalisation method for quantum field theory. *Commun. Math. Phys.* **97**, 91–110.

[2] P. Antal and A. Pisztora (1996). On the chemical distance in supercritical Bernoulli percolation. *Ann. Probab.* **24** 1036–1048.

[3] K. Athreya and P. Ney (1972). *Branching Processes.* Springer-Verlag, New York.

[4] A. Benassi (1996). Arbres et grandes déviations. In *Trees*, B. Chauvin, S. Cohen, A. Rouault (Editors), Birkhäuser.

[5] I. Benjamini, R. Pemantle and Y. Peres (1995). Martin capacity for Markov Chains. *Ann. Probab.* **23**, 1332–1346.

[6] I. Benjamini, R. Pemantle and Y. Peres (1998). Unpredictable paths and percolation. *Ann. Probab.* **26**, 1198–1211.

[7] I. Benjamini and Y. Peres (1992). Random walks on a tree and capacity in the interval. *Ann. IHP Probab. et. Statist.* **28**, 557–592.

[8] I. Benjamini and Y. Peres (1994a). Markov chains indexed by trees. *Ann. Probab.* **22**, 219–243.

[9] I. Benjamini and Y. Peres (1994b). Tree-indexed random walks and first passage percolation. *Probab. Th. Rel. Fields* **98**, 91–112.

[10] I. Benjamini and O. Schramm (1997) Every graph with a positive Cheeger constant contains a tree with a positive Cheeger constant, *Geom. Funct. Anal.* **7**, 403–419.

[11] J. D. Biggins (1977). Chernoff's theorem in the branching random walk. *J. Appl. Prob.* **14**, 630–636.

[12] P. M. Bloher, J. Ruiz and V. A. Zagrebnov (1995). On the purity of the limiting Gibbs state for the Ising model on the Bethe lattice. *J. Stat. Phys.* **79**, 473–482.

[13] J. Cavender (1978). Taxonomy with confidence. *Math. BioSci.* **40**, 271–280.

[14] J. T. Chayes, L. Chayes, J. P. Sethna and D. J. Thouless (1986). A mean field spin glass with short range interactions, *Comm. Math. Phys.* **106**, 41–89.

[15] T. M. Cover and J. A. Thomas (1991). *Elements of Information Theory.* Wiley, New York.

[16] K. L. Chung (1973). Probabilistic approach in potential theory to the equilibrium problem. *Ann. Inst. Fourier, Grenoble* **23**, 313–322.

[17] T. Cox and R. Durrett (1983). Oriented percolation in dimensions $d \geq 4$: bounds and asymptotic formulas. *Math. Proc. Camb. Phil. Soc.* **93**, 151–162.

[18] P. G. Doyle and E. J. Snell (1984). *Random walks and Electrical Networks.* Carus Math. Monographs **22**, Math. Assoc. Amer., Washington, D. C.

[19] L. Dubins and D. Freedman (1967). Random distribution functions. *Proc. fifth Berkeley Symp. on math. stat. and probab.* (LeCam and Neyman, Eds.), Vol. II, Part 1, 183–214.

[20] R. Durrett (1996). *Probability: Theory and Examples*, Second edition. Duxbury Press, Belmont, CA.

[21] A. Dvoretsky, P. Erdős and S. Kakutani (1950). Double points of paths of Brownian motion in n-space. *Acta Sci. Math. Szeged* **12**, 75–81.

[22] A. Dvoretsky, P. Erdős, S. Kakutani and S. J. Taylor (1957). Triple points of Brownian motion in 3-space. *Proc. Camb. Phil. Soc.* **53**, 856–862.

[23] W. Evans (1994). *Information Theory and Noisy Computation.* PhD thesis, Dept. of Computer Science, University of California at Berkeley.

[24] W. Evans, C. Kenyon, Y. Peres and L. J. Schulman (1995). A critical phenomenon in a broadcast process. *Unpublished manuscript.*

[25] W. Evans, C. Kenyon, Y. Peres and L. J. Schulman (1998). Broadcasting on trees and the Ising model. *Preprint.*

[26] W. Evans and L. J. Schulman (1993). Signal propagation, with application to a lower bound on the depth of noisy formulas. In *Proceedings of the 34th Annual Symposium on Foundations of Computer Science*, 594–603.

[27] S. Evans (1987). Multiple points in the sample paths of a Levy processes. *Probab. Th. Rel. Fields* **76**, 359–376.

[28] P.J. Fitzsimmons and T. Salisbury (1989). Capacity and energy for multi-parameter Markov processes. *Ann. IHP Probab. et. Statist.* **25**, 325–350.

[29] H. Furstenberg (1967). Disjointness in ergodic theory, minimal sets, and a problem in Diophantine approximation. *Math. Systems Theory* **1**, 1–49.

[30] H. O. Georgii (1988). *Gibbs Measures and Phase Transitions.* W. de Gruyter, Berlin.

[31] G. Grimmett (1989). *Percolation.* New York, Springer Verlag.

[32] G. R. Grimmett, Percolation and disordered systems, in *Lectures in Probability Theory and Statistics, Ecole d'Ete de Probabilites de Saint-Flour* **XXVI**-1996, P. Bernard (ed.), Springer Lecture Notes in Math **1665**, 153–300.

[33] G. R. Grimmett and H. Kesten (1984). *Unpublished.*

[34] G. R. Grimmett, H. Kesten and Y. Zhang (1993). Random walk on the infinite cluster of the percolation model. *Probab. Th. Rel. Fields* **96**, 33–44.

[35] G. R. Grimmett and J. M. Marstrand (1990). The supercritical phase of percolation is well behaved. *Proc. Royal Soc. London Ser. A* **430**, 439–457.

[36] O. Häggström and E. Mossel (1998). Nearest-neighbor walks with low predictability profile and percolation in $2 + \epsilon$ dimensions. *Ann. Probab.* **26**, 1212–1231.

[37] O. Häggström, Y. Peres and J. Steif (1997). Dynamical Percolation. *Ann. IHP Probab. et. Statist.* **33**, 497–528.

[38] J.M. Hammersley (1974). Postulates for subadditive processes. *Ann. Probab.* **2**, 652–680.

[39] T. Hara and G. Slade (1994). Mean field behavior and the lace expansion. In *Probability Theory and Phase Transitions* (ed. G. Grimmett), Proceedings of the NATO ASI meeting in Cambridge 1993, Klewer.

[40] J. Hawkes (1981). Trees generated by a simple branching process. *J. London Math. Soc.* **24**, 373–384.

[41] P. Hiemer (1998). Dynamical renormalisation in oriented percolation. *Preprint.*

[42] Y. Higuchi (1977). Remarks on the limiting Gibbs state on a (d+1)-tree. *Publ. RIMS Kyoto Univ.* **13**, 335–348.

[43] C. Hoffman (1998). Unpredictable nearest neighbor processes. *Preprint.*

[44] D. Ioffe (1996a). A note on the extremality of the disordered state for the Ising model on the Bethe lattice. *Lett. Math. Phys.* **37**, 137–143.

[45] D. Ioffe (1996b). A note on the extremality of the disordered state for the Ising model on the Bethe lattice. In *Trees*, B. Chauvin, S. Cohen, A. Rouault (Editors), Birkhäuser.

[46] J. P. Kahane and J. Peyrière (1976). Sur certaines martingales de B. Mandelbrot. *Advances Math.* **22**, 131–145.

[47] S. Kakutani (1944a). On Brownian motion in n-space. *Proc. Imp. Acad. Tokyo* **20**, 648–652.

[48] S. Kakutani (1944b). Two dimensional Brownian motion and harmonic functions. *Proc. Imp. Acad. Tokyo* **20**, 706–714.

[49] H. Kesten (1980). The critical probability of bond percolation on the square lattice equals $\frac{1}{2}$. *Commun. Math. Phys.* **109**, 109–156.

[50] H. Kesten and B. P. Stigum (1966a). A limit theorem for multidimensional Galton-Watson processes. *Ann. Math. Stat.* **37**, 1211–1223.

[51] H. Kesten and B. P. Stigum (1966b). Additional limit theorems for indecomposable multidimensional Galton-Watson processes. *Ann. Math. Stat.* **37**, 1463–1481.

[52] J.F.C. Kingman (1975). The first birth problem for an age-dependent branching process. *Ann. Probab.* **3**, 790–801.

[53] J.-F. Le Gall (1992) Some properties of planar Brownian motion. *École d'été de probabilités de Saint-Flour XX*, Lecture Notes in Math. **1527**, 111–235. Springer, New York.

[54] G. Lawler (1982). The probability of intersection of independent random walks in four dimensions. *Commun. Math. Phys.* **86**, 539–554.

[55] G. Lawler (1985). Intersections of random walks in four dimensions II. *Commun. Math. Phys.* **97**, 583–594.

[56] L. Le Cam (1974). *Notes on Asymptotic Methods in Statistical Decision Theory*. Centre de Rech. Math. Univ. Montréal.

[57] D. Levin and Y. Peres (1998). Energy and Cutsets in Infinite Percolation Clusters. To appear in *Proceedings of the Cortona Workshop on Random Walks and Discrete Potential Theory*, M. Picardello and W. Woess (editors).

[58] T. M. Liggett, R. H. Schonmann and A. M. Stacey (1996). Domination by product measures. *Ann. Probab.* **24**, 1711–1726.

[59] R. Lyons (1989). The Ising model and percolation on trees and tree-like graphs. *Comm. Math. Phys.* **125**, 337–352.

[60] R. Lyons (1990). Random walks and percolation on trees. *Ann. Probab.* **18**, 931–958.

[61] R. Lyons (1992). Random walks, capacity and percolation on trees. *Ann. Probab.* **20**, 2043–2088.

[62] R. Lyons (1995). Random walks and the growth of groups. *C.R. Acad. Sci. Paris* **320**, 1361–1366.

[63] R. Lyons and R. Pemantle (1992). Random walks in a random environment and first-passage percolation on trees. *Ann. Probab.* **20**, 125–136.

[64] R. Lyons, R. Pemantle and Y. Peres (1995). Conceptual proofs of $L \log L$ criteria for mean behavior of branching processes. *Ann. Probab.* **23**, 1125–1138.

[65] R. Lyons, R. Pemantle and Y. Peres (1996). Random walks on the lamplighter group, *Ann. Probab.* **24**, 1993–2006.

[66] R. Lyons and Y. Peres (1999). *Probability on Trees and Networks*, Cambridge University Press, in preparation. Current version available at http://php.indiana.edu/~rdlyons/.

[67] T. Lyons (1983) A simple criterion for transience of a reversible Markov chain, *Ann. Probab.* **11**, 393–402.

[68] P. Marchal (1998). The Best Bounds in a Theorem of Russell Lyons. *Elect. Comm. Probab.* **3**, 91–94. http://www.math.washington.edu/~ejpecp/

[69] T. Moore and J. L. Snell, (1979). A branching process showing a phase transition. *J. Appl. Probab.* **16**, 252–260.

[70] C. St. J.A. Nash-Williams (1959). Random walks and electric currents in networks. *Proc. Cambridge Philos. Soc.* **55**, 181-194

[71] R. Pemantle (1995). Tree-indexed processes. *Stat. Sci.* **10**, 200–213.

[72] R. Pemantle and Y. Peres (1994). Domination between trees and applications to an explosion problem. *Ann. Probab.* **22**, 180–194.

[73] R. Pemantle and Y. Peres (1995a). Critical random walk in random environment on trees. *Ann. Probab.* **23**, 105–140.

[74] R. Pemantle and Y. Peres (1995b). Galton-Watson trees with the same mean have the same polar sets. *Ann. Probab.* **23**, 1102–1124.

[75] R. Pemantle and Y. Peres (1996). Recursions on general trees and the Ising model at critical temperatures. *Unpublished manuscript.*

[76] Y. Peres (1996). Intersection-equivalence of Brownian paths and certain branching processes. *Commun. Math. Phys.* **177**, 417–434.

[77] Y. Peres and J. Steif (1998). The number of infinite clusters in dynamical percolation. *Probab. Th. Rel. Fields* **111**, 141–165.

[78] A. Pisztora (1996). Surface order large deviations for Ising, Potts and percolation models. *Probab. Th. Rel. Fields* **104**, 427–466.

[79] C. J. Preston (1974), *Gibbs States on Countable Sets,* Cambridge Univ. Press.

[80] A. Sidorenko (1994). An analytic approach to extremal problems for graphs and hypergraphs. *Extremal problems for finite sets* (Visegrad, 1991), 423–455, Bolyai Soc. Math. Stud. **3**, Janos Bolyai Math. Soc., Budapest.

[81] P. M. Soardi (1994). *Potential Theory on Infinite Networks.* Springer LNM, Berlin.

[82] F. Spitzer (1975). Markov random fields on an infinite tree. *Ann. Probab.* **3**, 387–394.

[83] M. Steel (1989). Distributions in bicolored evolutionary trees. *Ph.D. Thesis,* Massey University, Palmerston North, New Zealand.

[84] M. Steel and M. Charleston (1995). Five surprising properties of parsimoniously colored trees. *Bull. Math. Biology* **57**, 367–375.

[85] N. Tongring (1988). Which sets contain multiple points of Brownian motion? *Math. Proc. Camb. Phil. Soc.* **103**, 181–187.

[86] I. Vajda (1989). *Theory of Statistical Inference and Information.* Kluwer Academic Publishers, Dordrecht.

[87] B. Virag (1998). The speed of simple random walk on graphs. *Preprint.*

ECOLE D'ETE DE CALCUL DES PROBABILITES
SAINT-FLOUR

LISTE DES EXPOSES

ALILI Larbi
Reformulation de la Factorisation de Wiener-Hopf et quelques applications

ALOS Elisa
An extension of Itô's formula for anticipating processes

ARCHER Olivier
Comportement asymptotique de certains systèmes linéaires issus de modèles de fiabilité

BENAÏM Michel
Vertex-Reinforced Random Walks and a Conjecture of Pemantle

BENAROUS David

BENNIES Jürgen
A random walk approach to Galton-Watson trees

BRITTON Tom
Stochastic Fade-Outs of Epidemics

CHABOT Nicolas
Radial Limits for the Spead of Epidemics and Forest Fires with Constant Life-Time

DEBICKI Krzysztof
A large deviation result for the supremum from Gaussian processes

DELMAS Jean-François
Saucisses de serpent brownien. Equivalence de capacité pour l'image du serpent brownien

DHERSIN J.S. (avec J.F. LE GALL)
Test de Kolmogorov pour le super-mouvement brownien

FRANCOIS Olivier
Estimation de convergence à concentration fixée pour la dynamique de Glauber en temps discret

FRANZ Uwe
Processus de Lévy sur les bigèbres

GANTERT Nina
Large deviations for a Random Walk in a Random Environment

HIEMER Philip
Renormalisation in Oriented Percolation in Z^d

HOLROYD Alexander
   Rigidity Percolation

ISHIKAWA Yasushi
   Estimates of transition densities fo jump processes

LACHAL Aimé
   Temps de sortie d'un intervalle borné pour une certaine classe de fonctionnelles du mouvement brownien

LAGAIZE Sandrine
   Exposant de Hölder des champs de Lévy

LANJRI ZAIDI Noureddine
   Burgers equation perturbed by a white noise force

LEJAY Antoine
   Sur les opérateurs sous forme divergence

LEURIDAN Christophe
   Sur l'ensemble des instants de records d'un processus de Poisson ponctuel

MACCHI Cécile
   Grandes déviations en temps petit pour un processus de Markov à espace d'états dénombrable

MALOUCHE Dhafer
   Action Quadratique du Groupe des Homographies sur les familles exponentielles réelles

MARCHAL Philippe.
   Temps d'occupation pour les marches aléatoires

MARQUEZ-CARRERAS David.
   Développement de la densité en utilisant la décomposition en Chaos de Wiener

MARSALLE Laurence.
   Points lents des temps locaux

MAZET Olivier
   Caractérisation des semi-groupes de Markov sur R associés aux polynômes d'Hermite

MICLO Laurent
   Sur des relations entre isopérimétrie et trou spectral pour les chaînes de Markov finies

MORET Silvia
   Extension of Ito's formula for martingales

PHILIPPART Claudia
   Stability Conditions for Input-Output Processes

ROUSSEL Sandrine
   Marches aléatoires sur le groupe symétrique engendrées par les classes de conjugaison

ROVIRA Carles
   Large Deviations for stochastic Volterra equations in the plane

SAINT LOUBERT BIE Erwan
   Régularité de la solution d'une Equation aux Dérivées Partielles Stochastique conduite par un bruit poissonnien

SERLET Laurent
A large deviation principle for the Brownian snake

STEIF Jeff
The generalized TT¹ process

VIENS Frédéri (avec D. NUALART)
Evolution Equation of a Stochastic Semigroup with white-noise Drift

LISTE DES AUDITEURS

Mr.	ABRAHAM Romain	Université René Descartes, PARIS V
Melle	AKIAN Marianne	INRIA, Le Chesnay
Melle	AL-KHACH Rim	Université PARIS VI
Mr.	ALILI Larbi	University of Manchester
Melle	ALOS Elisa	Universitat de Barcelona
Mr.	ARCHER Olivier	Université de Toulouse
Mr.	AZEMA Jacques	Université PARIS VI
Mr.	BARDET Jean-Baptiste	Ecole Normale Supérieure, Lyon
Mr.	BARDINA Xavier	Universitat Autonoma de Barcelona
Mr.	BENAIM Michel	Université Paul Sabatier, Toulouse
Mr.	BENAROUS David	Université PARIS VI
Mr.	BENAROUS Gérard	Ecole Normale Supérieure, Paris
Mr.	BENASSI Albert	Université Blaise Pascal, Clermont-Fd
Mr.	BENNIES Jürgen	Universität Frankfurt
Mr.	BERNARD Pierre	Université Blaise Pascal, Clermont-Fd
Mr.	BODINEAU Thierry	Ecole Normale Supérieure, Paris
Mr.	BONACCORSI Stefano	University of Trento, Italie
Mr.	BOUFOUSSI Brahim	Université C. Ayyad, Marrakech, Maroc
Mr.	BOUGEROL Philippe	Université PARIS VI
Mr.	BRITTON Tom	Uppsala University, Suède
Mr.	BRUNAUD Marc	Université Denis Diderot, PARIS VII
Mme	CABALLERO Emilia	Université PARIS VI
Mr.	CHABOT Nicolas	CMI, Université de Provence, Marseille
Mme	CHALEYAT-MAUREL Mireille	Université René Descartes, PARIS V
Mr.	CHAUMONT Loïc	Université PARIS VI
Mme	CHAUVIN Brigitte	Université de Versailles, Saint Quentin
Mr.	COMETS Francis	Université Denis Diderot, PARIS VII
Mr.	DEBICKI Krzysztof	University of Wroclaw, Pologne
Mr.	DELMAS Jean-François	ENPC CERMICS, Marne-La-Vallée
Mr.	DHERSIN Stéphane	Université PARIS VI
Mme	DONATI-MARTIN Catherine	Université PARIS VI
Mr.	DONEY Ronald	University of Manchester, U.K.
Mr.	DUQUESNE Thomas	Ecole Normale Supérieure, Paris
Melle	ESTRADE Anne	Université d'Orléans
Mr.	FLEURY Gérard	Université Blaise Pascal, Clermont-Fd
Mme	FOURATI Sonia	Université PARIS VI
Mr.	FRANCOIS Olivier	Institut J.Fourier, Saint Martin d'Hères
Mr.	FRANZ Uwe	Institut Elie Cartan, Nancy
Melle	GANTERT Nina	Technion Haifa, Israël
Mr.	GARCIA DOMINGO Josep Lluis	Estudis Universitaris de Vic, Espagne
Mr.	GARNIER Josselin	Ecole Polytechnique, Palaiseau
Mr.	GATZOURAS Dimitris	University of Cambridge, U.K.
Mr.	GRORUD Axel	CMI, Université de Provence, Marseille
Mr.	GROSSMANN Steffen	Fachbereich Mathematik, Frankfurt, Allemagne
Mr.	GUIMIER Alain	Université des Sciences et Techniques, Gabon
Mr.	HIEMER Philipp	University of Cambridge, U.K.
Mr.	HOLROYD Alexander	University of Cambridge, U.K.
Mr.	ISHIKAWA Yasushi	University of Tsukuba, Japon
Mr.	KLEIN Etienne	Université Paris-Sud, Orsay
Mr.	LACHAL Aimé	INSA de Lyon, Villeurbanne

Melle	LAGAIZE Sandrine	Université d'Orléans
Mr.	LANJRIZAIDI Noureddine	Universitat de Barcelona, Espagne
Mr.	LE GALL Jean-François	Université PARIS VI
Mr.	LEJAY Antoine	CMI, Univrsité de Provence, Marseille
Mr.	LEURIDAN Christophe	Institut J. Fourier, Saint Martin d'Hères
Melle	MACCHI Cécile	Université Paul Sabatier, Toulouse
Mr.	MALOUCHE Dhafer	Université Paul Sabatier, Toulouse
Mr.	MARCHAL Philippe	Université PARIS VI
Mr.	MARDIN Arif	Institut National des Télécommunications, Evry
Mr.	MARQUEZ-CARRERAS David	Universitat de Barcelona, Espagne
Melle	MARSALLE Laurence	Université PARIS VI
Mr.	MAZET Olivier	Université Paul Sabatier, Toulouse
Mr.	MAZLIAK Laurent	Université PARIS VI
Mr.	MICLO Laurent	Université Paul Sabatier, Toulouse
Melle	MORET Silvia	Universitat de Barcelona, Espagne
Mr.	MOURRAGUI Mustapha	Université de Rouen
Mr.	NUALART David	Universitat de Barcelona, Espagne
Melle	PHILIPPART Claudia	Instituto Superior Tecnico, Lisboa, Portugal
Mr.	PICARD Jean	Université Blaise Pascal, Clermont-Fd
Mme	PONTIER Monique	Université d'Orléans
Melle	ROUSSEL Sandrine	Université Paul Sabatier, Toulouse
Mr.	ROUX Daniel	Université Blaise Pascal, Clermont-Fd
Mr.	ROVIRA Carles	Universitat de Barcelona, Espagne
Mr.	SAINT LOUBERT BIE Erwan	Université Blaise Pascal, Clermont-Fd
Melle	SAVONA Catherine	Université Blaise Pascal, C lermont-Fd
Mr.	SERLET Laurent	Université René Descartes, PARIS V
Mr.	STEIF Jeffrey	Chalmers University of Technology, Gothenburg, Suède
Melle	TALEB Marina	Université Denis Diderot, PARIS VII
Mr.	VIENS Frederi	University of North Texas, Denton, U.S.A.
Mr.	WU Liming	Université Blaise Pascal, Clermont-Fd
Mr.	ZEITOUNI Ofer	Israël Institute of Technology, Haifa, Israël

LIST OF PREVIOUS VOLUMES OF THE "Ecole d'Eté de Probabilités"

1978   R. AZENCOTT   (LNM 774)
"Grandes déviations et applications"
Y. GUIVARC'H
"Quelques propriétés asymptotiques des produits de
  matrices aléatoires"
R.F. GUNDY
"Inégalités pour martingales à un et deux indices :
  l'espace H^p"

1979   J.P. BICKEL   (LNM 876)
"Quelques aspects de la statistique robuste"
N. EL KAROUI
"Les aspects probabilistes du contrôle stochastique"
M. YOR
"Sur la théorie du filtrage"

1980   J.M. BISMUT   (LNM 929)
"Mécanique aléatoire"
L. GROSS
"Thermodynamics, statistical mechanics and
  random fields"
K. KRICKEBERG
"Processus ponctuels en statistique"

1981   X. FERNIQUE   (LNM 976)
"Régularité de fonctions aléatoires non gaussiennes"
P.W. MILLAR
"The minimax principle in asymptotic statistical theory"
D.W. STROOCK
"Some application of stochastic calculus to partial
  differential equations"
M. WEBER
"Analyse infinitésimale de fonctions aléatoires"

1982   R.M. DUDLEY   (LNM 1097)
"A course on empirical processes"
H. KUNITA
"Stochastic differential equations and stochastic
flow of diffeomorphisms"
F. LEDRAPPIER
"Quelques propriétés des exposants caractéristiques"

1983   D.J. ALDOUS   (LNM 1117)
"Exchangeability and related topics"
I.A. IBRAGIMOV
"Théorèmes limites pour les marches aléatoires"
J. JACOD
"Théorèmes limite pour les processus"

1984   R. CARMONA   (LNM 1180)
"Random Schrödinger operators"
H. KESTEN
"Aspects of first passage percolation"
J.B. WALSH
"An introduction to stochastic partial differential
  equations"

1985-87	S.R.S. VARADHAN	(LNM 1362)

1985-87   S.R.S. VARADHAN   (LNM 1362)
"Large deviations"
P. DIACONIS
"Applications of non-commutative Fourier
  analysis to probability theorems
H. FÖLLMER
"Random fields and diffusion processes"
G.C. PAPANICOLAOU
"Waves in one-dimensional random media"
D. ELWORTHY
Geometric aspects of diffusions on manifolds"
E. NELSON
"Stochastic mechanics and random fields"

1986   O.E. BARNDORFF-NIELSEN   (LNS M50)
"Parametric statistical models and likelihood"

1988   A. ANCONA   (LNM 1427)
"Théorie du potentiel sur les graphes et les variétés"
D. GEMAN
"Random fields and inverse problems in imaging"
N. IKEDA
"Probabilistic methods in the study of asymptotics"

1989   D.L. BURKHOLDER   (LNM 1464)
"Explorations in martingale theory and its applications"
E. PARDOUX
"Filtrage non linéaire et équations aux dérivées partielles
  stochastiques associées"
A.S. SZNITMAN
"Topics in propagation of chaos"

1990   M.I. FREIDLIN   (LNM 1527)
"Semi-linear PDE's and limit theorems for
  large deviations"
J.F. LE GALL
"Some properties of planar Brownian motion"

1991   D.A. DAWSON   (LNM 1541)
"Measure-valued Markov processes"
B. MAISONNEUVE
"Processus de Markov : Naissance,
  Retournement, Régénération"
J. SPENCER
"Nine Lectures on Random Graphs"

1992   D. BAKRY   (LNM 1581)
"L'hypercontractivité et son utilisation en théorie
des semigroupes"
R.D. GILL
"Lectures on Survival Analysis"
S.A. MOLCHANOV
"Lectures on the Random Media"

1993	P. BIANE "Calcul stochastique non-commutatif" R. DURRETT "Ten Lectures on Particle Systems"	(LNM 1608)
1994	R. DOBRUSHIN "Perturbation methods of the theory of Gibbsian fields" P. GROENEBOOM "Lectures on inverse problems" M. LEDOUX "Isoperimetry and gaussian analysis"	(LNM 1648)
1995	M.T. BARLOW "Diffusions on fractals" D. NUALART "Analysis on Wiener space and anticipating stochastic calculus"	(LNM 1690)
1996	E. GINE "Decoupling and limit theorems for U-statistics and U-processes" "Lectures on some aspects theory of the bootstrap" G. GRIMMETT "Percolation and disordered systems" L. SALOFF-COSTE "Lectures on finite Markov chains"	(LNM 1665)
1997	J. BERTOIN "Subordinators : examples and applications" F. MARTINELLI "Lectures on Glauber dynamics for discrete spin models" Y. PERES "Probability on Trees : an introductory climb"	(LNM 1717)

Lecture Notes in Mathematics

For information about Vols. 1–1525
please contact your bookseller or Springer-Verlag

Vol. 1568: F. Weisz, Martingale Hardy Spaces and their Application in Fourier Analysis. VIII, 217 pages. 1994.

Vol. 1569: V. Totik, Weighted Approximation with Varying Weight. VI, 117 pages. 1994.

Vol. 1570: R. deLaubenfels, Existence Families, Functional Calculi and Evolution Equations. XV, 234 pages. 1994.

Vol. 1571: S. Yu. Pilyugin, The Space of Dynamical Systems with the C^0-Topology. X, 188 pages. 1994.

Vol. 1572: L. Göttsche, Hilbert Schemes of Zero-Dimensional Subschemes of Smooth Varieties. IX, 196 pages. 1994.

Vol. 1573: V. P. Havin, N. K. Nikolski (Eds.), Linear and Complex Analysis - Problem Book 3 - Part I. XXII, 489 pages. 1994.

Vol. 1574: V. P. Havin, N. K. Nikolski (Eds.), Linear and Complex Analysis - Problem Book 3 - Part II. XXII, 507 pages. 1994.

Vol. 1575: M. Mitrea, Clifford Wavelets, Singular Integrals, and Hardy Spaces. XI, 116 pages. 1994.

Vol. 1576: K. Kitahara, Spaces of Approximating Functions with Haar-Like Conditions. X, 110 pages. 1994.

Vol. 1577: N. Obata, White Noise Calculus and Fock Space. X, 183 pages. 1994.

Vol. 1578: J. Bernstein, V. Lunts, Equivariant Sheaves and Functors. V, 139 pages. 1994.

Vol. 1579: N. Kazamaki, Continuous Exponential Martingales and BMO. VII, 91 pages. 1994.

Vol. 1580: M. Milman, Extrapolation and Optimal Decompositions with Applications to Analysis. XI, 161 pages. 1994.

Vol. 1581: D. Bakry, R. D. Gill, S. A. Molchanov, Lectures on Probability Theory. Editor: P. Bernard. VIII, 420 pages. 1994.

Vol. 1582: W. Balser, From Divergent Power Series to Analytic Functions. X, 108 pages. 1994.

Vol. 1583: J. Azéma, P. A. Meyer, M. Yor (Eds.), Séminaire de Probabilités XXVIII. VI, 334 pages. 1994.

Vol. 1584: M. Brokate, N. Kenmochi, I. Müller, J. F. Rodriguez, C. Verdi, Phase Transitions and Hysteresis. Montecatini Terme, 1993. Editor: A. Visintin. VII, 291 pages. 1994.

Vol. 1585: G. Frey (Ed.), On Artin's Conjecture for Odd 2-dimensional Representations. VIII, 148 pages. 1994.

Vol. 1586: R. Nillsen, Difference Spaces and Invariant Linear Forms. XII, 186 pages. 1994.

Vol. 1587: N. Xi, Representations of Affine Hecke Algebras. VIII, 137 pages. 1994.

Vol. 1588: C. Scheiderer, Real and Étale Cohomology. XXIV, 273 pages. 1994.

Vol. 1589: J. Bellissard, M. Degli Esposti, G. Forni, S. Graffi, S. Isola, J. N. Mather, Transition to Chaos in Classical and Quantum Mechanics. Montecatini Terme, 1991. Editor: 2S. Graffi. VII, 192 pages. 1994.

Vol. 1590: P. M. Soardi, Potential Theory on Infinite Networks. VIII, 187 pages. 1994.

Vol. 1591: M. Abate, G. Patrizio, Finsler Metrics - A Global Approach. IX, 180 pages. 1994.

Vol. 1592: K. W. Breitung, Asymptotic Approximations for Probability Integrals. IX, 146 pages. 1994.

Vol. 1593: J. Jorgenson & S. Lang, D. Goldfeld, Explicit Formulas for Regularized Products and Series. VIII, 154 pages. 1994.

Vol. 1594: M. Green, J. Murre, C. Voisin, Algebraic Cycles and Hodge Theory. Torino, 1993. Editors: A. Albano, F. Bardelli. VII, 275 pages. 1994.

Vol. 1595: R.D.M. Accola, Topics in the Theory of Riemann Surfaces. IX, 105 pages. 1994.

Vol. 1596: L. Heindorf, L. B. Shapiro, Nearly Projective Boolean Algebras. X, 202 pages. 1994.

Vol. 1597: B. Herzog, Kodaira-Spencer Maps in Local Algebra. XVII, 176 pages. 1994.

Vol. 1598: J. Berndt, F. Tricerri, L. Vanhecke, Generalized Heisenberg Groups and Damek-Ricci Harmonic Spaces. VIII, 125 pages. 1995.

Vol. 1599: K. Johannson, Topology and Combinatorics of 3-Manifolds. XVIII, 446 pages. 1995.

Vol. 1600: W. Narkiewicz, Polynomial Mappings. VII, 130 pages. 1995.

Vol. 1601: A. Pott, Finite Geometry and Character Theory. VII, 181 pages. 1995.

Vol. 1602: J. Winkelmann, The Classification of Three-dimensional Homogeneous Complex Manifolds. XI, 230 pages. 1995.

Vol. 1603: V. Ene, Real Functions - Current Topics. XIII, 310 pages. 1995.

Vol. 1604: A. Huber, Mixed Motives and their Realization in Derived Categories. XV, 207 pages. 1995.

Vol. 1605: L. B. Wahlbin, Superconvergence in Galerkin Finite Element Methods. XI, 166 pages. 1995.

Vol. 1606: P.-D. Liu, M. Qian, Smooth Ergodic Theory of Random Dynamical Systems. XI, 221 pages. 1995.

Vol. 1607: G. Schwarz, Hodge Decomposition - A Method for Solving Boundary Value Problems. VII, 155 pages. 1995.

Vol. 1608: P. Biane, R. Durrett, Lectures on Probability Theory. Editor: P. Bernard. VII, 210 pages. 1995.

Vol. 1609: L. Arnold, C. Jones, K. Mischaikow, G. Raugel, Dynamical Systems. Montecatini Terme, 1994. Editor: R. Johnson. VIII, 329 pages. 1995.

Vol. 1610: A. S. Üstünel, An Introduction to Analysis on Wiener Space. X, 95 pages. 1995.

Vol. 1611: N. Knarr, Translation Planes. VI, 112 pages. 1995.

Vol. 1612: W. Kühnel, Tight Polyhedral Submanifolds and Tight Triangulations. VII, 122 pages. 1995.

Vol. 1613: J. Azéma, M. Emery, P. A. Meyer, M. Yor (Eds.), Séminaire de Probabilités XXIX. VI, 326 pages. 1995.

Vol. 1614: A. Koshelev, Regularity Problem for Quasilinear Elliptic and Parabolic Systems. XXI, 255 pages. 1995.

Vol. 1615: D. B. Massey, Le Cycles and Hypersurface Singularities. XI, 131 pages. 1995.

Vol. 1616: I. Moerdijk, Classifying Spaces and Classifying Topoi. VII, 94 pages. 1995.

Vol. 1617: V. Yurinsky, Sums and Gaussian Vectors. XI, 305 pages. 1995.

Vol. 1618: G. Pisier, Similarity Problems and Completely Bounded Maps. VII, 156 pages. 1996.

Vol. 1619: E. Landvogt, A Compactification of the Bruhat-Tits Building. VII, 152 pages. 1996.

Vol. 1620: R. Donagi, B. Dubrovin, E. Frenkel, E. Previato, Integrable Systems and Quantum Groups. Montecatini Terme, 1993. Editors:M. Francaviglia, S. Greco. VIII, 488 pages. 1996.

Vol. 1621: H. Bass, M. V. Otero-Espinar, D. N. Rockmore, C. P. L. Tresser, Cyclic Renormalization and Auto-morphism Groups of Rooted Trees. XXI, 136 pages. 1996.

Vol. 1622: E. D. Farjoun, Cellular Spaces, Null Spaces and Homotopy Localization. XIV, 199 pages. 1996.

Vol. 1623: H.P. Yap, Total Colourings of Graphs. VIII, 131 pages. 1996.

Vol. 1624: V. Brınzanescu, Holomorphic Vector Bundles over Compact Complex Surfaces. X, 170 pages. 1996.

Vol.1625: S. Lang, Topics in Cohomology of Groups. VII, 226 pages. 1996.

Vol. 1626: J. Azéma, M. Emery, M. Yor (Eds.), Séminaire de Probabilités XXX. VIII, 382 pages. 1996.

Vol. 1627: C. Graham, Th. G. Kurtz, S. Méléard, Ph. E. Protter, M. Pulvirenti, D. Talay, Probabilistic Models for Nonlinear Partial Differential Equations. Montecatini Terme, 1995. Editors: D. Talay, L. Tubaro. X, 301 pages. 1996.

Vol. 1628: P.-H. Zieschang, An Algebraic Approach to Association Schemes. XII, 189 pages. 1996.

Vol. 1629: J. D. Moore, Lectures on Seiberg-Witten Invariants. VII, 105 pages. 1996.

Vol. 1630: D. Neuenschwander, Probabilities on the Heisenberg Group: Limit Theorems and Brownian Motion. VIII, 139 pages. 1996.

Vol. 1631: K. Nishioka, Mahler Functions and Transcendence.VIII, 185 pages.1996.

Vol. 1632: A. Kushkuley, Z. Balanov, Geometric Methods in Degree Theory for Equivariant Maps. VII, 136 pages. 1996.

Vol.1633: H. Aikawa, M. Essén, Potential Theory – Selected Topics. IX, 200 pages.1996.

Vol. 1634: J. Xu, Flat Covers of Modules. IX, 161 pages. 1996.

Vol. 1635: E. Hebey, Sobolev Spaces on Riemannian Manifolds. X, 116 pages. 1996.

Vol. 1636: M. A. Marshall, Spaces of Orderings and Abstract Real Spectra. VI, 190 pages. 1996.

Vol. 1637: B. Hunt, The Geometry of some special Arithmetic Quotients. XIII, 332 pages. 1996.

Vol. 1638: P. Vanhaecke, Integrable Systems in the realm of Algebraic Geometry. VIII, 218 pages. 1996.

Vol. 1639: K. Dekimpe, Almost-Bieberbach Groups: Affine and Polynomial Structures. X, 259 pages. 1996.

Vol. 1640: G. Boillat, C. M. Dafermos, P. D. Lax, T. P. Liu, Recent Mathematical Methods in Nonlinear Wave Propagation. Montecatini Terme, 1994. Editor: T. Ruggeri. VII, 142 pages. 1996.

Vol. 1641: P. Abramenko, Twin Buildings and Applications to S-Arithmetic Groups. IX, 123 pages. 1996.

Vol. 1642: M. Puschnigg, Asymptotic Cyclic Cohomology. XXII, 138 pages. 1996.

Vol. 1643: J. Richter-Gebert, Realization Spaces of Polytopes. XI, 187 pages. 1996.

Vol. 1644: A. Adler, S. Ramanan, Moduli of Abelian Varieties. VI, 196 pages. 1996.

Vol. 1645: H. W. Broer, G. B. Huitema, M. B. Sevryuk, Quasi-Periodic Motions in Families of Dynamical Systems. XI, 195 pages. 1996.

Vol. 1646: J.-P. Demailly, T. Peternell, G. Tian, A. N. Tyurin, Transcendental Methods in Algebraic Geometry. Cetraro, 1994. Editors: F. Catanese, C. Ciliberto. VII, 257 pages. 1996.

Vol. 1647: D. Dias, P. Le Barz, Configuration Spaces over Hilbert Schemes and Applications. VII. 143 pages. 1996.

Vol. 1648: R. Dobrushin, P. Groeneboom, M. Ledoux, Lectures on Probability Theory and Statistics. Editor: P. Bernard. VIII, 300 pages. 1996.

Vol. 1649: S. Kumar, G. Laumon, U. Stuhler, Vector Bundles on Curves – New Directions. Cetraro, 1995. Editor: M. S. Narasimhan. VII, 193 pages. 1997.

Vol. 1650: J. Wildeshaus, Realizations of Polylogarithms. XI, 343 pages. 1997.

Vol. 1651: M. Drmota, R. F. Tichy, Sequences, Discrepancies and Applications. XIII, 503 pages. 1997.

Vol. 1652: S. Todorcevic, Topics in Topology. VIII, 153 pages. 1997.

Vol. 1653: R. Benedetti, C. Petronio, Branched Standard Spines of 3-manifolds. VIII. 132 pages. 1997.

Vol. 1654: R. W. Ghrist, P. J. Holmes, M. C. Sullivan, Knots and Links in Three-Dimensional Flows. X, 208 pages. 1997.

Vol. 1655: J. Azéma, M. Emery, M. Yor (Eds.), Séminaire de Probabilités XXXI. VIII, 329 pages. 1997.

Vol. 1656: B. Biais, T. Björk, J. Cvitanic, N. El Karoui, E. Jouini, J. C. Rochet, Financial Mathematics. Bressanone, 1996. Editor: W. J. Runggaldier. VII, 316 pages. 1997.

Vol. 1657: H. Reimann, The semi-simple zeta function of quaternionic Shimura varieties. IX, 143 pages. 1997.

Vol. 1658: A. Pumarino, J. A. Rodrıguez, Coexistence and Persistence of Strange Attractors. VIII, 195 pages. 1997.

Vol. 1659: V. Kozlov, V. Maz'ya, Theory of a Higher-Order Sturm-Liouville Equation. XI, 140 pages. 1997.

Vol. 1660: M. Bardi, M. G. Crandall, L. C. Evans, H. M. Soner, P. E. Souganidis, Viscosity Solutions and Applications. Montecatini Terme, 1995. Editors: I. Capuzzo Dolcetta, P. L. Lions. IX, 259 pages. 1997.

Vol. 1661: A. Tralle, J. Oprea, Symplectic Manifolds with no Kähler Structure. VIII, 207 pages. 1997.

Vol. 1662: J. W. Rutter, Spaces of Homotopy Self-Equivalences – A Survey. IX, 170 pages. 1997.

Vol. 1663: Y. E. Karpeshina; Perturbation Theory for the Schrödinger Operator with a Periodic Potential. VII, 352 pages. 1997.

Vol. 1664: M. Väth, Ideal Spaces. V, 146 pages. 1997.

Vol. 1665: E. Giné, G. R. Grimmett, L. Saloff-Coste, Lectures on Probability Theory and Statistics 1996. Editor: P. Bernard. X, 424 pages. 1997.

Vol. 1666: M. van der Put, M. F. Singer, Galois Theory of Difference Equations. VII. 179 pages. 1997.

Vol. 1667: J. M. F. Castillo, M. González, Three-space Problems in Banach Space Theory. XII, 267 pages. 1997.

Vol. 1668: D. B. Dix, Large-Time Behavior of Solutions of Linear Dispersive Equations. XIV, 203 pages. 1997.

Vol. 1669: U. Kaiser, Link Theory in Manifolds. XIV, 167 pages. 1997.